VOLUME EIGHTY THREE

ADVANCES IN BOTANICAL RESEARCH

Phytoremediation

ADVANCES IN BOTANICAL RESEARCH

Series Editors

VOLUME EIGHTY THREE

ADVANCES IN BOTANICAL RESEARCH

Phytoremediation

Volume Editors

ANN CUYPERS
Hasselt University, Diepenbeek, Belgium

JACO VANGRONSVELD
Hasselt University, Diepenbeek, Belgium

ELSEVIER

ACADEMIC PRESS
An imprint of Elsevier

Academic Press is an imprint of Elsevier
125 London Wall, London EC2Y 5AS, United Kingdom
The Boulevard, Langford Lane, Kidlington, Oxford OX5 1GB, United Kingdom
50 Hampshire Street, 5th Floor, Cambridge, MA 02139, United States
525 B Street, Suite 1800, San Diego, CA 92101-4495, United States

First edition 2017

ISBN: 978-0-12-802853-7
ISSN: 0065-2296

For information on all Academic Press publications visit our website at https://www.elsevier.com/books-and-journals

Publisher: Zoe Kruze
Acquisition Editor: Kirsten Shankland
Editorial Project Manager: Thomas Van Der Ploeg
Production Project Manager: Magesh Kumar Mahalingam
Designer: Christian Bilbow

Typeset by TNQ Books and Journals

CONTENTS

CONTRIBUTORS

Vanessa Álvarez-López
Consejo Superior de Investigaciones Científicas (CSIC), Santiago de Compostela, Spain

Ilya A. Baklanov
University of Maryland, College Park, MD, United States

Cristina Becerra-Castro
Consejo Superior de Investigaciones Científicas (CSIC), Santiago de Compostela, Spain

Maribel Cabello-Conejo
Consejo Superior de Investigaciones Científicas (CSIC), Santiago de Compostela, Spain

Rufus L. Chaney
USDA-Agricultural Research Service, Crop Systems and Global Change Laboratory, Beltsville, MD, United States

Laura Coninx
Hasselt University, Diepenbeek, Belgium

Ann Cuypers
Hasselt University, Diepenbeek, Belgium

Sławomir Dresler
Maria Curie-Skłodowska University, Lublin, Poland

Nele Eevers
Hasselt University, Diepenbeek, Belgium

Helena Gawronska
Warsaw University of Life Sciences, Warsaw, Poland

Stanislaw W. Gawronski
Warsaw University of Life Sciences, Warsaw, Poland

Cristina Gonnelli
Università di Firenze, Florence, Italy

Sophie Hendrix
Hasselt University, Diepenbeek, Belgium

Christian Huber
Helmholtz Zentrum München, Neuherberg, Germany

Els Keunen
Hasselt University, Diepenbeek, Belgium

Petra S. Kidd
Consejo Superior de Investigaciones Científicas (CSIC), Santiago de Compostela, Spain

Tori Langill
Hasselt University, Diepenbeek, Belgium

Slawo Lomnicki
Louisiana State University, Baton Rouge, LA, United States

Veronika Martinova
Hasselt University, Diepenbeek, Belgium

Ángeles Prieto-Fernández
Consejo Superior de Investigaciones Científicas (CSIC), Santiago de Compostela, Spain

Francois Rineau
Hasselt University, Diepenbeek, Belgium

Adam Rostański
University of Silesia, Katowice, Poland

Peter Schröder
Helmholtz Zentrum München, Neuherberg, Germany

Federico Selvi
Università di Firenze, Florence, Italy

Arne Sæbo
Norwegian Institute of Bioeconomy Research (NIBIO), Ås, Norway

Sofie Thijs
Hasselt University, Diepenbeek, Belgium

Jaco Vangronsveld
Hasselt University, Diepenbeek, Belgium

Nele Weyens
Hasselt University, Diepenbeek, Belgium

Jason C. White
Connecticut Agricultural Experiment Station, New Haven, CT, United States

Małgorzata Wójcik
Maria Curie-Skłodowska University, Lublin, Poland

PREFACE

Phytoremediation, the use of plants and their associated microorganisms to remediate contaminated soils or (ground) water, concerns a range of *in situ*, solar-powered remediation technologies involving minimal site disturbance and maintenance with low implementation costs and high public acceptance. Since conventional (mostly civil-engineering based) remediation options for polluted soils and (ground) water currently available are frequently expensive (and at the same time environmentally invasive), phytoremediation often turns out to be a valuable alternative, especially for managing large areas with diffuse pollution. However, large-scale implementation of phytoremediation often faces a number of obstacles, including the levels of contaminants (being toxic for the organisms involved in remediation), the bioavailability of the contaminants (being too low) and, in certain cases, evapotranspiration of volatile organic pollutants from soil or groundwater to the atmosphere. The current volume contains several reviews addressing the potential for exploiting plants and their associated microorganisms to improve phytoremediation of trace metals and organic contaminants.

In general, remediation technologies, whether in place or *ex situ*, either remove the contaminants ('decontamination or clean-up'), or reduce the risk posed by the contaminants by reducing exposure ('immobilization or stabilization'). One 'gentle' plant-based site stabilization approach suitable for heavily contaminated sites is phytostabilization, aimed to decrease soil pollutant availability and spreading using plants or a combination of plants and soil amendments. Indeed, although some soils may be completely bare, due to extreme metal phytotoxicity, spontaneous development of metal tolerant vegetations has been seen in many cases (Chapter 1). Naturally occurring metalliferous soils are known to harbour unique vegetation types. Human-created metalliferous sites may become colonized by similar types of metal-tolerant plant communities. Such metal-tolerant communities often contain metal hyperaccumulators and constitute an excellent resource of species for rehabilitation or phytoremediation of degraded land (Chapters 1 and 5). Hyperaccumulators are excellent candidates for metal phytoextraction and phytomining (Chapter 5), which are based on trace element removal from the polluted substrate by concentrating them in harvestable plant parts. In general, a trace element phytoextraction protocol consists of (1) cultivation of the appropriate plant/crop species on the contaminated site; (2) removal of harvestable trace element–enriched biomass from the site; and (3) post harvest treatments (*i.e.*, composting, compacting, thermal treatment, etc.) to reduce volume and/or weight of biomass for disposal as a hazardous waste, or for its recycling to recover the trace elements that may have an economic value.

Metal uptake in plants does not only depend on the availability of the metals in soil and the expression of detoxification genes and metal transporters by the plant itself, but also on the plant-associated microbiota. Like other plant species, hyperaccumulator plants harbour a high diversity of bacteria and fungi in their rhizosphere and endosphere (Chapters 2 and 4), some of which were shown to have potential for assisting plants to grow in metal-contaminated soil and regulating plant metal uptake (Chapter 3). Some unique microbial groups that are highly abundant in heavily contaminated sites have been identified using next generation sequencing techniques and might have great potential for enhancing the efficiency of metal phytoextraction (Chapter 2). Also a profound knowledge of molecular and cellular mechanisms of toxicity and tolerance against toxic concentrations of metals is crucial to optimize phytoremediation strategies to clean up contaminated soils (Chapter 6).

In contrast to metals, organic contaminants can efficiently be degraded by the cooperation of plants and their associated microorganisms, generally termed rhizo- and phytodegradation. Since recently several reviews were published on rhizo- and phytodegradation of organic solvents and fuels, the review presented here concentrates on organic pesticides (Chapter 7).

More recently, the potential of air phytoremediation has received increased attention. Air pollution has become a global problem. Most of the pollution is of anthropogenic origin. Plants with their enormous biologically active surface area can be considered as crucial partners to combat this type of pollution. Plants themselves developed very efficient defence mechanisms against atmospheric pollutants, which can also be exploited to improve air quality. It is thus possible to exploit higher plants with their microbiome for the removal of airborne pollutants (including particulate matter) in outdoor and often also indoor locations (Chapter 8). Some of these pollutants are included by plants in their own metabolism, while others are sequestered. In some plant species, the processes of removing pollutants from the air are conducted in a very efficient way.

In conclusion, use of plants and their associated microbiome harbours great potential for remediation of polluted soils, (ground) water and air. Nevertheless, local environmental (and other) conditions need to be carefully investigated to define the most optimal implementation of phytoremediation.

Jaco Vangronsveld and Ann Cuypers

CHAPTER ONE

Metallophytes of Serpentine and Calamine Soils – Their Unique Ecophysiology and Potential for Phytoremediation

Małgorzata Wójcik*[,1], Cristina Gonnelli[§], Federico Selvi[§], Sławomir Dresler*, Adam Rostański[¶] and Jaco Vangronsveld[||]
*Maria Curie-Skłodowska University, Lublin, Poland
[§]Università di Firenze, Florence, Italy
[¶]University of Silesia, Katowice, Poland
[||]Hasselt University, Diepenbeek, Belgium
[1]Corresponding author: E-mail: mwojcik@umcs.pl

Contents

Advances in Botanical Research, Volume 83
ISSN 0065-2296
http://dx.doi.org/10.1016/bs.abr.2016.12.002

Abstract

For years, metallophytes of both natural and human-influenced metalliferous soils have focussed considerable attention due to their unique appearance and ability to colonize often extremely harsh habitats. A majority of metal-contaminated areas comprise serpentine (ultramafic, rich in Ni, Cr and Co) and calamine (rich in Zn, Pb and Cd) soils hosting characteristic serpentine and calamine flora, which is the focus of this review. Through microevolution, the plants inhabiting metalliferous habitats have developed a range of intriguing adaptive traits, demonstrated as characteristic morphological, behavioural and physiological alterations that enable them to avoid and/or tolerate metal toxicity. The mechanisms responsible for protection of the plant cell from metals entering the protoplast as well as for detoxification of toxic metal ions inside the cell by chelation, vacuolar sequestration and exclusion from the protoplast are reviewed. These mechanisms have resulted in highly specialized plants able to hyperaccumulate or avoid metals in the shoots. Potential applications of both kinds of metallophytes in rehabilitation and phytoremediation of metal-polluted sites are briefly discussed. Moreover, other beneficial applications of metal-rich plant biomass are mentioned, e.g., as a bio-ore for precious metal recovery (phytomining, agromining), a by-product for eco-catalyst production or a natural source of micronutrients that are essential for human diet and health (biofortification). The need of active protection of metalliferous sites and conservation of metallophyte biodiversity is pointed out.

1. INTRODUCTION

Heavy metals are present in all soils due to the mineral composition of the lithosphere. The background metal concentrations considered as natural in soils vary depending on the element content and geochemistry of the underlying local rocks and oscillate in a range from 0.1 to 150 mg/kg (e.g., for cobalt, Co; arsenic, As; cadmium, Cd; copper, Cu; nickel, Ni; chromium, Cr; lead, Pb; zinc, Zn; in increasing order) (Kabata-Pendias & Mukherjee, 2007) or from 200 to 2000 mg/kg (for manganese, Mn) (Reeves, 2006). However, due to natural processes or anthropogenic activity, these concentrations may become several thousand-fold higher, thereby significantly influencing the structure of ecosystems and the development of vegetation (Baker, Ernst, van der Ent, Malaisse, & Ginocchio, 2010).

Although some soils may be completely bare and devoid of vegetation due to their extreme phytotoxicity, spontaneous development of metal-tolerant flora has been observed in many cases (Tordoff, Baker, & Willis, 2000; Wójcik, Sugier, & Siebielec, 2014). Naturally occurring metalliferous soils are known to foster unique assemblages of vegetation; however, human-influenced metalliferous habitats may also support development of similar types of metal-tolerant plant populations (Baumbach & Hellwig, 2007; Szarek-Łukaszewska, 2009; Wierzbicka & Rostański, 2002). The latter habitats are sometimes species-poor. For instance, monocultures of a few grass species only, including *Agrostis stolonifera, Agrostis capillaris* or *Agropyron repens* are found in a close vicinity of some metal smelters in Poland, England, Belgium and the Netherlands (Brej, 1998; Dueck, Ernst, Faber, & Pasman, 1984; Vangronsveld, Colpaert, & Van Tichelen, 1996). In some newly heaped deposits, sparse vegetation patches dominated by grassland and ruderal vascular plants usually connected with shallow soils and dry and warm habitats are found (Szarek-Łukaszewska, 2009; Wójcik et al., 2014). However, the longer the process of colonization of metalliferous substrates, the bigger the number of plant species is usually established there. Many metalliferous sites are characterized not only by a remarkable diversity of higher plant species (several dozens or hundreds of species) but they also harbour a range of protected, rare or threatened taxa (Baker et al., 2010; Jędrzejczyk-Korycińska, 2009). Additionally, such habitats provide niches for bryophytes, lichens and a plethora of animal species, mainly insects, allowing them to persist and produce offspring.

Due to specific and harsh environmental conditions, plant communities developing on metalliferous sites represent a paradigmatic case of edaphic uniqueness that provides outstanding opportunities and model systems for ecological, physiological and evolutionary studies of adaptation and speciation. Moreover, they constitute an excellent reservoir of species for degraded land rehabilitation or phytoremediation. It is thus not surprising that metallicolous plants have fascinated many botanists for a long time and have been an object of studies across the Earth's continents.

2. NATURAL AND ANTHROPOGENIC METALLIFEROUS SITES

Metalliferous sites may be classified by the origin of metal contamination or by the mineral composition of the substrate. Baker et al. (2010) distinguished three types of metalliferous habitats based on the genesis of metal enrichment. 'Primary sites' have arisen naturally due to metal ore outcropping or weathering without any anthropogenic impact, creating the so-called metalliferous or orogenic soils. Such habitats are extremely rare today and are scattered as relics of the Late Glacial epoch, e.g., in Central Europe, the Pyrenees, the Alps and the Balkan region. Almost all of these habitats were destroyed or modified by mining activities dating back to the Bronze Age and intensified in the Middle Ages up to present time. In addition to soil-forming processes, other natural phenomena, such as volcanic eruptions and associated gaseous fumaroles or hot-spring activity, may significantly contribute to local heavy metal contamination (Kabata-Pendias & Mukherjee, 2007; Nagajyoti, Lee, & Sreekanth, 2010). The two other metalliferous habitats distinguished by Baker et al. (2010) are created by anthropogenic activity. The 'secondary sites', resulting from ore mining and processing, deposition of spoil and slug heaps, or other derived soil alterations, originally replaced the primary sites but afterwards they covered large areas located often at big distances from the primary sites. It is estimated that approximately 1% of the global land surface has been influenced by mining activities (Prach & Tolvanen, 2016). Finally, the 'tertiary metalliferous sites' originate by an input of metals usually in a nonmetal-enriched environment due to atmospheric deposition of metal pollutants or by alluvial deposition of metal-rich substrates in river floodplains or riverbanks. Important sources of metals to these habitats include all kinds of industrial emissions, e.g., from metal smelting, burning of fossil fuels, incineration of municipal wastes, production of batteries and other metal products, and ceramic and chemical industries. Also agricultural practices such as application of fertilizers and pesticides as well as sewage sludge and municipal waste disposal, transport and urbanization have a significant input into metal pollution (Ali, Khan, & Sajad, 2013; Kabata-Pendias & Mukherjee, 2007; Nagajyoti et al., 2010).

With regard to the mineral composition, both naturally and secondarily metal-enriched soils can be classified as serpentine, calamine, seleniferous or copper/cobalt-containing soils. Serpentine (ultramafic) soils are derived from ferromagnesian-rich mantle rocks and are characterized by anomalous Ni, Cr and Co concentrations, accompanied by high concentrations of iron (Fe) and magnesium (Mg) (more than 70%) and low concentrations of calcium (Ca), phosphorus (P) and potassium (K) (Kruckeberg, 2002; Martin & Coughtrey, 1982). Such soils have often a blue-green hue from serpentinite minerals (like lizardite, antigorite or chrysotile) and are the most abundantly represented metalliferous soils on all continents. Huge areas of ultramafic soils of tens of hundreds of square kilometres are found, e.g., in New Caledonia, Cuba or Turkey (Reeves, 2006). The vegetation inhabiting such soils is called 'serpentine/ultramafic flora' and the component species are called 'serpentinophytes'. Calamine soils, with their specific 'calamine flora', derived naturally from rock outcrops with galena (PbS), zinc blende (ZnS) and other calamine deposits (nonsulphide Zn minerals), are enriched mainly with Zn, Pb and Cd, sometimes with a significant admixture of sulphur and/or copper (Reeves, 2006; Tordoff et al., 2000). The richest and the most intensively studied patches of calamine vegetation are found in a few countries in Europe (Belgium, France, Germany and Poland) and in South Asia (China, Iran, Afghanistan and Pakistan) (Bothe, 2011; Ghaderian, Hemmat, Reeves, & Baker, 2007; Wierzbicka & Rostański, 2002). Soils rich in Cu and/or Co develop from argillites and dolomites containing sulphides of these metals; they occur mainly in Central Africa, in the DR Congo and Zambia (Reeves, 2006). Seleniferous soils (bearing more than 0.5 mg Se/kg) are derived from various Se-rich rock types, e.g., from Cretaceous shales, and are found in Australia, southern Asia, Ireland and the United States (Dhillon & Dhillon, 2001). Due to the abundance of natural and human-influenced ultramafic and calamine soils, serpentine and calamine vegetation has been the subject of the most intense studies and will be the focus of the present review.

Both the mineral composition and the level and origin of metal contamination determine the development and composition of vegetation in metalliferous habitats. However, not only high metal concentrations but also other unfavourable habitat conditions seriously hinder the development of vegetation. At primary sites, metal surplus is often accompanied by low nutrient availability (Baker et al., 2010). Anthropogenically influenced sites are usually characterized additionally by depleted organic matter concentrations, severe nutrient (especially phosphorous) deficiency, low water retention capacity, strong insolation and wind, acidification, poorly developed soil structure and sometimes by steep slopes (Dechamps, Noret, Mozek, Draye, & Meerts, 2008; Szarek-Łukaszewska, 2009; Wierzbicka & Rostański, 2002; Wójcik et al., 2014).

3. METALLOPHYTES

Plants that are adapted to live on metal-enriched soils and able to survive and reproduce there without suffering from toxicity are termed 'metallophytes' (Baker et al., 2010; Whiting et al., 2004). In general, metallophytes can be classified as obligate (absolute, strict or eu-) metallophytes, i.e., species found exclusively on metalliferous soils, or facultative (pseudo-) metallophytes, comprising species occurring both on metalliferous and nonmetalliferous, 'normal' soils (Baker et al., 2010; Kruckeberg, 2002; Whiting et al., 2004). There are also plants existing solely on nonmetalliferous soils; such species are designated as 'obligate nonmetallophytes'.

Obligate metallophytes are usually endemic to specific ecological niches provided by metalliferous soils with a well-defined metal composition. Due to their strict dependence upon occurrence of these metals in the substrate, such species have often been used as geobotanical indicators in mineral exploration (Brooks, 1998; Martin & Coughtrey, 1982). Classic examples of geobotanical indicators, successfully used for prospecting Cu deposits, are the 'copper flowers' of Katanga Province of Congo (formerly Shaba Province in Zaire) – *Becium homblei* and *Haumaniastrum katangense* (Brooks, Baker, & Malaisse, 1992). In general, the serpentine flora is globally the richest in endemic metallophytes with over 1000 higher plant species described to date (Kruckeberg, 2002; Pollard, Reeves, & Baker, 2014). Some genera are particularly rich in ultramafic obligate endemics, such as *Buxus* and *Leucocroton* (Euphorbiaceae) in the Caribbean region, *Phyllanthus* and *Psychotria* (Rubiaceae) in New Caledonia and the Caribbean region, *Homalium* (Salicaceae), *Xylosma* (Salicaceae), *Hybanthus* and *Rinorea* (Violaceae) in New Caledonia and tropical SE Asia, *Turnera* in South America, and *Alyssum*, *Odontarrhena* and *Noccaea* (Brassicaceae) and *Centaurea* (Asteraceae) in the Euro-Mediterranean and Irano-Turanian regions. *Clarkia* and *Epilobium* (Onagraceae), *Linum* (Linaceae) and *Streptanthus* (Brassicaceae) include several obligate endemics in western North America (Kruckeberg, 2002; Safford, Viers, & Harrison, 2005). In New Caledonia, i.e., the major hotspot of serpentine plant diversity in the world, endemic genera associated with ultramafic soils represent more than 37% of all the endemic genera, (39% of the indigenous flora), many of which are monotypic (Isnard, L'Huillier, Rigault, & Jaffré, 2016; Kruckeberg, 2002). A similar percentage is also found on Cuba, the other major centre of diversity of ultramafic flora at the global scale (32%–33% of all endemic genera; Borhidi, 1991; Kruckeberg, 2002). The Balkans are the major centre of the diversity of serpentine plants in Europe, with approximately 335 taxa of which 123 are obligate specialists of both neoendemic or paleoendemic type (Stevanović, Tan, & Iatrou, 2003), among which the two monotypic genera *Halacsya* and *Paramoltkia* of family Boraginaceae (Cecchi & Selvi, 2009; Coppi et al., 2014).

In contrast to serpentinophytes, species endemic to calamine soils are in fact a rare phenomenon. *Viola lutea* subsp. *calaminaria* or *Viola lutea* subsp. *westphalica* are examples of obligate metallophytes associated with this type of soil in Belgium, the Netherlands and Germany (Bizoux, Brevers, Meerts, Graitson, & Mahy, 2004; Bothe, 2011).

Facultative metallophytes are by far more numerous and widely taxonomically distributed in comparison with obligate metallophytes. They are usually recruited from adjacent nonmetalliferous environments and, though often not distinguishable at even the varietal rank, they develop metal-tolerant genotypes or ecotypes in response to the presence of specific metals in the soil. They occur, as species, in various metalliferous and nonmetalliferous soil types but when adapted to the metalliferous substrate their metallicolous ecotypes become dependent on the presence of metals. The most frequently studied examples of facultative metallophytes on calamine soils are *Noccaea* (formerly *Thlaspi*) *caerulescens*, *Noccaea goesingense*, *Arabidopsis* (formerly *Cardaminopsis*) *halleri*, *Silene vulgaris*, *Minuartia verna* s.l. and *Armeria maritima* (Bothe, 2011; Pollard et al., 2014 and references therein). Some authors distinguish a term 'pseudometallophytes' or 'bodenvag plants' ('soil wanderers') within facultative metallophytes to refer specifically to the group of plants indifferent to the parent rock material, moderately tolerant to heavy metals, but never dependent on their occurrence (Baker et al., 2010; Kruckeberg, 2002). Such plants belong to common monocot and dicot species with a wide geographic distribution range and often accompany metallophyte vegetation. Calamine pseudometallophytes are represented, for instance, by *Agrostis capillaris*, *Agrostis stolonifera*, *Deschampsia caespitosa*, *Festuca ovina* (*halleri*), *Festuca rubra*, *Holcus lanatus*, *Phragmites australis*, *Achillea millefolium*, *Campanula rotundifolia*, *Plantago lanceolata*, *Ranunculus acris*, *Rumex acetosella*, *Thymus pulegioides* and *Viola tricolor* (Baker et al., 2010; Bothe, 2011; Wójcik et al., 2014). Examples of bodenvags on serpentine soils include *Achillea lanulosa*, *Gilia capitata*, *Senecio pauperculus*, *Fragaria virginiana*, *Prunella vulgaris* or *Pinus contorta* (Kruckeberg, 2002). However, the term 'pseudometallophytes' is often indiscriminately used as synonymous with 'facultative metallophytes'. Even less popular and rather awkward is the term 'incidental pseudometallophytes' (Pollard et al., 2014) referring to 'associated nonmetal-tolerant' species (Baker et al., 2010) used for some weeds and ruderal plants appearing sporadically and showing reduced vigour in metalliferous habitats.

4. (MICRO)EVOLUTIONARY DYNAMICS OF METALLOPHYTES

Species constituting plant communities on metalliferous soils have adapted to the adverse environmental conditions through (micro)evolutionary processes and often represent genetically altered ecotypes. Normally, evolution of metal tolerance is achieved in plants over thousands or even millions of years (Whiting et al., 2004). However, under strong selection pressure imposed by extreme environmental conditions, it may evolve in a relatively short time of 40–150 years or even within less than a decade, i.e., in only a few generations (Ernst, 2006). The degree of specialization of the metal resistance trait is governed by the length of exposure to metals. At anthropogenically created metalliferous sites, e.g., spoil heaps or areas around metal smelters usually inhabited by pseudometallophytes, a limited level of metal tolerance is developed after a few years of metal emergence at the site. In plants facing the presence of mineral deposits at natural metal outcrops for centuries, more specialized mechanisms of tolerance are built up, resulting in their full dependence on metals (obligate metallophytes).

Many reports provide evidence that metal tolerance has evolved independently in different taxonomically unrelated plant affinities (within over 34 different plant families) and in geographically distant conspecific populations, being thus a clear-cut example of convergence (Bizoux, Daïnou, Raspé, Lutts, & Mahy, 2008). Moreover, colonization of metalliferous habitats can take place through multiple independent events from nonmetallicolous populations. For example, Patterson and Givnish (2003) showed that serpentine tolerance in *Calochortus* had evolved seven times independently across mountainous ranges in California and Oregon, while in *Streptanthus* this ability originated four to five times (Cacho, Burrell, Pepper, & Strauss, 2014). A further remarkable case of polyphyletic origin with multiple events of adaptation to ultramafic substrates is provided by the genus *Onosma* (Cecchi, Coppi, & Selvi, 2011) or *Odontarrhena*, until recently included in *Alyssum*, where serpentine and Ni hyperaccumulator endemics from southern Europe were scattered in three main evolutionary lineages without geographical cohesion (Cecchi et al., 2010; Cecchi, Colzi, Coppi, Gonnelli, & Selvi, 2013).

Rapid development of molecular techniques in the last few decades has provided excellent tools to study evolutionary changes in metalliferous habitats. Besides selection pressure, the mutation rate, reproductive success, gene flow from neighbouring populations and successive colonization events contribute to evolutionary processes (Babst-Kostecka, Parisod, Godé, Vollenweider, & Pauwels, 2014; Bickham, Sandhu, Hebert, Lounes, & Athwal, 2000). Both epigenetic processes and DNA changes may be the source of genetic divergence (Abratowska, Wąsowicz, Bednarek, Jelka, & Wierzbicka, 2012; Dresler et al., 2015 and references therein). Abratowska et al. (2012) pointed out that, more than the type of the substrate, geographic isolation had a crucial influence on the genetic structure of calamine populations of *Armeria maritima*; a similar effect may not be excluded in *Dianthus carthusianorum* (Wójcik, Dresler, Jawor, Kowalczyk, & Tukiendorf, 2013) and *Echium vulgare* (Dresler et al., 2015). Similarly, Nyberg Berglund and Westerbergh (2001) showed that serpentine populations of *Cerastium alpinum* in Scandinavia are genetically more similar to nonserpentine populations within the same geographic region than with distant serpentine populations. Gustafson, Romano, Latham, and Morton (2003) found close genetic relationships between serpentine and nonserpentine populations of *Cerastium velutinum* from North America and similar evidence was provided by Mengoni, Selvi, Cusimano, Galardi, and Gonnelli (2006) in a comparison of serpentine and limestone populations of *Onosma echioides* in Italy. On the other hand, in their study of several calamine and nonmetallicolous populations of *Biscutella laevigata*, Babst-Kostecka et al. (2014) showed that edaphic conditions and heavy metal exposition rather than geographic distance determined genetic divergence. Evidence for ecotypic differentiation on serpentine soil has been reported in numerous transplanting and cultivation experiments (Burgess, Szlavecz, Rajakaruna, & Swan, 2015; Chathuranga, Dharmasena, Rajakaruna, & Iqbal, 2015; O'Dell & Rajakaruna, 2011; Wright & Stanton, 2011), suggesting that the directional selection driven by the substrate can overcome the homogenization of populations by gene flow.

It is commonly accepted that strong selection pressure leads to reduced population size and genetic diversity. Accordingly, decreased genetic variability has been found in several metallicolous populations of *Deschampsia cespitosa* (Bush & Barrett, 1993), *Armeria maritima* (Vekemans & Lefèbvre, 1997), *Silene paradoxa* (Mengoni et al., 2001), *Sedum alfredii* (Deng et al., 2007) and *Biscutella laevigata* (Babst-Kostecka et al., 2014). However, in populations of *Viola calaminaria* and *Dianthus carthusianorum*, the level of genetic diversity was similar in metallicolous and nonmetallicolous populations (Bizoux et al., 2008; Wójcik et al., 2013). Moreover, other studies showed higher genetic diversity within the metallicolous compared to nonmetallicolous populations of *Viola tricolor* (Słomka, Sutkowska, et al., 2011) or

Echium vulgare (Dresler et al., 2015). The conflicting results obtained in the aforementioned studies might be related to the use of different molecular markers (PCR-based fingerprinting) (Dresler et al., 2015). However, they may also result from high heterogeneity of metalliferous sites in terms of different ecological factors and, therefore, genotypic variation and flexibility within a population may be crucial to survival.

5. PREADAPTIVE AND CONCOMITANT TRAITS

Besides evolving metal tolerance, plants growing on metal-enriched substrates were coselected for tolerance to other adverse site conditions. Therefore, many metallophytes are characterized by specific adaptations ensuring a high degree of resistance to drought and nutrient deficit or an ability to grow on a loose substrate (Baker et al., 2010). Preexisting adaptation, being most likely a plesiomorphic trait, seems a crucial prerequisite for the evolution of specialized ecotypes and/or taxa (Meyer et al., 2016). Which morphological and/or physiological features convey preadaptation is still not well understood, but drought resistance is undoubtedly a major trait due to the extremely xeric nature of most metalliferous environments (Brady, Kruckeberg, & Bradshaw, 2005; Proctor, 1999). Actually, drought resistance has been experimentally documented in several metallophytes, e.g., in serpentine *Mimulus* sp. (Hughes, Bachmann, Smirnoff, & Macnair, 2001) and *Cerastium* sp. (Nyberg Berglund, Dalgren, & Westerberg, 2004) as well as in calamine *Silene vulgaris* or *Biscutella laevigata* (Wierzbicka & Rostański, 2002). Recently, Armbruster (2014) commented that preexisting tolerance to open habitats, by themselves ecologically stressful, can explain the multiple origins of serpentine-soil endemism in Californian genera such as *Collinsia* and *Streptanthus*, and this trait is thought to have evolved before adaptation to serpentine soils (Cacho & Strauss, 2014). In fact, many metallophytes are sensitive to shade (Szarek-Łukaszewska, 2009). One of the most shade-sensitive species *Minuartia verna* disappears during succession of arborescent vegetation (Baker et al., 2010 and references therein). Therefore, as a means of active protection of unique calamine metallophytes inhabiting over 130 years old Zn–Pb waste spoil in Bolesław, Poland, as well as similar sites elsewhere, trees and shrubs spontaneously spreading there are being removed (Jędrzejczyk-Korycińska, Szarek-Łukaszewska, Kapusta, & Zagórna, 2014).

Due to their complex adaptive nature, metallophytes are usually restricted to metalliferous soils. This may be explained by the fact that they require relatively high levels of metals for optimal growth and reproduction or because they are poor competitors on 'normal' substrates in respect to herbivores and pathogens (Armbruster, 2014; Rascio & Navarri-Izzo, 2011). Their competitive weakness on nonmetalliferous soils may also result from a high energy cost incurred to development of adaptation mechanisms related to metal transport, storage and detoxification (Baker et al., 2010) or with their adaptation to drought or nutrient deficiency.

6. STRATEGIES OF METAL ACCUMULATION

There is clear evidence that different plant species take up different metals to varying degrees, even when grown on the same soil. Moreover, the propensity of a species to accumulate one metal does not necessarily imply that other metals will also be accumulated in superior concentrations and that the metals will be similarly distributed over roots and shoots. Plants growing on metalliferous soils may exclude, indicate or accumulate metallic elements in their shoots, being thus classified as 'excluders', 'indicators' or '(hyper)accumulators' (Baker, 1981).

A majority of metallophytes are 'excluders' since they are able to restrict either the entry of metals into the roots or their transport to the shoots over a wide range of soil metal concentrations (Wójcik et al., 2014). The shoot/soil metal concentration coefficient in such plants is always lower than unity. Metal excluders retain and detoxify most of the metals in the roots; however, when the control mechanisms break down, unrestricted metal transport to the shoots occurs, which is usually deleterious to the plant. The majority of serpentinophytes adapted to elevated Ni soil concentrations restrict element accumulation in their leaves and consequently show lower Ni concentrations in their tissues than nonserpentine conspecifics when grown in the presence of the metal (Nagy & Proctor, 1997; Pakdaman, Ghaderian, Ghasemi, & Asemaneh, 2013). This is also the case for numerous calamine metallophytes, including *Silene vulgaris* (Mohtadi, Ghaderian, & Schat, 2012) or *Dianthus carthusianorum* (Wójcik, Dresler, & Tukiendorf, 2015).

In 'indicators', shoot metal concentrations reflect soil metal concentrations (shoot/soil metal concentration coefficient is usually close to unity). Therefore, such plants have been proposed as biomonitors for the assessment of environment pollution (Mertens, Luyssaert, & Verheyen, 2005). In practice, the situation is more complicated since most metalliferous sites exhibit polymetallic contamination and, to date, no plant species are known to respond with a strong positive correlation between soil and shoot metal concentration to a wide range of elements.

Metal 'accumulators' are characterized by an efficient metal uptake and translocation to the shoots without showing any toxicity symptoms. The shoot/root and shoot/soil metal concentration ratios are higher than unity, although these criteria may sometimes be misleading in recognising such plants, as explained by Reeves (2006). Some accumulators, referred to as hyperaccumulators, have an intriguing strategy to concentrate elements in leaves at levels that are at least an order of magnitude higher than in other metallophytes from metalliferous soils or a few orders of magnitude higher than in ordinary plants from nonmetal-enriched soils (Pollard et al., 2014; Rascio & Navarri-Izzo, 2011). The threshold values of metal concentrations designating a plant as a hyperaccumulator have changed since the early reports defining hyperaccumulators and vary depending on the metal (Baker & Brooks, 1989; Reeves, 2006). The threshold criteria proposed currently are 100 mg for Cd, Se and thallium (Tl); 300 mg for Co, Cr and Cu; 1000 mg for Ni, Pb and As; 3000 mg for Zn and 10,000 mg for Mn determined in kg of dry leaf tissue of a plant growing in a natural habitat (Van der Ent, Baker, Reeves, Pollard, & Schat, 2013; Pollard et al., 2014). Based on these criteria, the innate ability to hyperaccumulate trace elements has been documented in approximately 500 species and is primarily attributed to metallophytes of serpentine soils accumulating Ni (over 90% of all hyperaccumulators identified so far; Pollard et al., 2014). Interestingly, the highest ever-recorded metal concentration present in a plant tissue (26%) is exactly for Ni, specifically in the latex of the *Sebertia acuminata* tree which is endemic to serpentine soils in New Caledonia (Sagner et al., 1998). Only approximately 50 species have been recognized as hyperaccumulators of other metals (Rascio & Navarri-Izzo, 2011). *Noccaea caerulescens* is probably the best-known hyperaccumulator of Zn, Cd or Ni, and together with *Arabidopsis halleri*, a Zn/Cd hyperaccumulator, used as a model species in studies on metal homeostasis and detoxification (Assunção, Schat, & Aarts, 2003; Meyer & Verbruggen, 2012). A majority of hyperaccumulators (approximately 85%–90%) are obligate metallophytes; however, some of them, including *Noccaea caerulescens* and *Arabidopsis halleri*, also occur on normal, nonmetalliferous soils.

Several hypotheses to explain metal hyperaccumulation have been proposed (Boyd, 2007). There is evidence that extreme metal concentrations in the leaves protect plants against herbivores (Zn) and pathogens (Ni). Accumulated metals may also be used as allelopathics against nearby competitors or serve as osmolytes in drought resistance. Finally, extraordinary metal concentrations may be a means of efficient metal tolerance and disposal mechanisms; alternatively, they may result from inadvertent metal uptake, which seems rather improbable in view of the large amount of metals accumulated without imposing toxicity symptoms.

7. SPECIFIC CHARACTERISTICS OF METALLOPHYTES

As already mentioned, metallophytes endemic to metalliferous sites may be considered as geobotanical indicators of mineral deposits. However, not only the presence of particular species but also floristic composition (assemblages of species) or some characteristic physiological or morphological changes (plant appearance) may be indicative of soil metal enrichment or potential mineral deposits. Typically, the vegetation of highly mineralized areas appears to be sparse and stunted, relative to surrounding areas, and is usually devoid of trees. Since heavy metals accumulate gradually in cells, it seems that short living herbs can cope better with the metal toxicity than long-lived plants (shrubs and trees) (Dickinson, Turner, & Lepp, 1991). However, shortening of the life cycle is apparently not an adaptive strategy on serpentine soils where perenniality seems an advantageous trait, though this lacks clear explanatory hypotheses (Selvi, 2007).

7.1 Morphological Variations

Many plant species exhibit abnormal appearance when growing on metal-enriched substrates. In comparison with individuals of the same species from unpolluted areas, they are characterized, among others, by reduction in stature (shrubbiness of arborescent species; dwarfism and plagiotropism in herbaceous species) although sometimes gigantisms may occur. Additionally, increased root system, xeromorphic foliage (with stenophylly, glaucescence), chlorosis or purpling of leaves, changes in the colour and pattern of flowers, and abnormally shaped fruits can be observed (Kruckeberg, 2002; Martin & Coughtrey, 1982; Wierzbicka & Rostański, 2002). In the case of serpentinophytes, such xerophytic and stunted appearance has been termed the 'serpentine syndrome' or 'serpentinomorphoses' (Menezes de Sequeira & Pinto da Silva, 1992). Reduced growth is a typical symptom on metalliferous substrates. Metallicolous plants are smaller, have more numerous but thinner and lower/shorter or sometimes trailing shoots, with less numerous, smaller and narrower leaves. This was observed, for instance, in calamine ecotypes of *Silene vulgaris*, *Dianthus carthusianorum* or *Arabidopsis halleri* in Poland (Bothe, 2011; Fiałkiewicz & Rostański, 2006; Wierzbicka & Rostański, 2002; Wójcik et al., 2013) and serpentine ecotypes of *Trifolium* sp. or *Jasione crispa* in Portugal (Menezes de Sequeira & Pinto da Silva, 1992). A significant increase in the plant size, manifested by, e.g., greater rosette diameter, larger stem and rosette leaves, or numerous lateral stem branches in metallicolous individuals of *Arabidopsis* (synonym *Cardaminopsis*) *arenosa* (Rostański, Myśliwiec, & Siwińska, 2005), is a rather rare phenomenon among metallophytes.

Plants occurring on metalliferous substrates exhibit various root morphology in terms of root length, depth and surface area. In general, no special root size or architecture distinguishes metallicolous from nonmetallicolous plants. Different root systems have been described in metallophytes, from short, thin, marginally branched primary roots through abundant fine root bundles to long and thick primary roots with or without well-developed side roots (Alford, Pilon-Smits, & Paschke, 2010; Bothe, 2011; Menezes de Sequeira & Pinto da Silva, 1992). However, what is striking at a comparison of individuals from noncontaminated and metalliferous soils, the latter often produce deeper roots and dense root hairs, probably as an adaptation to the dry substrate rather than the presence of metals (Menezes de Sequeira & Pinto da Silva, 1992; Wierzbicka & Rostański, 2002). Chlorosis of leaves is a very common symptom in plants from metal-enriched areas, although it may also result from accompanying deficiency of essential nutrients, e.g., nitrogen or iron in the soil. Similarly, the frequently observed anthocyanin accumulation manifested as intense purple/violet stem and leaf discolorations may at least partially be due to phosphorous deficiency on calamine and serpentine soils (Kabata-Pendias & Mukherjee, 2007; Menezes de Sequeira & Pinto da Silva, 1992). Variation in the colour of flowers has also been reported in plants exposed to heavy metals. For instance, *Epilobium angustifolium* has rose-purple coloured flowers which only exceptionally are white or pink; however, populations associated with uranium deposits in Canada or Alaska are distinguished by almost pure white sepals and petals (Martin & Coughtrey, 1982).

7.2 Behavioural Variations

Alterations in the rhythm of growth and flowering periods are other common symptoms found in plants growing in metalliferous habitats (Martin & Coughtrey, 1982; Wierzbicka & Rostański, 2002). To increase a species chance of survival, many metallophytes tend to shorten their vegetative phase and increase fertility by earlier and longer flowering, producing more flowers per plant, and delivering more seeds. A short life cycle and large reproductive effort at the expense of individual parental investment is typical for the 'r' strategy, where the ability to reproduce quickly and abundantly is crucial to survive under strong environmental pressure. Such behaviour was observed, among others, in calamine populations of *Silene vulgaris*, *Dianthus carthusianorum* (Wierzbicka & Rostański, 2002), *Arabidopsis arenosa* (Przedpelska-Wasowicz & Wasowicz, 2013) or *Minuartia verna* (Martin & Coughtrey, 1982).

Most plants avoid root proliferation in metal-enriched soil niches. However, some hyperaccumulating plant ecotypes, e.g., these of *Noccaea caerulescens* (Dechamps et al., 2008; Whiting, Leake, McGrath, & Baker, 2000) or *Sedum alfredii* (Li, Yang, He, & Yang, 2005) respond positively to trace elements in the soil by directing their root growth towards spots/patches with high metal content and increasing root biomass, length and hair root number while reaching them. This is, however, not a species-specific feature since other less metal-tolerant ecotypes of the same species do not show such foraging behaviour.

7.3 Physiological Variations

Intraspecific variation in the degree of metal tolerance has been found in most studied metallophytes (and specifically in facultative metallophytes) and the level of the tolerance has been consistently higher in ecotypes/populations growing at high metal concentrations in the substrate (Dresler, Bednarek, & Wójcik, 2014; Schat, Llugany, Vooijs, Hartley-Whitaker, & Bleeker, 2002; Wierzbicka & Rostański, 2002; Wójcik, Dresler, & Tukiendorf, 2015). To identify mechanisms responsible for enhanced metal tolerance, comparative methods are frequently employed in hydroponic experiments where natural metallicolous versus nonmetallicolous ecotypes or mutants/transgenic versus wild type plants are exposed to acute short-term single metal stress. Although such experiments are useful for fast screening of metal resistance and provide valuable information on the molecular and genetic background, they may not accurately reflect the response of plants grown in metalliferous habitats under chronic polymetallic stress (Wójcik, Dresler, & Tukiendorf, 2015). Therefore, a greater emphasis should be placed on mimicking more complex metal–plant–microorganism interactions in field trials or soil experiments to understand fully the complex nature of adaptation to metalliferous habitats.

8. MECHANISMS OF METAL UPTAKE, TRANSLOCATION AND TOLERANCE/DETOXIFICATION

Some metals, such as Zn, Cu, Fe, Mn, Mo, Co, Ni and vanadium (V) are required in minute quantities for plant growth and development; however, excessive amounts of these elements can become harmful to organisms. Other heavy metals or metalloids, such as Pb, Cd, mercury (Hg) or As do not have any beneficial effects on plants and they become toxic already at relatively low concentrations in the growth medium (Ernst, Kraus, Verkleij, & Wesenberg, 2008; Singh, Parihar, Singh, Singh, & Prasad, 2016). The level of the metal concentration at which toxicity symptoms appear (or adverse effects are recorded) strongly depends on the plant species and the mechanisms of metal tolerance developed.

The physiological mechanisms and genetic background of metal uptake, transport, sequestration and homeostasis in metal-accumulating and metal-nonaccumulating metallophytes have been extensively studied, yet are still not well understood/elucidated (Singh et al., 2016). Enhanced metal tolerance results from changes in the metal homeostasis network virtually shared by all higher plants. The differences between tolerant and nontolerant plants might possibly rely on different expression of genes found in plants, both hyperaccumulators and excluders, as already reported for Zn and Cd by Verbruggen, Hermans, and Schat (2009). Indeed, the mechanisms of metal tolerance in nonaccumulators seem to be under the control of relatively few genes (Schat & Vooijs, 1997), in contrast to hyperaccumulators, where the genetic background is much more complex (Pollard, Powell, Harper, & Smith, 2002; Van de Mortel et al., 2006).

Ernst et al. (2008) distinguished three degrees of plant metal tolerance: hypotolerance, basal tolerance and hypertolerance. In fact, metal-hypotolerant (hypersensitive) plants do not occur in nature; this term refers to mutants and transgenic plants with increased metal sensitivity in comparison with wild type plants. A great majority of plants, namely all pseudometallophytes, exhibit a basal level of metal tolerance (constitutive tolerance), which enables them to live on nonmetalliferous soils but also to cope with low, nontoxic concentrations of metals alternatively appearing in the environment (Wójcik, Dresler, & Tukiendorf, 2015). Basal metal tolerance is specific for various metals, plant species and ecotypes and is a starting point to evolve adaptive metal tolerance or hypertolerance in species or ecotypes growing and coping on highly metal-enriched soils (Ernst, 2006). A number of studies have demonstrated that plants develop enhanced tolerance specifically to the metal actually present in surplus in their natural environment; to all other metals, they have a basal level of tolerance. However, some cases of cotolerance have also been reported (Colzi, Rocchi, Rangoni, Del Bubba, & Gonnelli, 2014; Gonnelli, Galardi, & Gabbrielli, 2001; Remon, Bouchardon, & Faure, 2007; Schat & Vooijs, 1997).

8.1 Avoidance or Increased Metal Uptake

Only metal ions that are present in the form of soluble components in the soil solution or those that are easily solubilized, e.g., by root-mediated acidification of the rhizosphere, are available for plant uptake. Root exudates comprise the first line of defence against metal uptake. Many plants, predominantly monocots, excrete into the soil phytosiderophores, which chelate metal ions forming large complexes not taken up by the root cells. Exclusively, grass species use this strategy for acquisition of iron since they possess membrane transporters for Fe(III)—phytosiderophore complexes (Zhao & McGrath, 2009). Similarly, some carboxylic acids, including malate, citrate or oxalate as well as amino acids, e.g., histidine, excreted by roots bind metal ions in the soil solution thus preventing uptake thereof. This phenomenon is well documented and is crucial in the case of aluminium (Al) tolerance (Watanabe & Osaki, 2002); it has also been shown to reduce Ni uptake by *Thlaspi arvense* (Salt, Kato, Krämer, Smith, & Raskin, 2000). Alternatively, root exudates may selectively increase metal uptake by plants (Luo, Shen, & Li, 2008; Mench & Martin, 1991). However, in metal hyperaccumulators, such as Zn-accumulating *Nocceaa caerulescens* (Zhao, Hamon, & McLaughlin, 2001) or Ni-accumulating *Noccaea goesingense* (Salt et al., 2000), no increase in secretion of specific metal-chelating compounds has been reported.

Restricted metal uptake may also result from their binding to extracellular carbohydrates, such as callose or mucilage (Siedlecka et al., 2001). Furthermore, root cell walls play a key role in preventing toxic metal intake into the cytosol by immobilization thereof in lignopolysaccharide (cellulose, hemicelluloses and pectins) matrix. Depending on the kind of metal, its concentration, and the plant species and organ, the cell wall can retain a great majority of the metal taken up, also as a sink for metals excluded from the protoplast (Krzesłowska, 2011). It is also noteworthy that the cell wall is actively modified under metal exposure mainly by increasing the level of all compounds and particularly the low-methyl esterified fraction of pectins in order to increase its thickness and metal binding capacity (Krzesłowska, 2011). An interesting phenomenon has recently been found in *Noccaea caerulescens* with its additional peri-endodermal layer of cells (not found in its metal-sensitive relative *Thlaspi arvense*) characterized by irregularly thickened and lignin-impregnated cell walls (Zelko, Lux, & Czibula, 2008). Whether these cells act as an additional physical barrier for metal immobilization in extracellular metal movement across the root should still be elucidated.

The plasma membrane plays an important role in preventing or reducing metal uptake into the cell. The difference in the metal uptake and accumulation rate between excluders and hyperaccumulators results from different expression levels of the same plasma membrane transporters, although the existence of transporters with different kinetic properties cannot be excluded. Plants possess various families of transporters on the plasma membrane and tonoplast and these have been extensively reviewed recently (Guerinot, 2000; Haydon & Cobbett, 2007; Krämer, Talke, & Hanikenne, 2007; Manara, 2012; Singh et al., 2016). They belong mainly to the ZIP (Zinc-regulated transporter Iron-regulated transporter Protein), CDF (Cation Diffusion Facilitator also known as MTP, Metal Tolerance Protein), NRAMP (Natural Resistance Associated Macrophage Protein) and P-type ATPase (also known as HMAs, Heavy Metal ATPases) protein families of transporters (Krämer et al., 2007; Singh et al., 2016). Many metal transporters exhibit low ion selectivity and, in addition to essential elements, they may competitively transport nonessential and toxic ions with similar oxidation states and ionic radii (Alford et al., 2010). However, in the Ganges ecotype of *Noccaea caerulescens*, existence of a transporter in the root cell plasma membrane specific for Cd and independent on Zn transport has been suggested (Lombi, Zhao, McGrath, Young, & Sacchi, 2001). The entrance of Ni into the root cells seems to be mediated by a low-affinity uptake system with a high preference for Zn over Ni in *Noccaea caerulescens*

(Assunção et al., 2001), with genes *ZNT1*, *ZNT2* (zinc-regulated transporter belonging to ZIP family), *NRAMP3* and *NRAMP4* showing specific upregulation in this plant grown in the presence of Ni (Visioli, Gullì, & Marmiroli, 2014). Moreover, also IRT1 (iron-regulated transporter of ZIP family) and ZIP10 (ZRT, IRT-like protein) proteins have been indicated as candidates for Ni accumulation in the same plant (Halimaa et al., 2014). Enhanced Zn uptake by the Zn hyperaccumulators *Noccaea caerulescens* and *Arabidopsis halleri* is attributed to the constitutive overexpression of several genes for ZIP transporters (e.g., *ZNT1*, *ZNT2* and *ZIP6*, *ZIP9*, respectively), which in nonhyperaccumulating relatives are expressed only under Zn deficiency (Assunção et al., 2001). Recently it has been reported that rapid membrane voltage changes and diminution of the membrane depolarization may be the cause of higher metal tolerance of *Arabidopsis halleri* and *Arabidopsis arenosa* in comparison with *Arabidopsis thaliana* (Singh et al., 2016).

8.2 Root-to-Shoot Metal Translocation

There is a huge interspecies and intraspecies diversity in metal distribution within a plant. For instance, serpentine populations can show higher root-to-shoot translocation of Ni than nonserpentine ones (Chathuranga et al., 2015; Gonnelli et al., 2001). On the other hand, when grown in hydroponics, a metal-sensitive ecotype of *Dianthus carthusianorum* translocated more Zn and Pb to the shoots than a metal-tolerant one; however, the tendency was just opposite in plants cultivated on polluted soil (Wójcik, Dresler, & Tukiendorf, 2015). In fact, a majority of plants inhabiting metalliferous habitats preferentially retain metals in the roots, whereas efficient root-to-shoot translocation is typical for hyperaccumulators. This implies more efficient cytoplasmic detoxification and vacuolar sequestration of metal ions in the root cells of excluders, while in hyperaccumulators more metals are available for xylem loading due to either a low sequestration rate in the vacuole or efficient efflux out of the vacuole at the root level (Assunção et al., 2003; Rascio & Navarri-Izzo, 2011). Indeed, Lasat, Pence, Garvin, Abbs, and Kochian (2000) and Yang et al. (2006) found much lower amounts of Zn accumulated in the vacuole of root cells and faster Zn efflux from the vacuole in Zn hyperaccumulators *Noccaea caerulescens* and *Sedum alfredii* compared with their nonaccumulating relatives/accessions.

Several types of transporters have been recognized to be involved in metal transport to the xylem (Singh et al., 2016). It is now evident that the P_{1B}-type ATPases HMA4 and HMA2 play a crucial role in Zn and

Cd ion transport from the pericycle cells to the xylem vessels (Wong & Cobbett, 2009). In *Noccaea caerulescens*, the Ni–nicotianamine complex seems to be transported by proteins YSL3 (yellow stripe like) for the vascular loading and translocation (Gendre et al., 2007; Halimaa et al., 2014).

Metal ions loaded into the xylem are transported to the shoots in the form of complexes with various chelators, primarily with organic acids and amino acids or their derivatives (Haydon & Cobbett, 2007) but also as free hydrated cations (Krämer, Cotter-Howwells, Charnock, Baker, & Smith, 1996). Malate was supposed to be involved in Ni translocation in the hyperaccumulator *Stackhousia tryonii* (Bhatia, Walsh, & Baker, 2005), whereas citrate and other organic acids complexed almost all Ni in the latex of *Sebertia acuminata* (Callahan et al., 2008). Histidine and nicotianamine were reported as the major xylem chelators for Ni in the Ni-hyperaccumulating *Alyssum* (Krämer et al., 1996) and *Noccaea caerulescens* (Mari et al., 2006), respectively. The same compounds have been found to complex Zn, Cd, Pb, Mn, Cu, etc. for long distance transport (Álvarez-Fernández, Díaz-Benito, Abadía, López-Millán, & Abadia, 2014). It should also be noted that metals may be distributed within the plant via the phloem, being largely complexed with the same ligands. In the case of Cd, Cd–phytochelatin complexes have been found both in xylem and phloem saps (Álvarez-Fernández et al., 2014).

8.3 Metal Tolerance Mechanisms

Rapid removal of free metal ions from metabolically active cell compartments to prevent physiological damage is at the core of metal tolerance mechanisms. This can be achieved by metal chelation in the cytoplasm by a variety of ligands, metal sequestration in the vacuole or exclusion from the protoplast. It cannot be definitely declared that these processes are metal-specific; the same mechanisms are usually employed to cope with different metals, although their efficiency is different from metal to metal, species to species and even between the ecotypes/populations of a given species. They also depend on the metal concentration in the growth medium and time of exposure (acute or chronic stress) (Wójcik, Dresler, & Tukiendorf, 2015).

8.3.1 Metal Binding Ligands

The production of metal binding ligands to detoxify and facilitate transport and storage of metals is crucial to cope with the metal surplus. Up to date, no specific metabolites have been detected in metal-tolerant ecotypes. The most important ligands comprise thiol peptides (phytochelatins, PC;

metallothioneins, MT; glutathione, GSH), organic acids (malate, citrate), amino acids (histidine, proline) and phytate (Anjum et al., 2015; Manara, 2012; Rauser, 1999). Metallothioneins are cysteine-rich metal binding peptides that have hardly been isolated from plants although their genes are abundant and expressed (or even overexpressed in some metallophytes, including *Noccaea caerulescens*) upon metal exposure virtually in all plant tissues (Cobbett & Goldsbrough, 2002; Manara, 2012). At present, however, there is no direct evidence in favour of their role in metal tolerance or hyperaccumulation. Glutathione is the most abundant thiol compound in plants serving a broad range of biochemical functions, including its role in metal detoxification and tolerance (Jozefczak, Remans, Vangronsveld, & Cuypers, 2012; Wójcik & Tukiendorf, 2011). It is a rather poor metal chelator and its role in enhanced, adaptive metal tolerance but not basal metal tolerance is rather doubtful (Ernst et al., 2008; Schat et al., 2002; Wójcik, Dresler, & Tukiendorf, 2015); therefore, it will not be further addressed in this paper.

Phytochelatins are the best-characterized metal chelators (Cobbett, 2000; Cobbett & Goldsbrough, 2002; Pal & Rai, 2010). These thiol peptides are synthetized from GSH and have a general structure (γ-Glu-Cys)$_n$-Gly ($n = 2-11$, usually 2–5). They chelate metal ions, and this mechanism is the best recognized for Cd ions, through thiol groups of cysteine to form the so-called LMW (low molecular weight) PC–metal complexes in the cytoplasm for further transport thereof to the vacuole (Ernst et al., 2008; Pal & Rai, 2010; Rauser, 1999). This transport is mediated by ATP-dependent ABC transporters in the tonoplast. Once in the vacuole, LMW PC–metal complexes incorporate acid labile sulphur (S^{2-}), yielding HMW (high molecular weight) PC–metal complexes with higher stability under acidic vacuolar conditions (Wójcik, 2009). Phytochelatin synthesis is rapidly activated by different heavy metal ions, mainly Cd but also Cu, Zn, Pb, silver (Ag), gold (Au) and Hg, although stable in vivo complexes were only detected with Cd, Cu and Au (Cobbett & Goldsbrough, 2002). In addition to PCs found in all plants studied so far, PC homologues varying in the kind of the carboxy terminal amino acid (βAla, Ser, Glu, Gln instead of Gly) or its absence have been distinguished in some plant species (Rauser, 1999; Siedlecka et al., 2001); however, their role in metal detoxification and tolerance is not well recognized.

Similarly, contrasting evidence has been reported for the role of PCs in metal tolerance of metallophytes. Many studies, including these on mutants, transgenic plants or based on other modifications of the PC synthesis pathway, have confirmed that these peptides are the main factor for basal tolerance, especially Cd tolerance (Verbruggen et al., 2009; Wójcik & Tukiendorf, 2011). There are also a few examples of higher PC accumulation in metallicolous versus nonmetallicolous populations, e.g., in *Echium vulgare* or *Dittrichia viscosa* (Dresler et al., 2014; Fernández, Fernández-Fuego, Bertrand, & González, 2014). Nevertheless, in a great majority of studies, more intense PC synthesis was found in nonmetal-adapted populations in comparison with tolerant ones, both within calamine metal hyperaccumulators, e.g., *Sedum alfredii* or *Noccaea caerulescens*, and nonaccumulators, among others in *Silene vulgaris* or *Dianthus carthusianorum* (Schat et al., 2002; Sun, Ye, Wang, & Wong, 2007; Wójcik, Dresler, & Tukiendorf, 2015). Further evidence against the role of PC in adaptive metal tolerance is provided by the fact that hardly ever have these peptides been detected in plants growing in metal-enriched soils, either in controlled experiments or under natural conditions (Ernst et al., 2008 and references therein; Wójcik, Dresler, & Tukiendorf, 2015; Wójcik, Vangronsveld, & Tukiendorf, 2005). This is most probably related to the high metabolic costs of sulphur assimilation necessary for their synthesis (Ernst et al., 2008; Rascio & Navarri-Izzo, 2011).

Thiol peptides have a role in homeostasis and detoxification of metal ions with high affinity to thiol groups, such as Cd, Cu, Hg, Ag and As; detoxification of other metals, e.g., Ni or Zn, is more related to hydroxyl and amino groups (Rauser, 1999; Ernst et al., 2008). Organic acids, such as malate, citrate and oxalate, chelate metal ions in the cytosol but mainly participate in storage thereof in the vacuole (Haydon & Cobbett, 2007; Wójcik, 2009). Citrate has a higher capacity for metal ions than the other organic acids and exhibits strong affinity for Ni and Cd (Anjum et al., 2015), whereas malate was proposed as the main ligand for Zn (Mathys, 1977). However, the existence of both citrate and malate complexes with all these metal ions was confirmed *in planta* (Krämer, Pickering, Prince, Raskin, & Salt, 2000; Sarret et al., 2002). In the case of the increase in the root mitochondrial synthesis of malic and citric acids, Agrawal, Czymmek, Sparks, and Bais (2013) reported a fast and transient influx of Ni in mitochondria in *Alyssum murale* prior to sequestration thereof in vacuoles. In turn, in the case of metal storage in cell walls, Halimaa et al. (2014) showed that the Ni—malate complex was transported to the apoplast by malate anion channel ALMT12 (aluminium-activated malate transporter) of the root stele plasma membrane.

Although higher constitutive levels of organic acids have been reported in some hyperaccumulators and nonaccumulating metal-tolerant metallophytes (Assunção et al., 2003; Wójcik, Dresler, & Tukiendorf, 2015), a clear relationship between metal accumulation and the level of organic acids has not been established. A positive correlation between metal treatment and the concentrations of these compounds was found in *Arabidopsis halleri* (Zhao, Lombi, Breedon, & McGrath, 2000), *Noccaea caerulescens* (Wójcik, Skórzyńska-Polit, & Tukiendorf, 2006) and Ni-tolerant accessions of *Arabidopsis thaliana* (Agrawal, Lakshmanan, Kaushik, & Bais, 2012). On the other hand, increasing Zn, Pb or Cd concentrations did not affect malate or citrate concentrations in metallicolous and nonmetallicolous ecotypes of *Dianthus carthusianorum* (Wójcik, Dresler, Plak, & Tukiendorf, 2015; Wójcik, Dresler, & Tukiendorf, 2015; Wójcik & Tukiendorf, 2014), *Silene vulgaris* (Harmens, Koevoets, Verkleij, & Ernst, 1994) or *Armeria maritima* (Olko, Abratowska, Żyłkowska, Wierzbicka, & Tukiendorf, 2008). Similarly, Ni did not affect malate concentrations in Ni-susceptible accessions of *Arabidopsis thaliana* (Agrawal et al., 2012). Probably the constitutively high tissue content of organic acids, which are involved, besides metal detoxification, in a plethora of physiological functions, may be sufficient for effective chelation of the increasing concentrations of metals. Nonetheless, their role in hypertolerance and hyperaccumulation is questionable (Callahan, Baker, Kolev, & Wedd, 2006; Verbruggen et al., 2009).

Amino acids (histidine, proline) and derivatives (nicotianamine) have also been reported to chelate metal ions in the cytoplasm before loading thereof to the xylem. Histidine is considered the most important ligand of this group of compounds, especially for Ni detoxification, translocation, tolerance and hyperaccumulation (Callahan et al., 2006). Complexation of Ni to histidine has been suggested to play a role in preventing metal entrapment in root cell vacuoles and in keeping the metal in the cytosol in a detoxified form available to be released into the xylem in *Alyssum* species and *Noccaea caerulescens* (Krämer et al., 1996; Kerkeb & Krämer, 2003; Richau et al., 2009). For generation of such a large pool of histidine, the gene for the first step of its biosynthesis encoding an ATP-phosphoribosyl transferase was reported to be constitutively overexpressed in the roots of the Ni hyperaccumulator *Alyssum lesbiacum* (Ingle, Mugford, Rees, Campbell, & Smith, 2005). In contrast, increased histidine concentrations in response to Ni do not appear to be a feature of the Ni hyperaccumulators *Noccaea goesingense* (Persans, Yan, Patnoe, Krämer, & Salt, 1999) and *Noccaea caerulescens* (Richau et al., 2009). Another ligand involved in the Ni and Zn chelation in the cytosol in their path towards the xylem is nicotianamine, showing metal-triggered accumulation in roots in *Noccaea caerulescens* and *Arabidopsis halleri* (Mari et al., 2006; Weber, Harada, Vess, Roepenack-Lahaye, & Clemens, 2004). Nicotianamine also chelates Zn, Cu and Fe for storage thereof in the vacuole (Manara, 2012).

8.3.2 Metal Exclusion from the Protoplast and Plant Organism

Compartmentalization of metals in the vacuole is a key mechanism to remove excess metal ions from the cytosol and seems to be crucial for metal tolerance. Indeed, the vacuole, which in plants may account for up to 90% of the cell volume, is generally considered the main metal storage compartment in plant cells (Wójcik, 2009). A number of tonoplast transporters are responsible for metal transport into the vacuole, both in the form of free metal ions and in complexes with ligands (Manara, 2012; Wójcik, 2009). The enhanced ability of Ni vacuolar storage in shoots seems to be mediated by high expression of *MTP1* (metal transport protein of the CDF family) in *Noccaea goesingense* (Persans, Nieman, & Salt, 2001). This protein appears to be involved also in the Ni efflux from the cytoplasm to the cell wall (Kim et al., 2004). Another tonoplast located metal transporter, i.e., a ferroportin transporter PgIREG1, showed higher expression in *Psychotria gabriellae* and was suggested as a candidate for Ni tolerance and hyperaccumulation by Merlot et al. (2014). Zinc and Cd sequestration into the shoot vacuole is probably mediated by transporters belonging to tonoplast P_{1B}-ATPase and CAX (cation exchanger) families, respectively, whose genes are overexpressed in the Zn/Cd hyperaccumulators *Noccaea caerulescens* and *Arabidopsis halleri*, probably contributing to their hypertolerance (Rascio & Navarri-Izzo, 2011; Wójcik, 2009 and references therein).

Active metal efflux from the cell is well documented in bacterial and animal cells but only a few plasma membrane efflux transporters have been recognized in plants so far (Manara, 2012). Based on sequence similarity to microbial and animal proteins, the most likely candidates for active metal ion pumping out of the cytoplasm are the P_{1B}-ATPases and the CDF families of transporters. Heavy metal ATPases (HMAs): HMA2 and HMA4 belonging to the first family export selectively Zn and Cd (Krämer et al., 2007). The AtPDR8 protein of the ABC transporter family localized in the plasma membrane of *Arabidopsis thaliana* root hairs and epidermal cells probably acts as an efflux pump for Cd and Pb, thus conferring tolerance to these metals (Kim, Bovet, Maeshima, Martinoia, & Lee, 2007).

When excluded from the protoplast, heavy metals may be removed from the organism by excretion or deposited in the cell wall, inside idioblasts, trichomes, or in some plant organs, the latter being often shed when overburdened with metals (Siedlecka et al., 2001). Active excretion of metals from leaf cells can occur via multicellular salt glands, e.g., in *Armeria maritima* ssp. *halleri* or via hydathodes, e.g., in *Minuartia verna*, when grown on calamine soils (Olko et al., 2008). Nickel and Cd have also been reported to be excluded from stomatal cells (Broadhurst, Chaney, Angle, Erbe, & Maugel, 2004; Cosio, De Santis, Frey, Diallo, & Keller, 2005). Many plants allocate metals in particular epidermal cells, including trichomes, and even in the cuticle (Asemaneh, Ghaderian, Crawford, Marshall, & Baker, 2006; Krämer, Grime, Smith, Hawes, & Baker, 1997; Mesjasz-Przybylowicz, Balkwill, Przybylowicz, & Annegarn, 1994; Słomka, Kuta, et al., 2011). In some halophytes, such leaf hairs are discarded when overloaded (Bothe, 2011); whether such behaviour is exhibited by metallophytes should still be elucidated. Instead, the phenomenon of seasonal intensive metal deposition in older leaves and other plant parts followed by shedding these organs is well documented. It occurs, among others, in Zn/Cd excluding *Armeria maritima*, *Minuartia verna*, *Anthyllis vulneraria* or *Biscutella laevigata*, in Cu excluding *Becium homblei* and in Ni excluding *Indigofera setiflora* (Baker & Brooks, 1989; Bothe, 2011; Siedlecka et al., 2001).

If the defence mechanisms discussed above fail, the plant will suffer metal toxicity expressed as growth inhibition, chlorosis and necrosis, senescence and ultimately death. These phytotoxicity symptoms may result from alterations in numerous physiological processes, such as photosynthesis, respiration, mineral nutrition, water relations, hormone balance, cell division and elongation (Siedlecka et al., 2001; Singh et al., 2016). Another common consequence of metal toxicity is the enhanced production of reactive oxygen species (ROS) leading to oxidative stress resulting in lipid peroxidation, membrane dismantling, ion leakage, deterioration of macromolecules and DNA-strand cleavage (Jozefczak et al., 2012). Therefore, a further defence mechanism adopted by metal-exposed plants relies on activation of the cell antioxidant system, including induction of enzymatic and nonenzymatic free radical scavengers (Jozefczak et al., 2012; Manara, 2012). Among the few studies performed on metallicolous versus nonmetallicolous ecotypes, Gonnelli et al. (2001) reported that a serpentine population of *Silene paradoxa* seemed able to counteract Ni-induced oxidative stress by limiting the metal-inhibiting effect on peroxisomal H_2O_2 scavenging enzymes. This problem has been comprehensively analysed in chapter "Molecular and Cellular Aspects of Contaminant Toxicity in Plants: The Importance of Sulphur and Associated Signalling Pathways" by Hendrix et al. (2017).

A better understanding of the processes involved in metal tolerance and adaptations to metalliferous habitats may be important as a basis for more targeted strategies for developing plants suitable for phytoremediation.

9. POTENTIAL OF METALLOPHYTES FOR REHABILITATION AND PHYTOREMEDIATION

Metalliferous areas, and especially the recent ones of anthropogenic origin and usually devoid of vegetation, are a source of secondary metal pollution via water and wind erosion, posing a threat to adjacent ecosystems and human health (Tordoff et al., 2000; Wójcik et al., 2014). Some plant species occur spontaneously on degraded areas due to natural succession; however, this process is very slow and may last decades, often merely resulting in sparse patches of vegetation. Establishing the plant cover is one of the best ways to prevent metal migration and to increase the aesthetical value of degraded areas. To that end, it is important to use native species growing in the immediate neighbourhood or not far from the area to be reclaimed, because these plants are often more efficient in terms of survival, growth and reproduction under environmental stress than plants introduced from other environments (Tordoff et al., 2000). Indigenous metallophytes may be used as a source of seedlings or seeds as well as fragments of natural turf with its soil, plants, microorganisms, mycorrhizas, seed bank etc. to increase the persistence of the plant cover and its biodiversity (Wójcik et al., 2014).

Various environmental approaches using metal-tolerant plants are recommended, such as restoration, rehabilitation or phytoremediation of metal-contaminated sites (Prach & Tolvanen, 2016; Whiting et al., 2004). The term 'restoration' refers to a process where the recovery of a degraded or destroyed ecosystem to its preexisting conditions is designed and prosecuted, which implies that only metallophytes originally present at the site can/should be used (Whiting et al., 2004). Since the metallophyte populations in most degraded mine sites have not been catalogued, such sites cannot be ecologically restored. On the other hand, 'rehabilitation' (or 'reclamation') means establishing a vegetative cover that is not representative of the preexisting conditions and therefore any plant species able to tolerate the harsh site conditions can be exploited. Two approaches can be distinguished based on the level of human intervention to the environment. Assisted reclamation aims at speeding up the natural regeneration of ecosystems by improving the site conditions and/or introduction of target species, whereas technical reclamation often involves drastic restructuring of the habitat, including importing soil followed by sowing or planting some plants (Prach & Tolvanen, 2016). Vegetation can be introduced to a contaminated area for either phytostabilization or phytoextraction. Phytostabilization relies on stabilizing contaminants in the soil by plant imposed accumulation and precipitation of toxic elements in the rhizosphere or adsorption thereof on root surface. In contrast, phytoextraction assumes removal of metals and metalloids from soils through plants. Both approaches represent phytoremediation technologies and have been a subject of many reviews recently (Ali et al., 2013; Cundy et al., 2016; Vangronsveld et al., 2009). They will also be extensively addressed in another chapter of this volume (chapter: Phytoremediation and Phytomining: Status and Promise by Chaney and Baklanov (2017)); therefore, they will be only briefly characterized here. The mode of management of metalliferous areas depends upon the level and area of pollution and the final purpose of the site utilization/exploitation.

9.1 Phytostabilization

There are plenty of species of metal-tolerant plants that may be considered for phytostabilization. These plants should be characterized by restricted metal uptake and/or transfer to the shoots (excluders) to avoid entry of metals into the food chain, dense rooting system, rapid growth to ensure adequate ground coverage, ease of establishment and maintenance under field conditions, and tolerance to high concentrations of metals and other adverse soil conditions (Vangronsveld et al., 2009; Whiting et al., 2004).

Metal-tolerant plant populations/ecotypes (mainly grasses but also some trees) have been commonly used to stabilize and revegetate waste spoils and other severely contaminated sites (Cundy et al., 2016; Johnson, Cooke, & Stevenson, 1994; Stuczynski et al., 2007; Szarek-Łukaszewska, 2009; Tordoff et al., 2000). Phytostabilization by grasses has been performed with success, among others, in Pb/Zn/Cd-contaminated sites in Poland (Daniels, Stuczynski, Chaney, Pantuck, & Pistelok, 1998; Stuczynski, Siebielec, Daniels, McCarty, & Chaney, 2007), Belgium (Vangronsveld et al., 1996), France (Cundy et al., 2016), United States (Li, Chaney, Siebielec, & Kershner, 2000) and China (Shu, Xia, Zhang, Lan, & Wong, 2002). The most widely used grass species include *Agrostis capillaris*, *Festuca rubra*, *Festuca arundinacea*, *Poa pratensis*, *Vetiveria zizanioides* and *Deschampsia cespitosa*. Zn/Pb mining waste deposits near Bukowno-Bolesław in Poland as well as other metalliferous remnants of mining and smelting activities elsewhere have been and are still being afforested by planting trees and shrubs, which yields stable woodlands characteristic for the landscape of these regions (Szarek-Łukaszewska, 2009). Trees that are most often introduced into reclaimed areas include pine (*Pinus sylvestris*), birch (*Betula pendula*), larch (*Larix decidua*) and black locust (*Robinia pseudoacaccia*). To improve the vegetation capacity and increase the persistence of a plant cover, it is recommended to combine the introduced plants with some metal-tolerant ecotypes of legume species, e.g., *Trifolium repens* or *Lotus corniculatus*, which can naturally maintain a nitrogen supply through symbiotic fixation. Successful attempts of such reclamation approaches have been reported, e.g., in the Parc Pb/Zn mine in North Wales (Whiting et al., 2004) or in abandoned metalliferous fluorspar dams in Derbyshire, UK (Tordoff et al., 2000). There are several examples of using metallicolous populations of herbaceous plants for revegetation of metal-contaminated sites. For instance, introduction of *Arabidopsis halleri* and *Armeria maritima* subsp. *halleri* originating from Central European calaminarian grassland into the Auby smelter area in France produced a positive effect (Bert, Macnair, De Laguerie, Saumitou-Laprade, & Petit, 2000; Dahmani-Müller, van Oort, Gélic, & Balabane, 2000). Seeds of native metallophytes were collected and successfully used for revegetation of Australian mineral wastes (Whiting et al., 2004). Similarly, seeds of a calamine population of *Biscutella laevigata* were used with a big success on another Zn—Pb spoil heap, in Upper Silesia, Poland creating a numerous and still increasing population just after four years (Rostański, 2014).

9.2 Phytoextraction

In contrast to organic pollutants, heavy metals do not undergo biodegradation and therefore persist in soils for thousands of years. For instance, Pb, one of the most toxic metals, has soil retention time of 150—5000 years and its high concentration may be maintained in the soil for 150 years (Singh et al., 2016). The only effective way of removal of metallic elements is extraction thereof using conventional physical or chemical methods (e.g., in situ vitrification, soil incineration, washing or flushing, but usually soil excavation and replacement — the so-called 'dig and dump'), which are usually very expensive and destructive to the soil ecosystem, or using phytoextraction (Ali et al., 2013; Vangronsveld et al., 2009). Phytoextraction represents a cost-effective, efficient, environment-friendly and eco-friendly alternative to the conventional methods but is also the most challenging task among all phytoremediation approaches/technologies.

The idea of phytoextraction is to concentrate high amounts of metals in aboveground plant parts that are subsequently harvested and incinerated or processed in another way (Sas-Nowosielska et al., 2004). Therefore, ideally, it should combine hyperaccumulation with high biomass production; however, no such plant has been recognized so far. Hyperaccumulators are usually slow-growing low-biomass plants, often with restricted and local occurrence, whilst fast-growing and high-biomass plants are predominantly poor metal accumulators with a low to moderate metal tolerance level. Nevertheless, several successful applications of both kinds of plants for phytoextraction on a field scale have been noted. For instance, the Ni hyperaccumulators *Alyssum bertolonii* and *Strephanthus polygaloides* have been proposed for Ni phytoextraction (Li et al., 2003). Small-scale field trials showed a potential of some ecotype of the Zn/Cd hyperaccumulator *Noccaea caerulescens* for phytoextraction of Cd from polluted soils (Maxted et al., 2007; McGrath et al., 2006). On the other hand, high biomass plants such as willow (*Salix* sp.) (Janssen et al., 2015), poplar (*Populus* sp.) (Ruttens et al., 2011), maize (*Zea mays*) (Van Slycken et al., 2013), sunflower (*Helianthus annuus*) (Nehnevajova, Herzig, Bourigault, Bangerter, & Schwitzguébel, 2009), *Brassica juncea* (Tlustoš et al., 2006) or *Jatropha curcas* (Papazoglou, 2014) appeared to be good candidates for decontamination of moderately polluted soils.

In phytoremediation treatment, only rarely can metal-tolerant vegetation be established directly on the metalliferous substrate (like in Rostański, 2014). Such soils are often too phytotoxic and usually some physical or chemical modifications (mostly using some amendments) are applied prior to introduction of plants to improve soil fertility, render the soil less hostile and, depending on the aim, to decrease or increase metal bioavailability and accumulation for successful site management (Vangronsveld et al., 2009).

Phytoextraction efficiency of plants can also be substantially improved through natural cultivation for some desirable traits, mutagenesis, hybridization or genetic engineering (Ali et al., 2013; Singh et al., 2016; Vangronsveld et al., 2009). However, due to the complex genetic background of mechanisms involved in metal uptake, translocation and sequestration as well as poor legal and societal acceptance of application of genetically modified plants, this strategy is still far from introduction in practice.

Successful phytoextraction requires decreasing the pollutant level to that complying with environmental regulations. From the practical point of view, phytoextraction is only feasible on low to moderately contaminated soils, like these at tertiary human-influenced sites. In the case of many severely metal-enriched soils influenced by industrial or mining activities, it would be very difficult or even impossible to carry out phytoextraction within a reasonable period; then, stabilization using metal-tolerant plants is a more logical alternative to initiate a healthy ecosystem (Whiting et al., 2004; Vangronsveld et al., 2009; Zhao & McGrath, 2009).

10. OTHER POTENTIAL BENEFICIAL APPLICATIONS OF METALLOPHYTES

In addition to the rehabilitation or phytoremediation of metal-contaminated areas, several other benefits from application of metallophytes can be mentioned. Metal-rich plant biomass from phytoextraction may be combusted to get energy and the remaining ash may be in some cases used as 'bio-ore' for recovery of precious metals (Ni; Tl; Au; palladium, Pd and platinum, Pt) in a process called phytomining or agromining (Van der Ent et al., 2015; chapter: Phytoremediation and Phytomining: Status and Promise by Chaney and Baklanov (2017)). The difference between these two applications is associated with the type of soil used for obtaining plant biomass. Phytomining relates to degraded or mined lands being under the reclamation process, whereas agromining assumes using low-productivity agricultural metal-contaminated soils to generate economic profits to farmers. Up to date, phytomining has been commercially applied for Ni using *Alyssum murale* and *Alyssum corsicum* and it appeared to be more eco-effective and cost-effective than the conventional metal extraction methods with reference to many ores (Chaney et al., 2007).

Another novel and interesting way of processing with metal-rich plant biomass is to use it for eco-catalyst production (Escande, Olszewski, & Grison, 2015). A polymetallic solid obtained after thermal and chemical treatment of plant material can serve as a catalyst or reagent in synthesis of valuable products for the fine chemical, e.g., fragrances and cosmetics, industry.

Metal hyperaccumulators may also have an alternative use — in biofortification (Clemens, 2016; Guerinot & Salt, 2001; Zhao & McGrath, 2009). Biofortification aims at production of crops enriched with trace elements that are essential for human diet in edible plant parts through cultivation, biotechnological approaches or the use of appropriate fertilizers.

Zinc, Se, Fe and iodine (I) are the major target micronutrients since their deficiencies affect more than half of the world population.

11. CONCLUDING REMARKS

Metallophytes are widespread throughout the world, forming an integral part of the biodiversity on metalliferous soils. Due to their amazing ecology and physiology as well as multiplicity of potential applications, there is a need to continue screening metallicolous vegetation for its ability to tolerate metals, both in terms of hyperaccumulation and exclusion. Severely contaminated sites bearing well-developed metal-resistant flora are peculiar ecological islands that should be protected as gene reservoirs. These include not only the natural outcrops or tracts of metalliferous areas (e.g., serpentine soils), being often a cradle for endemic vegetation, but also anthropogenically created metalliferous waste deposits (e.g., from mining and smelting of Zn and Pb ores) becoming niches for specific and unique vegetation. The latter artificial habitats, although often considered dull scars and eyesores of the landscapes, may in fact represent very interesting biocenoses exceeding the preexisting habitats with their biodiversity. Therefore, integrated efforts should be focused on protecting the native environments of metallophytes, both of natural and human origin. Several strategies have been proposed for preserving the biodiversity of metallophytes: active protection of metalliferous habitats in situ, establishing collections of metallophytes in botanical gardens, arboreta, and seed or germplasm banks, etc. (Baker et al., 2010; Whiting et al., 2004).

Considerable attention has been given worldwide to the possibility of using metallophytes, either alone or in combination with microorganisms for rehabilitation or phytoremediation of contaminated sites. Moreover, exploitation of metallophytes as bioores coupled with bioenergy production or, alternatively, as a natural source of crucial micronutrients for improving human diet and public health is an attractive idea. However, scientific information on the mechanisms of adaptation to adverse environmental conditions is a basis for planning effective reclamation of contaminated sites using metallophytes or other ways of beneficial or even profitable utilization of metallophytes. Thus, it is of pivotal importance to understand the interspecies and intraspecies variation in mineral uptake, distribution, metabolism and tolerance, and the molecular mechanisms responsible for the process. Many details of these processes have already been presented; yet, there is still a lot to be elucidated and recognized, especially under natural site-specific field conditions.

ACKNOWLEDGEMENTS

M. Wójcik and S. Dresler would like to dedicate this article to Prof. Anna Tukiendorf, their long-time supervisor, who inspired them to work on metallophytes, as well as their mentor and friend. J. Vangronsveld acknowledges support from the Hasselt University Methusalem project 08M03VGRJ.

REFERENCES

Abratowska, A., Wąsowicz, P., Bednarek, P. T., Jelka, J., & Wierzbicka, M. (2012). Morphological and genetic distinctiveness of metallicolous and non-metallicolous populations of *Armeria maritima* s.l. (Plumbaginaceae) in Poland. *Plant Biology, 14*, 586–595.

Agrawal, B., Czymmek, K. J., Sparks, D. L., & Bais, H. P. (2013). Transient influx of nickel in root mitochondria modulates organic acid and reactive oxygen species production in nickel hyperaccumulator *Alyssum murale*. *The Journal of Biological Chemistry, 288*, 7351–7362.

Agrawal, B., Lakshmanan, V., Kaushik, S., & Bais, H. P. (2012). Natural variation among *Arabidopsis* accessions reveals malic acid as a key mediator of nickel (Ni) tolerance. *Planta, 236*, 477–489.

Alford, É. R., Pilon-Smits, E. A. H., & Paschke, M. W. (2010). Metallophytes – A view from the rhizosphere. *Plant and Soil, 337*, 33–50.

Ali, H., Khan, E., & Sajad, M. A. (2013). Phytoremediation of heavy metals – Concepts and applications. *Chemosphere, 91*, 869–881.

Álvarez-Fernández, A., Díaz-Benito, P., Abadía, A., López-Millán, A.-F., & Abadia, J. (2014). Metal species involved in long distance metal transport in plants. *Frontiers in Plant Science, 5*, 105.

Anjum, N. A., Hasanuzzaman, M., Hossain, M. A., Thangavel, P., Roychoudhury, A., Gill, S. S., … Ahmad, I. (2015). Jacks of metal/metalloid chelation trade in plants – An overview. *Frontiers in Plant Sciences, 6*, 192.

Armbruster, W. S. (2014). Multiple origins of serpentine-soil endemism explained by preexisting tolerance to open habitats. *Proceedings of the National Academy of Sciences of the United States of America, 111*, 14968–14969.

Asemaneh, T., Ghaderian, S. M., Crawford, S. A., Marshall, A. T., & Baker, A. J. M. (2006). Cellular and subcellular compartmentation of Ni in the Eurasian serpentine plants *Alyssum bracteatum, Alyssum murale* (Brassicaceae) and *Cleome heratensis* (Capparaceae). *Planta, 225*, 193–290.

Assunção, A. G. L., Da Costa Martins, P., De Folter, S., Vooijs, R., Schat, H., & Aarts, M. G. M. (2001). Elevated expression of metal transporter genes in three accessions of the metal hyperaccumulator *Thlaspi caerulescens*. *Plant, Cell and Environment, 24*, 217–226.

Assunção, A. G. L., Schat, H., & Aarts, M. G. M. (2003). *Thlaspi caerulescens*, an attractive model species to study heavy metal hyperaccumulation in plants. *New Phytologist, 159*, 351–360.

Babst-Kostecka, A. A., Parisod, C., Godé, C., Vollenweider, P., & Pauwels, M. (2014). Patterns of genetic divergence among populations of the pseudometallophyte *Biscutella laevigata* from southern Poland. *Plant and Soil, 383*, 245–256.

Baker, A. J. M. (1981). Accumulators and excluders – Strategies in the response of plants to heavy metals. *Journal of Plant Nutrition, 3*, 643–654.

Baker, A. J. M., & Brooks, R. R. (1989). Terrestrial higher plants which hyperaccumulate metallic elements: A review of their distribution, ecology and phytochemistry. *Biorecovery, 1*, 81–126.

Baker, A. J. M., Ernst, W. H. O., van der Ent, A., Malaisse, F., & Ginocchio, R. (2010). Metallophytes: The unique biological resource, its ecology and conservational status in Europe, central Africa and Latin America. In L. C. Batty, & K. H. Hallberg (Eds.), *Ecology of industrial pollution* (pp. 7–40). Cambridge: Cambridge University Press.

Baumbach, H., & Hellwig, F. H. (2007). Genetic differentiation of metallicolous and non-metallicolous *Armeria maritima* (Mill.) Willd. taxa (Plumbaginaceae) in Central Europe. *Plant Systematics and Evolution, 269*, 245–258.

Bert, V., Macnair, M. R., De Laguerie, F., Saumitou-Laprade, P., & Petit, D. (2000). Zinc tolerance and accumulation in metallicolous and nonmetallicolous populations of *Arabidopsis halleri* (Brassicaceae). *New Phytologist, 146*, 225–233.

Bhatia, N. P., Walsh, K. B., & Baker, A. J. M. (2005). Detection and quantification of ligands involved in nickel detoxification in a herbaceous Ni hyperaccumulator *Stackhousia tryonii* Bailey. *Journal of Experimental Botany, 56*, 1343–1349.

Bickham, J. W., Sandhu, S., Hebert, P. D. N., Lounes, C., & Athwal, R. (2000). Effects of chemical contaminants on genetic diversity in natural populations: Implications for biomonitoring ecotoxicology. *Mutation Research – Reviews in Mutation Research, 463*, 33–51.

Bizoux, J. P., Brevers, F., Meerts, P., Graitson, E., & Mahy, G. (2004). Ecology and conservation of Belgian populations of *Viola calaminaria*, a metallophyte with a restricted geographic distribution. *Belgian Journal of Botany, 137*, 91–104.

Bizoux, J.-P., Daïnou, K., Raspé, O., Lutts, S., & Mahy, G. (2008). Fitness and genetic variation of *Viola calaminaria*, an endemic metallophyte: Implication of population structure and history. *Plant Biology, 10*, 684–693.

Borhidi, A. (1991). *Phytogeography and vegetation ecology of Cuba*. Budapest: Akademiai Kiado.

Bothe, H. (2011). Plants in heavy metal soils. In I. Sherameti, & A. Varma (Eds.), *30. Detoxification of heavy metals, soil biology* (pp. 35–57). Berlin Heidelberg: Springer-Verlag.

Boyd, R. S. (2007). The defense hypothesis of elemental hyperaccumulation: Status, challenges and new directions. *Plant and Soil, 293*, 153–176.

Brady, K. U., Kruckeberg, A. R., & Bradshaw, H. D., Jr. (2005). Evolutionary ecology of plant adaptation to serpentine soils. *Annual Review of Ecology, Evolution and Systematics, 36*, 243–266.

Brej, T. (1998). Heavy metal tolerance in *Agropyron repens* (L.) P. Beauv. populations from the Legnica copper smelter area, Lower Silesia. *Acta Societatis Botanicorum Poloniae, 67*, 325–333.

Broadhurst, C. L., Chaney, R. L., Angle, J. S., Erbe, E. F., & Maugel, T. K. (2004). Nickel localization and response to increasing Ni soil levels in leaves of the Ni hyperaccumulator *Alyssum murale*. *Plant and Soil, 265*, 225–242.

Brooks, R. R. (1998). Geobotany and hyperaccumulators. In R. R. Brooks (Ed.), *Plants that hyperaccumulate heavy metals: Their role in phytoremediation, microbiology, archaeology, mineral exploration and phytomining* (pp. 55–94). Oxon: CAB International.

Brooks, R. R., Baker, A. J. M., & Malaisse, F. (1992). Copper flowers. *National Geographic Research and Exploration, 8*, 338–351.

Burgess, J., Szlavecz, K., Rajakaruna, N., & Swan, C. (2015). Ecotypic differentiation of mid-Atlantic *Quercus* species in response to ultramafic soils. *Australian Journal of Botany, 63*, 308–323.

Bush, E. J., & Barrett, S. C. H. (1993). Genetics of mine invasions by *Deschampsia cespitosa* (Poaceae). *Canadian Journal of Botany, 71*, 1336–1348.

Cacho, N. I., Burrell, A. M., Pepper, A. E., & Strauss, S. Y. (2014). Novel nuclear markers inform the systematics and the evolution of serpentine use in *Streptanthus* and allies (Thelypodieae, Brassicaceae). *Molecular Phylogenetics and Evolution, 72*, 71–81.
Cacho, N. I., & Strauss, S. Y. (2014). Occupation of bare habitats, an evolutionary precursor to soil specialization in plants. *Proceedings of the National Academy of Sciences of the United States of America, 111*, 15132–15137.
Callahan, D. L., Baker, A. J. M., Kolev, S. D., & Wedd, A. G. (2006). Metal ion ligands in hyperaccumulating plants. *Journal of Biological Inorganic Chemistry, 11*, 2–12.
Callahan, D. L., Roessner, U., Dumontet, V., Perrier, N., Wedd, A. G., O'Hair, R. A., ... Kolev, S. D. (2008). LC–MS and GC–MS metabolite profiling of nickel (II) complexes in the latex of the nickel hyperaccumulating tree *Sebertia acuminata* and identification of methylated aldaric acid as a new nickel (II) ligand. *Phytochemistry, 69*, 240–251.
Cecchi, L., Colzi, I., Coppi, A., Gonnelli, C., & Selvi, F. (2013). Diversity and biogeography of Ni-hyperaccumulators of *Alyssum* section *Odontarrhena* (Brassicaceae) in the Central Western Mediterranean: Evidence from karyology, morphology and DNA sequence data. *Botanical Journal of the Linnaean Society, 173*, 269–289.
Cecchi, L., Coppi, A., & Selvi, F. (2011). Evolutionary dynamics of serpentine adaptation in *Onosma* (Boraginaceae) as revealed by ITS sequence data. *Plant Systematics and Evolution, 297*, 185–199.
Cecchi, L., Gabbrielli, R., Arnetoli, M., Gonnelli, C., Hasko, A., & Selvi, F. (2010). Evolutionary lineages of Ni-hyperaccumulation and systematics in European Alysseae (Brassicaceae): Evidence from nrDNA sequence data. *Annals of Botany, 106*, 751–767.
Cecchi, L., & Selvi, F. (2009). Phylogenetic relationships of the monotypic genera *Halacsya* and *Paramoltkia* and the origins of serpentine adaptation in circummediterranean Lithospermeae (Boraginaceae): Insights from ITS and matK DNA sequences. *Taxon, 58*, 700–714.
Chaney, R. L., Angle, J. S., Broadhurst, C. L., Peters, C. A., Tappero, R. V., & Sparks, D. L. (2007). Improved understanding of hyperaccumulation yields commercial phytoextraction and phytomining technologies. *Journal of Environmental Quality, 36*, 1429–1443.
Chaney, R. L., & Baklanov, I. A. (2017). Phytoremediation and phytomining: Status and promise. In A. Cuypers, & J. Vangronsveld (Eds.), *Advances in botanical research* (Vol. 83, pp. 189–222).
Chathuranga, P. K. D., Dharmasena, S. K. A. T., Rajakaruna, N., & Iqbal, M. C. M. (2015). Growth and nickel uptake by serpentine and non-serpentine populations of *Fimbristylis ovata* (Cyperaceae) from Sri Lanka. *Australian Journal of Botany, 63*, 128–133.
Clemens, S. (2016). How metal hyperaccumulating plants can advance Zn biofortification. *Plant and Soil*. http://dx.doi.org/10.1007/s11104-016-2920-3.
Cobbett, C. S. (2000). Phytochelatin biosynthesis and function in heavy metal detoxification. *Current Opinion in Plant Biology, 3*, 211–216.
Cobbett, C. S., & Goldsbrough, P. (2002). Phytochelatins and metallothioneis: Roles in heavy metal detoxification and homeostasis. *Annual Review of Plant Biology, 53*, 159–182.
Colzi, I., Rocchi, S., Rangoni, M., Del Bubba, M., & Gonnelli, C. (2014). Specificity of metal tolerance and use of excluder metallophytes for the phytostabilization of metal polluted soils: The case of *Silene paradoxa* L. *Environmental Science and Pollution Research, 21*, 10960–10969.
Coppi, A., Cecchi, L., Mengoni, A., Pustahija, F., Tomović, G., & Selvi, F. (2014). Low genetic diversity and contrasting patterns of differentiation in the two monotypic genera *Halacsya* and *Paramoltkia* (Boraginaceae) endemic to the Balkan serpentines. *Flora, 209*, 5–14.
Cosio, C., De Santis, L., Frey, B., Diallo, S., & Keller, C. (2005). Distribution of cadmium in leaves of *Thlaspi caerulescens*. *Journal of Experimental Botany, 56*, 565–575.

Cundy, A. B., Bardos, R. P., Puschenreiter, M., Mench, M., Bert, V., Friesl-Hanl, W., ... Vangronsveld, J. (2016). Brownfields to green fields: Realising wider benefits from practical contaminant phytomanagement strategies. *Journal of Environmental Management, 184*, 67–77.

Dahmani-Müller, H., van Oort, E., Gélic, B., & Balabane, M. (2000). Strategies of heavy metal uptake by three plant species growing near a smelter site. *Environmental Pollution, 109*, 231–238.

Daniels, W. L., Stuczynski, T., Chaney, R. L., Pantuck, K., & Pistelok, F. (1998). Reclamation of Pb/Zn smelter wastes in Upper Silesia, Poland. In H. R. Fox, H. M. Moore, & A. D. McIntosh (Eds.), *Land reclamation: Achieving sustainable benefits* (pp. 269–276). Rotterdam: Balkema.

Dechamps, C., Noret, N., Mozek, R., Draye, X., & Meerts, P. (2008). Root allocation in metal-rich patch by *Thlaspi caerulescens* from normal and metalliferous soil – New insight into the rhizobox approach. *Plant and Soil, 310*, 211–224.

Deng, J., Liao, B., Ye, M., Deng, D., Lan, C., & Shu, W. (2007). The effects of heavy metal pollution on genetic diversity in zinc/cadmium hyperaccumulator *Sedum alfredii* populations. *Plant and Soil, 297*, 83–92.

Dhillon, K. S., & Dhillon, S. K. (2001). Distribution and management of seleniferous soils. *Advances in Agronomy, 79*, 119–184.

Dickinson, N. M., Turner, A. P., & Lepp, N. W. (1991). How do trees and other long-lived plants survive in polluted environments? *Functional Ecology, 5*, 5–11.

Dresler, S., Bednarek, W., & Wójcik, M. (2014). Effect of cadmium on selected physiological and morphological parameters in metallicolous and nonmetallicolous populations of *Echium vulgare* L. *Ecotoxicology and Environmental Safety, 104*, 332–338.

Dresler, S., Tyrka, M., Szeliga, M., Ciura, J., Wielbo, J., Wójcik, M., & Tukiendorf, A. (2015). Increased genetic diversity in the populations of *Echium vulgare* L. colonising Zn-Pb waste heaps. *Biochemical Systematics and Ecology, 60*, 28–36.

Dueck, T. A., Ernst, W. H. O., Faber, J., & Pasman, F. (1984). Heavy metal emission and genetic constitution of plant populations in the vicinity of two metal emission sources. *Angewandte Botanik, 58*, 47–59.

Ernst, W. H. O. (2006). Evolution of metal tolerance in higher plants. *Forest, Snow Landscape Research, 80*, 251–274.

Ernst, W. H. O., Kraus, G.-J., Verkleij, J. A. C., & Wesenberg, D. (2008). Interaction of heavy metals with the sulphur metabolism in angiosperms from an ecological point of view. *Plant, Cell and Environment, 31*, 123–143.

Escande, V., Olszewski, T., & Grison, C. (2015). From biodiversity to catalytic diversity: How to control the reaction mechanism by the nature of metallophytes. *Environmental Science and Pollution Research, 22*, 5653–5666.

Fernández, R., Fernández-Fuego, D., Bertrand, A., & González, A. (2014). Strategies for Cd accumulation in *Dittrichia viscosa* (L.) Greuter: Role of the cell wall, non-protein thiols and organic acids. *Plant Physiology and Biochemistry, 78*, 63–70.

Fiałkiewicz, B., & Rostański, A. (2006). Morphological variability of *Cardaminopsis halleri* (L.) Hayek from selected habitats in the Silesian Upland (southern Poland). *Biodiversity Research and Conservation, 1–2*, 34–40.

Gendre, D., Czernic, P., Conejero, G., Pianelli, K., Briat, J. F., Lebrun, M., & Mari, S. (2007). *TcYSL3*, a member of the *YSL* gene family from the hyper-accumulator *Thlaspi caerulescens*, encodes a nicotianamine-Ni/Fe transporter. *The Plant Journal, 49*, 1–15.

Ghaderian, S. M., Hemmat, G. R., Reeves, R. D., & Baker, A. J. M. (2007). Accumulation of lead and zinc by plants colonising a metal mining area in Central Iran. *Journal of Applied Botany and Food Quality, 81*, 145–150.

Gonnelli, C., Galardi, F., & Gabbrielli, R. (2001). Nickel and copper tolerance and toxicity in three Tuscan populations of *Silene paradoxa*. *Physiologia Plantarum, 113*, 507–514.

Guerinot, M. L. (2000). The ZIP family of metal transporters. *Biochimica and Biophysica Acta, 1465*, 190–198.

Guerinot, M. L., & Salt, D. E. (2001). Fortified foods and phytoremediation. Two sides of the same coin. *Plant Physiology, 125*, 164–167.

Gustafson, D. J., Romano, G., Latham, R. E., & Morton, J. K. (2003). Amplified fragment length polymorphism analysis of genetic relationships among the serpentine barrens endemic *Cerastium velutinum* Rafinesque var. *villosissimum* Pennel (Caryophyllaceae) and closely related *Cerastium* species. *Journal of Torrey Botanical Society, 130*, 218–223.

Halimaa, P., Lin, Y. F., Ahonen, V. H., Blande, D., Clemens, S., Gyenesei, A., ... Tervahauta, A. I. (2014). Gene expression differences between *Noccaea caerulescens* ecotypes help to identify candidate genes for metal phytoremediation. *Environmental Science and Technology, 48*, 3344–3353.

Harmens, H., Koevoets, P. L. M., Verkleij, J. A. C., & Ernst, W. H. O. (1994). The role of low molecular weight organic acids in the mechanism of increased zinc tolerance in *Silene vulgaris* (Moench) Garcke. *New Phytologist, 126*, 615–621.

Haydon, M. J., & Cobbett, C. S. (2007). Transporters of ligands for essential metal ions in plants. *New Phytologist, 174*, 499–506.

Hendrix, S., Schröder, P., Keunen, E., Huberm, C., & Cuypers, A. (2017). Molecular and cellular aspects of contaminant toxicity in plants: The importance of sulphur and associated signalling pathways. In A. Cuypers, & J. Vangronsveld (Eds.), *Advances in botanical research* (Vol. 83, pp. 227–318).

Hughes, R., Bachmann, K., Smirnoff, N., & Macnair, M. R. (2001). The role of drought tolerance in serpentine tolerance in the *Mimulus guttatus* Fischer ex DC. complex. *South African Journal of Science, 97*, 581–586.

Ingle, R. A., Mugford, S. T., Rees, J. D., Campbell, M. M., & Smith, J. A. C. (2005). Constitutively high expression of the histidine biosynthetic pathway contributes to nickel tolerance in hyperaccumulator plants. *The Plant Cell, 17*, 2089–2106.

Isnard, S., L'Huillier, L., Rigault, F., & Jaffré, T. (2016). How did the ultramafic soils shape the flora of the New Caledonian hotspot? *Plant and Soil*. http://dx.doi.org/10.1007/s11104-016-910-5.

Janssen, J., Weyens, N., Croes, S., Beckers, B., Meiresonne, L., Van Peteghem, P., ... Vangronsveld, J. (2015). Phytoremediation of metal contaminated soil using willow: Exploiting plant-associated bacteria to improve biomass production and metal uptake. *International Journal of Phytoremediation, 17*, 1123–1136.

Jędrzejczyk-Korycińska, M. (2009). Former zinc-lead ores mining areas — Their floristic diversity and chances of conservation. *Problemy Ekologii Krajobrazu, 24*, 71–80 (in Polish).

Jędrzejczyk-Korycińska, M., Szarek-Łukaszewska, G., Kapusta, P., & Zagórna, M. (2014). Changes in species composition of calamine grassland after the removal of woody plants — Effect of a conservation management. *Acta Biologica Cracoviensia, Series Botanica, 56*(Suppl. 2), 26.

Johnson, M. S., Cooke, J. A., & Stevenson, J. K. (1994). Revegetation of metalliferous wastes and land after metal mining. In R. E. Hester, & R. M. Harrison (Eds.), *Mining and its environmental impact* (pp. 31–48). Cambridge: Royal Society of Chemistry, Thomas Graham House.

Jozefczak, M., Remans, T., Vangronsveld, J., & Cuypers, A. (2012). Glutathione is a key player in metal-induced oxidative stress defences. *International Journal of Molecular Sciences, 13*, 3145–3175.

Kabata-Pendias, A., & Mukherjee, A. B. (2007). *Trace elements from soil to human*. Berlin Heidelberg: Springer-Verlag.

Kerkeb, L., & Krämer, U. (2003). The role of free histidine in xylem loading of nickel in *Alyssum lesbiacum and Brassica juncea*. *Plant Physiology, 131*, 716–724.

Kim, D., Gustin, J. L., Lahner, B., Persans, M. W., Baek, D., Yun, D. J., & Salt, D. E. (2004). The plant CDF family member TgMTP1 from the Ni/Zn hyperaccumulator *Thlaspi goesingense* acts to enhance efflux of Zn at the plasma membrane when expressed in *Saccharomyces cerevisiae*. *Plant Journal, 39*, 237–251.

Kim, D. Y., Bovet, L., Maeshima, M., Martinoia, E., & Lee, Y. (2007). The ABC transporter AtPDR8 is a cadmium extrusion pump conferring heavy metal resistance. *Plant Journal, 50*, 207–218.

Krämer, U., Cotter-Howwells, J. D., Charnock, J. M., Baker, A. J. M., & Smith, J. A. C. (1996). Free histidine as a metal chelator in plants that accumulate nickel. *Nature, 379*, 635–638.

Krämer, U., Grime, G. W., Smith, J. A. C., Hawes, C. R., & Baker, A. J. M. (1997). MicroPIXE as a technique for studying nickel localization in leaves of the hyperaccumulator plant *Alyssum lesbiacum*. *Nuclear Instruments and Methods in Physics Research Bulletin, 130*, 346–350.

Krämer, U., Pickering, I. J., Prince, R. C., Raskin, I., & Salt, D. E. (2000). Subcellular localization and speciation of nickel in hyperaccumulator and non-accumulator *Thlaspi* species. *Plant Physiology, 122*, 1343–1354.

Krämer, U., Talke, I. N., & Hanikenne, M. (2007). Transition metal transport. *FEBS Letters, 581*, 2263–2272.

Kruckeberg, A. R. (2002). *Geology and plant life: The effects of landforms and rock type on plants*. Seattle/London: University Washington Press.

Krzesłowska, M. (2011). The cell wall in plant cell response to trace metals: Polysaccharide remodeling and its role in defense strategy. *Acta Physiologiae Plantarum, 33*, 35–51.

Lasat, M. M., Pence, N. S., Garvin, D. F., Abbs, S. D., & Kochian, L. V. (2000). Molecular physiology of zinc transport in the Zn hyperaccumulator *Thlaspi caerulescens*. *Journal of Experimental Botany, 51*, 71–79.

Li, T. Q., Yang, X. E., He, Z. L., & Yang, J. Y. (2005). Root morphology and Zn^{2+} uptake kinetics of the Zn hyperaccumulator of *Sedum alfredii* Hance. *Journal of Integrative Plant Biology, 47*, 927–934.

Li, Y.-M., Chaney, R., Brewer, E., Roseberg, R., Angle, J. S., Baker, A., … Nelkin, J. (2003). Development of a technology for commercial phytoextraction of nickel: Economic and technical considerations. *Plant and Soil, 249*, 107–115.

Li, Y.-M., Chaney, R. L., Siebielec, G., & Kershner, B. A. (2000). Response of four turfgrass cultivars to limestone and biosolids-compost amendment of a zinc and cadmium contaminated soil at Palmerton, Pennsylvania. *Journal of Environmental Quality, 29*, 1440–1447.

Lombi, E., Zhao, F. J., McGrath, S. P., Young, S. D., & Sacchi, G. A. (2001). Physiological evidence for a high-affinity cadmium transporter highly expressed in a *Thlaspi caerulescens* ecotype. *New Phytologist, 149*, 53–60.

Luo, C. L., Shen, Z. G., & Li, X. D. (2008). Root exudates increase metal accumulation in mixed cultures: Implications for naturally enhanced phytoextraction. *Water, Air and Soil Pollution, 193*, 147–154.

Manara, A. (2012). Plant responses to heavy metal toxicity. In A. Furini (Ed.), *Plants and heavy metals. SpringerBriefs in biometals* (pp. 27–53). The Netherlands: Springer.

Mari, S., Gendre, D., Pianelli, K., Ouerdane, L., Lobinski, R., Briat, J. F., … Czernic, P. (2006). Root to shoot long-distance circulation of nicotianamine and nicotianamine–nickel chelates in the metal hyperaccumulator *Thlaspi caerulescens*. *Journal of Experimental Botany, 57*, 4111–4122.

Martin, M. H., & Coughtrey, P. J. (1982). Biological indicators of natural ore-bodies: Geobotanical and biogeochemical prospecting for heavy metal deposits. In M. H. Martin, & P. J. Coughtrey (Eds.), *Biological monitoring of heavy metal pollution. Land and air* (pp. 34–59). London, New York: Applied Science Publishers Ltd.

Mathys, W. (1977). The role of malate, oxalate, and mustard oil glucosides in the evolution of zinc resistance in herbage plants. *Physiologia Plantarum, 40*, 130–136.

Maxted, A. P., Black, C. R., West, H. M., Crout, N. M. J., McGrath, S. P., & Young, S. D. (2007). Phytoextraction of cadmium and zinc from arable soils amended with sewage sludge using *Thlaspi caerulescens*: Development of a predictive model. *Environmental Pollution, 150*, 363–372.

McGrath, S. P., Lombi, E., Gray, C. W., Caille, N., Dunham, S. J., & Zhao, F.-J. (2006). Field evaluation of Cd and Zn phytoextraction potential by the hyperaccumulators *Thlaspi caerulescens* and *Arabidopsis halleri*. *Environmental Pollution, 141*, 115–125.

Mench, M., & Martin, E. (1991). Mobilization of cadmium and other metals from two soils by root exudates of *Zea mays* L., *Nicotiana tabacum* L., and *Nicotiana rustica* L. *Plant and Soil, 132*, 187–196.

Menezes de Sequeira, E., & Pinto da Silva, A. R. (1992). Ecology of serpentinized areas of north-east Portugal. In B. A. Roberts, & J. Proctor (Eds.), *The ecology of areas with serpentinized rocks. A world view* (pp. 169–197). Dordrecht, Germany: Kluwer Academic Publishers.

Mengoni, A., Barabesi, C., Gonnelli, C., Galardi, F., Gabbrielli, R., & Bazzicalupo, M. (2001). Genetic diversity of heavy metal-tolerant populations in *Silene paradoxa* L. (Caryophyllaceae): A chloroplast microsatellite analysis. *Molecular Ecology, 10*, 1909–1916.

Mengoni, A., Selvi, F., Cusimano, N., Galardi, F., & Gonnelli, C. (2006). Genetic diversity inferred from AFLP fingerprinting in populations of *Onosma echioides* (Boraginaceae) from serpentine and nrDNA sequence data. *Plant Biosystems, 140*, 211–219.

Merlot, S., Hannibal, L., Martins, S., Martinelli, L., Amir, H., Lebrun, M., & Thomine, S. (2014). The metal transporter PgIREG1 from the hyperaccumulator *Psychotria gabriellae* is a candidate gene for nickel tolerance and accumulation. *Journal of Experimental Botany, 65*, 1551–1564.

Mertens, J., Luyssaert, S., & Verheyen, K. (2005). Use and abuse of trace metal concentrations in plant tissue for biomonitoring and phytoextraction. *Environmental Pollution, 138*, 1–4.

Mesjasz-Przybylowicz, J., Balkwill, K., Przybylowicz, W. J., & Annegarn, H. J. (1994). Proton microprobe and *X-ray* fluorescence investigations of nickel distribution in serpentine flora from South Africa. *Nuclear Instruments and Methods Bulletin, 89*, 208–212.

Meyer, C.-L., Pauwels, M., Briset, L., Godé, C., Bourceaux, A., Souleman, D., … Verbruggen, N. (2016). Potential preadaptation to anthropogenic pollution: Evidence from a common quantitative trait locus for zinc and cadmium tolerance in metallicolous and nonmetallicolous accessions of Arabidopsis halleri. *New Phytologist, 212*, 934–943.

Meyer, C.-L., & Verbruggen, N. (2012). The use of the model species *Arabidopsis halleri* towards phytoextraction of cadmium polluted soils. *New Biotechnology, 30*, 9–14.

Mohtadi, A., Ghaderian, S. M., & Schat, H. (2012). Lead, zinc and cadmium accumulation from two metalliferous soils with contrasting calcium contents in heavy metal-hyperaccumulating and non-hyperaccumulating metallophytes: A comparative studies. *Plant and Soil, 361*, 109–118.

Nagajyoti, P. C., Lee, K. D., & Sreekanth, T. V. M. (2010). Heavy metals, occurrence and toxicity for plants: A review. *Environmental Chemistry Letters, 8*, 199–216.

Nagy, L., & Proctor, J. (1997). Soil Mg and Ni as causal factors of plant occurrence and distribution at the Meikle Kilrannoch ultramafic site in Scotland. *New Phytologist, 135*, 561–566.

Nehnevajova, E., Herzig, R., Bourigault, C., Bangerter, S., & Schwitzguébel, J. P. (2009). Stability of enhanced yield and metal uptake by sunflower mutants for improved phytoremediation. *International Journal of Phytoremediation, 11*, 329–346.

Nyberg Berglund, A. B., Dalgren, S., & Westerberg, A. (2004). Evidence for parallel evolution and site-specific selection of serpentine tolerance in *Cerastium alpinum* during the colonization of Scandinavia. *New Phytologist, 161*, 199–209.

Nyberg Berglund, A. B., & Westerbergh, A. (2001). Two postglacial immigration lineages of the polyploid *Cerastium alpinum* (Caryophyllaceae). *Hereditas*, 134,171–183.

O'Dell, R. E., & Rajakaruna, N. (2011). Intraspecific variation, adaptation, and evolution. In S. P. Harrison, & N. Rajakaruna (Eds.), *Serpentine: The evolution and ecology of a model system* (pp. 97–137). Berkeley: University of California Press.

Olko, A., Abratowska, A., Żyłkowska, J., Wierzbicka, M., & Tukiendorf, A. (2008). *Armeria maritima* from a calamine heap – Initial studies on physiologic-metabolic adaptations to metal enriched soil. *Ecotoxicology and Environmental Safety, 69*, 209–218.

Pakdaman, N., Ghaderian, S. M., Ghasemi, R., & Asemaneh, T. (2013). Effects of calcium/magnesium quotients and nickel in the growth medium on growth and nickel accumulation in *Pistacia atlantica*. *Journal of Plant Nutrition, 36*, 1708–1718.

Pal, R., & Rai, J. P. N. (2010). Phytochelatins: Peptides involved in heavy metal detoxification. *Applied Biochemistry and Biotechnology, 160*, 945–963.

Papazoglou, E. (2014). *Jatropha curcas* I: A multiple Energy crop. *Fresenius Environmental Bulletin, 23*, 2695–2699.

Patterson, T. B., & Givnish, T. J. (2003). Geographic cohesion, chromosomal evolution, parallel adaptive radiations, and consequent floral adaptations in *Calochortus* (Calochortaceae): Evidence from a cpDNA phylogeny. *New Phytologist, 161*, 253–264.

Persans, M. W., Nieman, K., & Salt, D. E. (2001). Functional activity and role of cation-efflux family members in Ni hyperaccumulation in *Thlaspi goesingense*. *Proceedings of the National Academy of Sciences of the United States of America, 98*, 9995–10000.

Persans, M. W., Yan, X., Patnoe, J. M., Krämer, U., & Salt, D. E. (1999). Molecular dissection of the role of histidine in nickel hyperaccumulation in *Thlaspi goesingense* (Hálácsy). *Plant Physiology, 121*, 1117–1126.

Pollard, A. J., Powell, K. D., Harper, F. A., & Smith, J. A. C. (2002). The genetic basis of metal hyperaccumulation in plants. *Critical Reviews in Plant Sciences, 21*, 539–566.

Pollard, A. J., Reeves, R. D., & Baker, A. J. M. (2014). Facultative hyperaccumulation of heavy metals and metalloids. *Plant Science, 217–218*, 8–17.

Prach, K., & Tolvanen, A. (2016). How can we restore biodiversity and ecosystem services in mining and industrial sites? *Environmental Science and Pollution Research, 23*, 13587–13590.

Proctor, J. (1999). Toxins, nutrient shortages and droughts: The serpentine challenge. *Trends in Ecology and Evolution, 14*, 334–335.

Przedpelska-Wasowicz, E., & Wasowicz, P. (2013). Does zinc concentration in the substrate influence the onset of flowering in *Arabidopsis arenosa* (Brassicaceae)? *Plant Growth Regulation, 69*, 87–97.

Rascio, N., & Navarri-Izzo, F. (2011). Heavy metal hyperaccumulating plants: How and why do they do it? And what makes them so interesting? *Plant Science, 180*, 169–181.

Rauser, W. E. (1999). Structure and function of metal chelators produced by plants: The case for organic acids, amino acids, phytin and metallothioneins. *Cell Biochemistry and Biophysics, 31*, 19–48.

Reeves, R. D. (2006). Hyperaccumulation of trace elements by plants. In J.-L. Morel, G. Echevarria, & N. Goncharova (Eds.), *Phytoremediation of metal-contaminated soils* (pp. 25–52). Dordrecht: Springer.

Remon, E., Bouchardon, J. L., & Faure, O. (2007). Multi-tolerance to heavy metals in *Plantago arenaria* Waldst. & Kit.: Adaptive versus constitutive characters. *Chemosphere, 69*, 41–47.

Richau, K. H., Kozhevnikova, A. D., Seregin, I. V., Vooijs, R., Koevoets, P. L. M., Smith, J. A. C., ... Schat, H. (2009). Chelation by histidine inhibits the vacuolar sequestration of nickel in roots of the hyperaccumulator *Thlaspi caerulescens*. *New Phytologist, 183*, 106–116.

Rostański, A. (2014). Intentional introduction of *Biscutella laevigata* L. on post-flotation lead-zinc ore spoil heap in Piekary Śląskie (Upper Silesia, Southern Poland). *Environmental and Socio-Economic Studies, 2*, 24–32.

Rostański, A., Myśliwiec, I., & Siwińska, D. (2005). Variability of *Cardaminopsis arenosa* (L.) Hayek populations in areas polluted with heavy metals. In W. Prus-Głowacki (Ed.), *Variability and evolution – New perspectives* (pp. 547–561). Poznań: Wydawnictwo Naukowe UAM.

Ruttens, A., Boulet, J., Weyens, N., Smeets, K., Adriaensen, K., Meers, E., ... Vangronsveld, J. (2011). Short rotation coppice culture of willow and poplar as energy crops on metal contaminated agricultural soils. *International Journal of Phytoremediation, 13*, 194–207.

Safford, H. D., Viers, J. H., & Harrison, S. P. (2005). Serpentine endemism in the California flora: A database of serpentine affinity. *Madrono, 52*, 22–257.

Sagner, S., Kneer, R., Wanner, G., Cosson, J. P., Deus-Neumann, B., & Zenk, M. H. (1998). Hyperaccumulation, complexation and distribution of nickel in *Sebertia acuminata*. *Phytochemistry, 47*, 339–347.

Salt, D. E., Kato, N., Krämer, U., Smith, R. D., & Raskin, I. (2000). The role of root exudates in nickel hyperaccumulation and tolerance in accumulator and nonaccumulator species of *Thlaspi*. In N. Terry, & G. Banuelos (Eds.), *Phytoremediation of contaminated soil and water* (pp. 189–200). Boca Raton: CRC Press LLC.

Sarret, G., Saumitou-Laprade, P., Bert, V., Proux, O., Hazemann, J.-L., Traverse, A., ... Manceau, A. (2002). Forms of zinc accumulated in the hyperaccumulator *Arabidopsis halleri*. *Plant Physiology, 130*, 1815–1826.

Sas-Nowosielska, A., Kucharski, R., Małkowski, E., Pogrzeba, M., Kuperberg, J. M., & Kryński, K. (2004). Phytoextraction crop disposal – An unsolved problem. *Environmental Pollution, 128*, 373–379.

Schat, H., Llugany, M., Vooijs, R., Hartley-Whitaker, J., & Bleeker, P. M. (2002). The role of phytochelatins in constitutive and adaptive heavy metal tolerance in hyperaccumulator and nonhyperaccumulator metallophytes. *Journal of Experimental Botany, 53*, 2381–2392.

Schat, H., & Vooijs, R. (1997). Multiple tolerance and co-tolerance to heavy metals in *Silene vulgaris*: A co-segregation analysis. *New Phytologist, 136*, 489–496.

Selvi, F. (2007). Diversity, geographic variation and conservation of the serpentine flora of Tuscany (Italy). *Biodiversity and Conservation, 16*, 1423–1439.

Shu, W. S., Xia, H. P., Zhang, Z. Q., Lan, C. Y., & Wong, M. H. (2002). Use of vetiver and three other grasses for revegetation of Pb/Zn mine tailings: Field experiment. *International Journal of Phytoremediation, 4*, 47–57.

Siedlecka, A., Tukendorf, A., Skórzyńska-Polit, E., Maksymiec, W., Wójcik, M., Baszyński, T., & Krupa, Z. (2001). Angiosperms (Asteraceae, Convolvulaceae, Fabaceae and Poaceae; other than Brassicaceae). In M. N. V. Prasad (Ed.), *Metals in the environment. Analysis by biodiversity* (pp. 171–217). New York, Basel: Marcel Dekker, Inc.

Singh, S., Parihar, P., Singh, R., Singh, V. P., & Prasad, S. M. (2016). Heavy metal tolerance in plants: Role of transcriptomics, proteomics, metabolomics, and ionomics. *Frontiers in Plant Science, 6*, 1143.

Słomka, A., Kuta, E., Szarek-Łukaszewska, G., Godzik, B., Kapusta, P., Tylko, G., & Bothe, H. (2011). Violets of the section Melanium, their colonization by arbuscular mycorrhizal fungi and their occurrence on heavy metal heaps. *Journal of Plant Physiology, 168*, 1191–1199.

Słomka, A., Sutkowska, A., Szczepaniak, M., Malec, P., Mitka, J., & Kuta, E. (2011). Increased genetic diversity of *Viola tricolor* L. (Violaceae) in metal-polluted environments. *Chemosphere, 83*, 435–442.

Stevanović, V., Tan, K., & Iatrou, G. (2003). Distribution of the endemic Balkan flora on serpentine I. Obligate serpentine endemics. *Plant Systematics and Evolution, 242*, 149–170.

Stuczynski, T., Siebielec, G., Daniels, W. L., McCarty, G., & Chaney, R. L. (2007). Biological aspects of metal waste reclamation with biosolids. *Journal of Environmental Quality, 36*, 1154–1162.

Sun, Q., Ye, Z. H., Wang, X. R., & Wong, M. H. (2007). Cadmium hyperaccumulation leads to an increase of glutathione rather than phytochelatins in the cadmium hyperaccumulator *Sedum alfredii*. *Journal of Plant Physiology, 164*, 1489–1498.

Szarek-Łukaszewska, G. (2009). Vegetation of reclaimed and spontaneously vegetated Zn-Pb mine wastes in Southern Poland. *Polish Journal of Environmental Studies, 18*, 717–733.

Tlustoš, P., Száková, J., Hrubý, J., Hartman, I., Najmanová, J., Nedělník, J., … Batysta, M. (2006). Removal of As, Cd, Pb, and Zn from contaminated soil by high biomass producing plants. *Plant, Soil and Environment, 52*, 413–423.

Tordoff, G. M., Baker, A. J. M., & Willis, A. J. (2000). Current approaches to the revegetation and reclamation of metalliferous mine wastes. *Chemosphere, 41*, 219–228.

Van de Mortel, J. E., Villanueva, L. A., Schat, H., Kwekkeboom, J., Coughlan, S., Moerland, P. D., … Aarts, M. G. M. (2006). Large expression differences in genes for iron and zinc homeostasis, stress response, and lignin biosynthesis distinguish roots of *Arabidopsis thaliana* and the related metal hyperaccumulator *Thlaspi caerulescens*. *Plant Physiology, 142*, 239–260.

Van der Ent, A., Baker, A. J. M., Reeves, R. D., Chaney, R. L., Anderson, C. W. N., Meech, J. A., … Mulligan, D. R. (2015). Agromining: Farming for metals in the future? *Environmental Science and Technology, 49*, 4773–4780.

Van der Ent, A., Baker, A. J. M., Reeves, R. D., Pollard, A. J., & Schat, H. (2013). Hyperaccumulators of metal and metalloid trace elements: Facts and fiction. *Plant and Soil, 362*, 319–334.

Van Slycken, S., Witters, N., Meers, E., Peene, A., Michels, E., Adriaensen, K., … Tack, F. M. G. (2013). Safe use of metal-contaminated agricultural land by cultivation of energy maize (*Zea mays*). *Environmental Pollution, 178*, 375–380.

Vangronsveld, J., Colpaert, J., & Van Tichelen, K. (1996). Reclamation of a bare industrial area contaminated by non-ferrous metals: Physico-chemical and biological evaluation of the durability of soil treatment and revegetation. *Environmental Pollution, 94*, 131–140.

Vangronsveld, J., Herzig, R., Weyens, N., Boulet, J., Adriaensen, K., Ruttens, A., … Mench, M. (2009). Phytoremediation of contaminated soils and groundwater: Lesson from the field. *Environmental Science and Pollution Research, 16*, 765–794.

Vekemans, X., & Lefèbvre, C. (1997). On the evolution of heavy metal tolerant populations in *Armeria maritima*: Evidence from allozyme variation and reproduction barriers. *Journal of Evolutionary Biology, 10*, 175–191.

Verbruggen, N., Hermans, C., & Schat, H. (2009). Molecular mechanisms of metal hyperaccumulation in plants. *New Phytologist, 181*, 759–776.

Visioli, G., Gullì, M., & Marmiroli, N. (2014). *Noccaea caerulescens* populations adapted to grow in metalliferous and non-metalliferous soils: Ni tolerance, accumulation and expression analysis of genes involved in metal homeostasis. *Environmental and Experimental Botany, 105*, 10–17.

Watanabe, T., & Osaki, M. (2002). Mechanisms of adaptation to high aluminium conditions in native plant species growing in acidic soils: A review. *Communications in Soil Science and Plant Analysis, 33*, 1247–1260.

Weber, M., Harada, E., Vess, C., Roepenack-Lahaye, E. V., & Clemens, S. (2004). Comparative microarray analysis of *Arabidopsis thaliana* and *Arabidopsis halleri* roots identifies nicotianamine synthase, a ZIP transporter and other genes as potential metal hyperaccumulation factors. *Plant Journal, 37*, 269–281.

Whiting, S. N., Leake, J. R., McGrath, S. P., & Baker, A. J. M. (2000). Positive responses to Zn and Cd by roots of the Zn and Cd hyperaccumulator *Thlaspi caerulescens*. *New Phytologist, 145*, 199–210.

Whiting, S. N., Reeves, R. D., Richards, D., Johnson, M. S., Cooke, J. A., Malaisse, F., ... Baker, A. J. M. (2004). Research priorities for conservation of metallophyte biodiversity and their potential for restoration and site remediation. *Restoration Ecology, 12*, 106–116.

Wierzbicka, M., & Rostański, A. (2002). Microevolutionary changes in ecotypes of calamine waste heap vegetation near Olkusz, Poland: A review. *Acta Biologica Cracoviensia Series Botanica, 44*, 7–19.

Wójcik, M. (2009). Vacuole as a multifunctional compartment in plant responses to stress factors. In W. Maksymiec (Ed.), *Compartmentation of responses to stresses in higher plants, true or false* (pp. 91–123). Kerala: Transworld Research Network.

Wójcik, M., Dresler, S., Jawor, E., Kowalczyk, K., & Tukiendorf, A. (2013). Morphological, physiological, and genetic variation between metallicolous and nonmetallicolous populations of *Dianthus carthusianorum*. *Chemosphere, 90*, 1249–1257.

Wójcik, M., Dresler, S., Plak, A., & Tukiendorf, A. (2015a). Naturally evolved enhanced Cd-tolerance of *Dianthus carthusianorum* L. is not related to accumulation of thiol peptides and organic acids. *Environmental Science and Pollution Research, 22*, 7906–7917.

Wójcik, M., Dresler, S., & Tukiendorf, A. (2015). Physiological mechanisms of adaptation of *Dianthus carthusianorum* L. to growth on a Zn-Pb waste deposit – The case of chronic multi-metal and acute Zn stress. *Plant and Soil, 390*, 237–250.

Wójcik, M., Skórzyńska-Polit, E., & Tukiendorf, A. (2006). Organic acids accumulation and antioxidant enzyme activities in *Thlaspi caerulescens* under Zn and Cd stress. *Plant Growth Regulation, 48*, 145–155.

Wójcik, M., Sugier, P., & Siebielec, G. (2014). Metal accumulation strategies in plants spontaneously inhabiting Zn-Pb waste deposits. *Science of the Total Environment, 487*, 313–322.

Wójcik, M., & Tukiendorf, A. (2011). Glutathione in adaptation of *Arabidopsis thaliana* to cadmium stress. *Biologia Plantarum, 55*, 125–132.

Wójcik, M., & Tukiendorf, A. (2014). Accumulation and tolerance of lead in two contrasting ecotypes of *Dianthus carthusianorum*. *Phytochemistry, 100*, 60–65.

Wójcik, M., Vangronsveld, J., & Tukiendorf, A. (2005). Cadmium tolerance in *Thlaspi caerulescens*. I. Growth parameters, metal accumulation and phytochelatin synthesis in response to cadmium. *Environmental and Experimental Botany, 53*, 151–161.

Wong, C. K. E., & Cobbett, C. S. (2009). HMA P-type ATPases are the major mechanisms for root-to-shoot Cd translocation in *Arabidopsis thaliana*. *New Phytologist, 181*, 71–78.

Wright, J. W., & Stanton, M. L. (2011). Local adaptation in heterogeneous landscapes – Reciprocal transplant experiments and beyond. In S. P. Harrison, & N. Rajakaruna (Eds.), *Serpentine: The evolution and ecology of a model system* (pp. 155–180). Berkeley: University of California Press.

Yang, X., Li, T., Yang, J., He, Z., Lu, L., & Meng, F. (2006). Zinc compartmentation in root, transport into xylem, and absorption into leaf cells in the hyperaccumulating species of *Sedum alfredii* Hance. *Planta, 224*, 185–195.

Zelko, I., Lux, A., & Czibula, K. (2008). Difference in the root structure of hyperaccumulator *Thlaspi caerulescens* and non-hyperaccumulator *Thlaspi arvense*. *International Journal of Environment and Pollution, 33*, 123–132.

Zhao, F.-J., Hamon, R. E., & McLaughlin, M. J. (2001). Root exudates of the hyperaccumulator *Thlaspi caerulescens* do not enhance metal immobilization. *New Phytologist, 151*, 613–620.

Zhao, F. J., Lombi, E., Breedon, T., & McGrath, S. P. (2000). Zinc hyperaccumulation and cellular distribution in *Arabidopsis halleri*. *Plant Cell and Environment, 23*, 507–514.

Zhao, F.-J., & McGrath, S. P. (2009). Biofortification and phytoremediation. *Current Opinion in Plant Biology, 12*, 373–380.

CHAPTER TWO

The Bacterial and Fungal Microbiota of Hyperaccumulator Plants: Small Organisms, Large Influence

Sofie Thijs[1], Tori Langill and Jaco Vangronsveld
Hasselt University, Diepenbeek, Belgium
[1]Corresponding author: E-mail: sofie.thijs@uhasselt.be

Contents

Abstract

Hyperaccumulator plants can take up large amounts of metals in their shoots without showing significant signs of toxicity. This makes hyperaccumulators ideal candidates for metal phytoremediation and phytomining. Hyperaccumulation of metals does not only depend on the bioavailability of the metals in soil and the expression of detoxification genes and metal transporters provided by the plant, but also on the plant-associated microbiota. Studies using culture-dependent strategies have shown that hyperaccumulator plants carry a high diversity of bacteria and fungi in the rhizosphere and endosphere, some of which are shown to have potential for assisting

Advances in Botanical Research, Volume 83
ISSN 0065-2296
http://dx.doi.org/10.1016/bs.abr.2016.12.003

plants to grow in metal-contaminated soil and regulating plant metal uptake. However, there is yet little information available about the total microbial communities and their functions associated with the majority of hyperaccumulators known to date. Culture-independent molecular techniques and next generation sequencing allow to uncover a broader diversity of microbial species than the classical 1% cultivable fraction and can provide unprecedented insights in microbial community functions. Some unique microbial groups that are highly abundant in heavy contaminated sites have been detected in this way and have great potential for improving the efficiency of metal phytoextraction. This chapter provides the latest insights in the plant-associated microbiota of common hyperaccumulator plant species and discusses on the implications, and future prospects, of exploiting the microbiome for enhancing metal uptake by plants.

1. INTRODUCTION

Environmental contamination with trace metals is a serious problem worldwide as it reduces the yield and quality of crops on agricultural soils and threatens human, plant and wildlife. Hot spots of soil contamination are located in areas of large industrial activities (ore smelters), and surrounding agricultural areas can be affected by atmospheric deposition of these trace metals (Vangronsveld et al., 2009). In addition, agricultural practices such as the application of sewage sludge, pesticides or phosphate fertilisers has led to increased metal concentrations in soil (Puschenreiter, Horak, Friesl, & Hartl, 2005). The most common trace metals found in contaminated soils include cadmium (Cd), zinc (Zn), nickel (Ni), lead (Pb), arsenic (As) and copper (Cu). In China, more than 1.5 million sites of trace metal exposure are reported (Hu, Jin, & Kavan, 2014), with 19.4% of the investigated farmland soils polluted and in need of remediation (Zhang, Zhong, Liu, & Ouyang, 2015). In Europe, an estimated 137,000 km^2 agricultural land is polluted with metals and needs remediation actions (Toth, Hermann, Da Silva, & Montanarella, 2016). Besides anthropogenic sources of metal pollution, metal contamination in soils may also derive from geogenic sources e.g., ore-rich soils at mining sites. Mine sites with high concentrations of Cu, Cr, Fe or Ni are primary resources for recovering valuable metals for both clean energy technologies, smartphones and in the medicinal industry. However, mining takes a negative toll to the environment, with waste tailings, acid run-off water and seepage ponds of toxic metals, causing long-term pressure to the environment. These sites thus require restoration.

The majority of the trace metals interfere with numerous biochemical and physiological processes thus adversely affecting organisms and environment (Jaishankar, Tseten, Anbalagan, Mathew, & Beeregowda, 2014). Besides their toxic effects, several trace metals are classified as carcinogenic and mutagenic (Hengstler et al., 2003). For example, long-term exposure to cadmium can increase the incidence of cancer, cardiovascular disease, hormonal disturbance and neurodegenerative effects (Hogervorst et al., 2007). Arsenic exposure induces cancers including skin, lung and kidney (Singh, Goel, & Kaur, 2011). Arsenic is also a neurotoxin, and damages DNA, contributing genetic alteration (O'Bryant, Edwards, Menon, Gong, & Barber, 2011). So it is important to reduce the environmental impact of trace metal contaminated sites.

Phytoremediation using hyperaccumulator plants to clean up contaminated soils has a great potential (Baker & Smith, 2000). Hyperaccumulator plants have the capability of growing in soils with very high concentrations of metals and accumulate higher levels of metals than normal plants in their aerial tissues (Reeves, 2003). Physicochemical strategies to remove metals from contaminated soil often include the addition of chemical chelates such as ethylenediaminetetraacetic acid (EDTA), glycoletherdiamine tetraacetic acid (EDGA) and citric acid, but these expensive compounds can increase the metal-leaching risk and show negative effects on plant growth or soil structures (Kidd et al., 2015). Physicochemical remediation techniques are in general more expensive to remove metals from contaminated soils compared to phytoextraction and are less feasible for large areas. Phytoremediation is both more sustainable and cheap and creates additional advantages such as biomass for biofuel and increasing local biodiversity (Weyens, van der Lelie, Taghavi, Newman, & Vangronsveld, 2009; Weyens, van der Lelie, Taghavi, & Vangronsveld, 2009). As an example, according to the European Environment Agency, the total cost for clean-up of polluted site requires expenses of about 59–109 billion euros (Mahmood, Malik, & Hussain, 2010). While it was estimated that for the removal of lead through phytoextraction from 1.4 to 0.4 g/kg of soil in 10 years is cheaper in comparison to conventional methods (Evangelou, Ebel, & Schaeffer, 2007). Besides phytoremediation, metal phytomining is also a popular strategy to combine the need for site remediation coupled with the recovery of highly valuable metals (Echevarria et al., 2015). In this case, hyperaccumulator plants are used that can accumulate the metal of interest from metal-rich soils and transport them to the shoots (>1%), followed by harvesting the shoots as a bioore (Echevarria et al., 2015).

However, some major drawbacks limit the application of hyperaccumulators for phytoextraction and phytomining, including slow growth, low biomass production, variable metal uptake under field conditions and thus variation in the accumulation of trace metals in the tissues (Gerhardt, Huang, Glick, & Greenberg, 2009). Indeed, metal-tolerant plants alone cannot sustain in contaminated areas. These plants find constraints to perform their action under extreme conditions such as high contaminant level, lack of essential nutrients and a reduced microflora. These factors considerably reduce the efficiency of hyperaccumulators for metal phytoextraction. Therefore, it is necessary to update phytoextraction by increasing the growth of hyperaccumulator plants to yield higher biomass, increase plant tolerance, metal accumulation, survival under field uncertainties and to sustain a healthy microflora under the physicochemical soil conditions in the course of time.

There is increasing evidence that plant—microbe interactions play a critical role in metal phytoextraction and metal-mining, as they mediate different physicochemical and biological activities to facilitate ecological performances of the host plant (Muehe et al., 2015). The root environment is a very dynamic microsystem in which microorganisms, roots and soil interact, and roots can get access to metals and nutrients (Berendsen, Pieterse, & Bakker, 2012). In the rhizosphere, bacteria and fungi can in some cases enhance metal uptake by the plant, reduce (White, Sayer, & Gadd, 1997) or show no effect (Rajkumar, Nagendran, Lee, Lee, & Kim, 2006). Among the rhizosphere microorganisms involved in plant interactions with the soil environment, the plant growth-promoting rhizobacteria (PGPR) have received special attention for their unique abilities to promote plant growth (Lugtenberg & Kamilova, 2009). Certain of these PGPR can increase solubility of metals and change speciation of metals through producing organic ligands, exuding metabolites (organic acids, siderophores) or reducing soil pH (Sessitsch et al., 2013). Similarly, different trace metal-tolerance mechanisms have been found in symbiotic fungi colonizing the plant root, with exclusion, biosorption, precipitation and complexation of metals in different extracellular or intracellular spaces (see chapter: Mycorrhiza-Assisted Phytoremediation by Coninx, Martinova, & Rineau, 2017; Miransari, 2011a,b). These processes can influence the solubility and the bioavailability of the metal to the plant in the rhizosphere, thus modifying the toxic effects of the metals. Some microorganisms are competent of entering inside plants and colonize the interior of plants as endophytes (Hardoim et al., 2015; Hardoim, van Overbeek,

& Elsas, 2008). Endophytes have been shown to increase plant resistance to pathogens, enhance plant growth and alleviate plants from trace metal toxicity, and moreover they enjoy a more favourable, less competitive environment (Ma, Prasad, Rajkumar, & Freitas, 2011; Weyens, van der Lelie, Taghavi, & Newman, 2009; Weyens, van der Lelie, Taghavi, & Vangronsveld, 2009).

The beneficial effects of both rhizospheric and endophytic microorganisms on plant growth, physiology, functional attributes and impact on the efficiency of metal extraction by the plant made them very attractive for biotechnological applications (Ma et al., 2011; Rajkumar, Ae, & Freitas, 2009; Weyens, van der Lelie, Taghavi, & Newman, 2009; Weyens, van der Lelie, Taghavi, & Vangronsveld, 2009). Characterization of the hyperaccumulator rhizosphere and endosphere microorganisms is of prime importance to facilitate these applications. First of all, identification of potential microbial cultures for bioaugmentation, single or consortia, is important for amelioration of the rhizosphere environment for metal uptake. Next, microbial characteristics for plant growth and survival are important strategies towards achieving better plant performance. Finally, enhanced metal tolerance and plant growth in the trace metal contaminated soils by application of inoculants would minimize application of other ameliorating agents such as fertiliser, compost and thereby reduce the cost of treatment system (Alvarez-Lopez, Prieto-Fernandez, Cabello-Conejo, & Kidd et al., 2016).

While there has been a lot of research in characterizing the culturable bacteria and fungi from hyperaccumulator plants (Visioli, D'Egidio, & Sanangelantoni, 2014), there has been very little research on identifying the total rhizospheric and endophytic microbiota and their roles in metal tolerance and transformation. Nevertheless, microorganisms have a large impact on metal phytoremediation and phytomining. In this chapter, we summarize the latest insights in the bacterial and fungal communities associated with common hyperaccumulator plants, together with future prospects, and we provide suggestions to contribute research in this exciting field.

2. HYPERACCUMULATOR PLANT SPECIES

Hyperaccumulator plants are able to take up large amounts of metals in their tissues without showing any symptoms of toxicity (Baker & Smith, 2000; Goolsby & Mason, 2015). For example plant species accumulating

>1000 mg/kg of Cu, Co, Cr, Ni or Pb or >10,000 mg/kg of Mn or Zn in their tissues have been defined as hyperaccumulator species (Baker & Brooks, 1989), and these concentrations are typically up to four order of magnitude higher than in other adjacent plants (Baker & Smith, 2000). These unusually higher concentrations occur even when the metal is present at low external concentrations, suggesting that plants may benefit from metal accumulation rather than being injured by it (Rascio & Navari-Izzo, 2011). The accumulation of metals in the aboveground tissues is therefore sometimes suggested to provide a protection against herbivores (Freeman et al., 2007). Though, the evolution of hyperaccumulator plants is still under investigation, and we have not yet fully understood the physiology (van der Ent, Baker, Reeves, Pollard, & Schat, 2013; Goolsby & Mason, 2015; Verbruggen, Hermans, & Schat, 2009). Molecular mechanisms of metal hyperaccumulation suggest that hyperaccumulators have probably evolved through adaption of metal homoeostasis processes (van der Ent et al., 2013; Verbruggen et al., 2009). Some of these internal tolerance mechanisms that have been identified include (1) sequestration of the metals in root tissues, (2) transport to cell components not involved in physiological processes such as vacuole, cell wall, apoplast (Leitenmaier & Kupper, 2013) and (3) complexation of metals by organic acids and amino acids such as metallothioneines and phytochelatins (Azevedo et al., 2016; Verbruggen et al., 2009). Further, to accumulate and sequestrate trace metals, metals must be taken up for which high expression of metal transporter genes, secretion of chelating agents and active root proliferation towards contaminated soil spots have been reported as important mechanisms (Pence et al., 2000).

Hyperaccumulator plants comprize currently a collection of more than 500 plant species of which the majority are Ni-hyperaccumulators growing on serpentine (ultramafic) soils (Goolsby & Mason, 2015). Although hyperaccumulation of trace metals is found throughout the whole plant kingdom and climate conditions, hyperaccumulator plant species are typically restricted to native species growing on mineralized soils and related rock types (Baker & Smith, 2000). The most important types of metalliferous soil colonized by hyperaccumulator plants are serpentine soils which are enriched in Co, Cr and Ni, 'calamine' soils enriched in Cd, Pb and Zn, Se-rich soils and Cu- and Co-containing soils (Reeves & Baker, 2000). Although we focus on metal accumulating plant species that can be used for phytoremediation, some of the plants can also be used in phytomining (growing plants to harvest metals) (Chaney et al., 2007) or for food fortification (Zhu, Pilon-Smits, Zhao, Williams, & Meharg, 2009). We can classify hyperaccumulators related to their soil habitat, into accumulators of Ni; Cd, Pb and Zn; Se; Co and Cu; and other elements (Al, As, Cr, Mn, Tl) (Reeves & Baker, 2000). *Pteris vittata* (PV. Chinese brake fern) was the first known As-hyperaccumulator which can accumulate up to 2%–3% As in its shoots (Ma et al., 2001). *Pteris multifidi* (PM, spider brake fern) is a more recent discovered arsenic hyperaccumulator, which can also accumulate >1000 mg/kg As in its fronds (Slonecker, Haack, & Price, 2009). Several plant species have been proposed for phytoextraction of Ni, Cd, Zn and Pb., e.g., *Alyssum bertolonii*, *Arabidopsis halleri*, *Solanum nigrum*, *Eichhornia crassipes* and *Thlaspi caerulescens* (McGrath et al., 2006). *A. bertolonii* Desv. (Brassicaceae) is a nickel hyperaccumulator endemic to the serpentine soils of central Italy and belonging to a genus that has been used commonly for phytoremediation (Chaney, Angle, Baker, & Li, 1998). It can contain in its aerial parts from 7000 to 12,000 μg of Ni/g dry weight and can be used for phytomining (Galardi et al., 2007). *A. halleri* naturally accumulates and tolerates leaf concentrations as high as 2.2% Zn and 0.028% Cd in dry biomass (Dahmani-Muller, Van Oort, Gelie, & Balabane, 2000), a constitutive trait as has been found in a large number of European populations (Bert, Bonnin, Saumitou-Laprade, De Laguérie, & Petit, 2002). *S. nigrum* L. is an annual or perennial herb which has been characterized as a Cd-hyperaccumulator with a remarkable ecological adaptability throughout the world (Khan et al., 2014; Wei, Zhou, & Koval, 2006). *Sedum plumbizincicola* is a Zn/Cd hyperaccumulator from lead-zinc mining areas native to Zhejiang Province, China (Wu, Li, & Luo, 2008). *Phytolacca americana* is a Mn-hyperaccumulator widely distributed in Mn mining wastelands and Mn-contaminated soils (Wei et al., 2014). *Sedum alfredii* (SA) has been identified as a Zn and Cd cohyperaccumulator plant species, also native to China (Yang et al., 2004).

Hyperaccumulator plants have been studied for a long time mostly from the plant perspective (Verbruggen et al., 2009). This means, hyperaccumulation is seen as an intrinsic ability of a plant based on the presence of appropriate physiological adaptations (Rascio & Navari-Izzo, 2011). Since the major revolution in cultivability of microbes and molecular tools to study them, the hypothesis that metal hyperaccumulation is only related to plant genetic traits has been revised, and it has become more and more clear that also the plant-associated microorganisms and their genes are very likely to explain together the observed hyperaccumulator plant phenotype. The high bacterial and fungal diversity in the rhizosphere is due to the high level of nutrients such as amino acids, organic acids and sugars exuded from the plant roots and capable of supporting microbial growth (Lopez-Guerrero, Ormeno-Orrillo, Rosenblueth, Martinez-Romero, & Martinez-Romero, 2013). On the other hand, bacteria and fungi can promote plant growth by providing compounds or by facilitating nutrient uptake from the environment (Segura, Rodríguez-Conde, Ramos, & Ramos, 2009). They can also improve plant competitiveness and response to external stress factors and promote plant health by the production of plant hormones (Pilon-Smits, 2005). In this complex and dynamic microenvironment, the microflora may increase metal mobility and availability to the plant or else decrease metal solubility due to precipitation. Currently however, it is estimated that less than 10% of hyperaccumulator plant species have been investigated for the diversity of rhizospheric and endophytic microorganisms and their role in metal uptake (Alford, Pilon-Smits, & Paschke, 2010).

3. RHIZOSPHERIC AND ENDOPHYTIC MICROORGANISMS

The plant rhizosphere and plant endophytic tissues – the internal plant including root, stem, leaf apoplast spaces and transport vessels – are ubiquitous habitats for microorganisms (Compant, Clément, & Sessitsch, 2010; Hardoim et al., 2015). Plant roots provide inorganic nutrients and water to the plant, whereas the green shoots fix carbon through photosynthesis, and part of the organic compounds are transported back to the roots. Roots excrete these organic compounds back into the surrounding soil environment, considered the rhizosphere, a soil biologically and biochemically influenced by the living roots and soil organisms. The rhizosphere is a very dynamic environment hosting a high number and wide diversity of (micro)organisms (Berendsen et al., 2012). For instance, it is estimated that the rhizosphere contains up to 10^{11} microbial cells per gram soil and thousands of different species (Berendsen et al., 2012; Mendes, Garbeva, & Raaijmakers, 2013). These rhizosphere-associated microorganisms, including bacteria, fungi and archaea, are very important to their host and can influence plant growth in many ways (Mendes et al., 2013).

While some plant-associated microorganisms may be detrimental to plant health and cause disease, the majority of plant-associated microorganisms support the plant or even increase plant fitness by one or various mechanisms. Specific bacterial populations called PGPR can competitively colonize plant roots and can promote plant growth and/or reduce the incidence of soilborne diseases (Lugtenberg & Kamilova, 2009). PGPR can directly promote plant growth by production of bacterial metabolites, phytohormones, that positively affect the plant (Spaepen & Vanderleyden, 2011). For example, many plant-associated bacteria synthesize the plant hormone and growth regulator indole-3-acetic acid (IAA) (Spaepen, Vanderleyden, & Remans, 2007; Spaepen & Vanderleyden, 2011), and enhanced production of IAA by a rhizosphere bacterium can improve plant growth (Mohite, 2013). In addition, the production of cytokinin, gibberellins and acetoin by rhizobacteria has been suggested to enhance plant growth (Arkhipova et al., 2007; Bottini, Cassán, & Piccoli, 2004). Another mechanism to improve plant growth includes increasing the availability, mobility and uptake of nutrients (e.g., phosphate, iron, magnesium, potassium, nitrogen...) and water (Glick, 2012). Some free-living and symbiotic rhizobacteria can fix atmospheric nitrogen (De Bruijn, 2015, pp. 215–225; Dobbelaere, Vanderleyden, & Okon, 2003). The production of siderophores by PGPR plays an important role in iron competition and has been identified as a mechanism contributing to biocontrol activity (Miethke & Marahiel, 2007; Ponraj, Shankar, Ilakkiam, & Gunasekaran, 2012; Rajkumar, Ae, Prasad, & Freitas, 2010). Furthermore, it has been suggested that siderophores might be involved in the induction of systemic resistance (Bakker, Pieterse, & van Loon, 2007; Kloepper, Ryu, & Zhang, 2004). PGPR can also regulate plant indigenous hormone profiles by the production of 1-amino-cyclopropane-1-carboxylate (ACC)-deaminase (Glick, Penrose, & Li, 1998; Saleem, Arshad, Hussain, & Bhatti, 2007). ACC-deaminase producing bacteria play an important role in the alleviation of different types of stress in plants including trace metals, by lowering ethylene concentrations in plants (Arshad, Saleem, & Hussain., 2007; Glick, 2014). Many PGPR can also suppress phytopathogens by mechanisms such as antibiosis or by inducing systemic resistance (Compant, Duffy, Nowak, Clement, & Barka, 2005). A variety of biocontrol strains simply outcompete pathogens by efficiently colonizing plants and consuming resources (Whipps, 2001).

Hyperaccumulator plants, because they accumulate large amounts of trace metals in their tissues, provide also a specific environment for the plant-associated endophytic bacteria that must adapted to survive in these high metal concentrations. Bacterial endophytes live inside plant tissues for all or part of their life cycle and cause no symptoms of disease (Rosenblueth & Martinez-Romero 2006; Schulz & Boyle, 2006). Endophytic bacteria first colonize the plant surface and may subsequently gain entry into plants via tissue wounds, root hairs, stomata, lenticels or germinating radicles (Compant et al., 2010). Bacteria may also actively invade intact plants by penetrating root hair cells or by producing cell wall–degrading enzymes (Prieto et al., 2011). The mechanisms for biocontrol and growth promotion by an endophyte are very similar to those exhibited by rhizosphere bacteria, which makes sense as they often exist part of their life as soil or rhizosphere bacteria (Hardoim et al., 2008). Endophytes enjoy also a more favourable environment and experience less competition due to their close interaction with the host (Brader, Compant, Mitter, Trognitz, & Sessitsch, 2014). Systemic endophytes or those species that are able to colonize the xylem and phloem transport vessels can even be transmitted via the seeds to next generations, an important mechanism of microbe-mediated plant adaptation to metal tolerance (Mastretta et al., 2009; Truyens et al., 2015; Truyens, Weyens, Cuypers, & Vangronsveld, 2015).

Fungi are also known to inhabit the rhizosphere and engage in a mutually beneficial symbiosis with the plant (Bonfante & Genre, 2008; see chapter: Mycorrhiza-Assisted Phytoremediation by Coninx et al., 2017). The symbiotic mycorrhizal fungi, including the ectomycorrhizal (EM) fungi, arbuscular mycorrhizal (AM) fungi and ericoid mycorrhiza are known to play an important role in plant health through various ways. Important functions exerted by mycorrhiza are for example nutrient supply, water house holding and protection against pathogens (Bonfante & Anca, 2009). In the case of hyperaccumulators, arbuscular mycorrhiza are by far the most studied (Gohre & Paszkowski, 2006; Leung et al., 2013). AM fungi help hyperaccumulators by alleviating the negative effects that are associated with hyperaccumulation (Miransari, 2011a,b). EM fungi can induce the immobilization of trace metals on the surface of living hyphae using chemical bonding groups in the cell wall or by complexation of the trace metal with small molecular organic compounds exuded in the rhizosphere (Luo et al., 2014).

4. EFFECT OF TRACE METALS ON MICROBIAL COMMUNITIES

Trace metal contamination in soil is known to impact plant quality and yield, but it also causes changes in the size, composition, growth and activity of microbial communities in soil (Hassan Sel, Boon, St-Arnaud, & Hijri, 2011; Op De Beeck et al., 2015; Sessitsch & Puschenreiter, 2008). Together with other environmental factors, pH, moisture, temperature, nutrient status, trace metals thus influence microbial community structure. Numerous studies have shown the effect of trace metals, in inorganic or organic forms, on microbial diversity, biomass and activity (Sessitsch &

Puschenreiter, 2008). In most cases, high metal pollution has shown to reduce bacterial and fungal diversity in the bulk soil (Azarbad et al., 2013; Rajapaksha, Tobor-Kaplon, & Baath, 2004; Stefanowicz, Niklinska, & Laskowski, 2008). On the other hand, it has been found that bacteria isolated from polluted soils are more tolerant to higher levels of metals than those isolated from unpolluted soils (Almås, Bakken, & Mulder, 2004; Azarbad et al., 2013; Croes et al., 2013; Kandeler et al., 2000). That is because metal-sensitive bacteria die or are outcompeted by the tolerant strains. Metal-contaminated soils are also a hotspot of horizontal gene-transfer of plasmids containing metal resistance genes which may greatly contribute to the adaption of surviving bacteria in metal-contaminated soils (Sentchilo et al., 2013).

The toxicity of metals can be different in the rhizosphere because of the influence of the living root and consequent interactions with the microorganisms. Studies have shown that in response to trace metal pollution, plants alter the secretion of roots exudates (amount and composition), which can significantly alter the concentrations of bioavailable metals, their toxicity and consequently effect on the rhizosphere microbial communities (Chen et al., 2003; Qin, Hirano, & Brunner, 2007; Zhou, Liu, Wang, & Xu, 2011). Root exudates may for instance either complex metals, making them unavailable to microbes, or enhance metal release, for example, by altering the pH (Puschenreiter et al., 2005). Also in the rhizosphere a higher number of trace metal-tolerant bacteria are found on contaminated soil compared to unpolluted soil (Luo et al., 2011; Ma et al., 2015; Mirete, de Figueras, & Gonzalez-Pastor, 2007). Because plants accumulate high concentrations of trace metals in shoot tissues, they also provide a specific environment to endophytic bacteria that could be adapted to survive in high metal concentrations (Luo et al., 2011; Ma et al., 2015; Rajkumar et al., 2009).

5. STRUCTURE AND DIVERSITY OF THE HYPERACCUMULATOR RHIZOSPHERE MICROBIOME

Attempts made to characterize and describe the rhizospheric and endophytic microbial communities associated with hyperaccumulators have focussed for a large part on the culturable isolates (Visioli et al., 2014). Though the number of soil bacteria that can be cultivated is estimated to be less than 1%–10% (Pham & Kim, 2012), thus we have overlooked a significant part of the microbiota. Culture-independent techniques relying on the total isolated DNA from soil have also been used to characterize the microbial community structure and dynamics, with denaturating gradient gel electrophoresis, amplified rDNA restriction analyses (ARDRA), terminal restriction fragment length polymorphism (T-RFLP) and automated ribosomal intergenic spacer analyses. To examine the phylogenetic structure of metal-resistant communities associated with the host plant, primers were used that specifically amplify the bacterial 16S rRNA gene or fungal internal transcribed spacer (ITS) region and further analyzed by Sanger sequencing (Mengoni, Pini, Huang, Shu, & Bazzicalupo, 2009). Alternatively, primers that are specific for metal transporters or tolerance genes have been used for polymerase chain reaction (PCR)-amplification, and further analyzed by clone libraries and Sanger sequencing (Mirete et al., 2007).

Nowadays, with the advent of next generation sequencing technologies such as llumina, Ion Torrent, Pacific Biosciences and Oxford Nanopore sequencing, addressing microbial community structure has become much more achievable and at tremendously increased resolution (Caporaso et al., 2012; Rhoads & Au, 2015). These technologies allow the massive parallel sequencing of complex environmental DNA, RNA or protein sequences at unprecedented depth. For example, the total isolated microbial DNA can be sheared in smaller fragments and directly used for sequencing, termed shotgun metagenomics, to obtain insight in the potential active genes and microbial species involved. Although these technologies are state-of-the-art in human health research, the implementation of these technologies in the field of hyperaccumulators, and soil remediation, has lagged behind but a rapid catch up race is expected (Bell, Joly, Pitre, & Yergeau, 2014).

The ability to quantify microbial community structure and diversity in depth using these high-throughput massive parallel sequencing technologies has already led to increasing insights in the genomic content and functional diversity of microbial communities of various crop species (Sessitsch et al., 2012), and the human gut microbiome (Arumugam et al., 2011). Human and plant microbiome studies have shown that the microbiome extends the phenotype of the host microorganisms and plays a very important role in the host fitness and functioning. Moreover, manipulation of the microbiome has resulted in successful new therapies to treat human diseases (Rooks et al., 2014) or control plant pests (Berg, Krause, & Mendes, 2015). Similarly, it is well known that the establishment of plant species in metal-contaminated soils depends greatly on the symbiotic association with rhizospheric and endophytic microorganisms which provide tolerance to different environmental stresses (Bell et al., 2015; Muehe et al., 2015). For example, the presence of indigenous rhizosphere microflora may increase plant metal accumulation capacity and increase shoot metal content and plant chlorophyll values (Farinati et al., 2009; Farinati, DalCorso, Panigati, & Furini, 2011; Muehe et al., 2015). Due to their importance for practical applications, the desire to model and manage microbial communities, exploit host–microbiome interactions and understand how the microbiota assist the host is receiving increased interest in various research fields. A first step towards integrating microbial community functions into bioremediation strategies requires a detailed insight in the microbial community structure and diversity of different plant species and the correlation with functional traits of the plant hosts in the natural environment.

5.1 Bacteria

The rhizosphere bacterial community colonizing the roots of the **Ni-hyperaccumulator *Alyssum murale*** when grown in ultramafic soil in Italy with high levels of Ni was determined by pyrosequencing (S. Lopez, pers. communication). Analyses of the generated 16S rRNA gene sequences showed that a diverse number of taxa, genera and species were able to colonize *Alyssum* rhizosphere, despite the high Ni concentrations in soil. A high proportion of Chloroflexi was found (greater than 50%), and the higher the soil Ni contents, the higher the relative abundance of Proteobacteria and Actinobacteria. On the contrary, the abundance of TM7 decreased with increasing levels of bioavailable Ni.

Similarly, a study using T-RFLP analysis and culture-dependent methods showed that Proteobacteria dominated in the rhizosphere of the **Ni-hyperaccumulator *A. bertolonii***, endemic to serpentine soils in central Italy (Mengoni et al., 2004). The plant rather than the locality shaped the microbial community structure (Mengoni et al., 2004). In addition, the cultivation-dependent analyses showed that the culturable Ni resistant rhizosphere community was dominated by *Pseudomonas* whereas soil samples contained a high number of Ni resistant *Streptomyces* spp. (Mengoni, Barzanti, Gonnelli, Gabbrielli, & Bazzicalupo, 2001). Furthermore, for all soils sampled from three Ni containing serpentine sites in Italy, an increasing number of resistant strains in increasing proximity to the Ni hyperaccumulator plant were found indicating a possible important role of these microbes for plant growth (Mengoni et al., 2001). Barzanti et al. (2007) isolated 83 endophytic bacteria from the roots, stems and leaves of *A. bertolonii*, and they were classified by restriction analysis of 16S rDNA (ARDRA) and partial 16S rDNA sequencing into 23 different taxonomic groups, with the highest diversity in roots, and then stem and leaves. Most of the ARDRA types were represented by Gram-positive bacteria, in particular the genera *Bacillus, Paenibacillus, Leifsonia, Curtobacterium, Microbacterium, Micrococcus* and *Staphylococcus* (Barzanti et al., 2007). Overall, Firmicutes was the most dominant group (60%) followed by Actinobacteria (28%). Furthermore, most of the isolates were resistant to more than one trace metal, with coresistance to Ni, Cr, Zn and Cu the most frequent (Barzanti et al., 2007). The results pointed out that the diverse endophytic flora found in the tissues of *A. bertolonii* with a high level of resistance to trace metals, could potentially help plant growth and Ni hyperaccumulation. In a subsequent study, also the yet-to-be-cultured fraction of the leaf-associated community of this hyperaccumulator was investigated using T-RFLP fingerprinting (Mengoni et al., 2009). Fifty-nine unique terminal-restriction fragments (TRFs) were identified, with more than half of the taxonomically interpreted TRFs assigned to Alpha- and Gamma-proteobacteria and Clostridia (Mengoni et al., 2009). Results indicated a very high plant by plant variation of leaf-associated communities (93%), but also a core of plant specific TRFs was found. Future studies are certainly needed to obtain further insights in the association between bacteria and the Ni hyperaccumulator *Alyssum*.

A study involving the **Ni hyperaccumulator *Thlaspi goesingense*** showed a high percentage of *Methylobacteria, Rhodococcus* spp. and *Okibacterium* spp. in the rhizosphere as determined by T-RFLP, cloning and sequencing of the 16S rRNA genes (Idris, Trifonova, Puschenreiter, Wenzel, & Sessitsch, 2004). In addition, *T. goesingense* hosted members of the phylum Holophaga/Acidobacterium, high-GC (guanine-cytosine) Gram-positive bacteria, members of the Bacteroidetes phylum and Verrucomicrobia (Idris et al., 2004). Overall, the microbial community structure of the shoot-associated endophytes was highly different from the rhizosphere. *Methylobacteria* were identified as dominant endophytic colonizers of *T. goesingense* (Idris et al., 2004), but different strains of *Methylobacterium* were found inside the plant and in the rhizosphere, suggesting that these habitats provide distinct growth conditions for microorganisms. One strain fell into a newly described species, *Methylobacterium goesingense* (Idris et al., 2006). As many *Methylobacteria* strains displayed tolerance to high Ni concentration, this leads to the speculation that *Methylobacteria* could play a role in metal tolerance and metal hyperaccumulation in Brassicaceae. Actually, *Methylobacteria* are known to produce phytohormones that could also help plant growth (Idris et al., 2006; Kutschera, 2007). Similar to what was found for the Ni hyperaccumulator *T. goesingense*, most isolates from the shoots of **Zn hyperaccumulator *T. caerulescens*** were affiliated with the genera *Methylobacterium* and *Sphingomonas* (Lodewyckx, Mergeay, Vangronsveld, Clijsters, & Van Der Lelie, 2002). *T. caerulescens* was grown in soil from an abandoned Zn–Pb mine and smelter in Plombières, Belgium (Lodewyckx et al., 2002). The shoot endophytic bacteria showed higher resistance to Zn and Cd than those from the rhizoplane and roots, and the rhizosphere contained more metal-resistant strains than the surrounding bulk soil. These differences in the characteristics of the bacterial microflora associated with *T. caerulescens* were speculated to reflect altered metal speciation in the different soils and plant compartments studied (Lodewyckx et al., 2002). In another study, a comparative 16S rDNA and 16S rRNA-based analysis was used to characterize bacterial diversity in trace metal contaminated bulk soil and in the rhizosphere of *T. caerulescens* (Gremion, Chatzinotas, & Harms, 2003). DNA-based analysis indicated the dominance of Alpha- and Beta-Proteobacteria, Acidobacteria and Planctomycetes in the rhizosphere. In contrast, analysis of the metabolically active population showed that members of the Rubrobacteria subdivision, belonging to the high-GC Gram-positive Actinobacteria, were highly dominating (Gremion et al., 2003). This is one of the first studies providing insights in the metabolically active populations in trace metal contaminated rhizosphere soil.

Muehe et al. (2015) studied the role of the rhizosphere microbiome on metal uptake in ***A. halleri***, a **hyperaccumulator of Zn and Cd**. In total 1449 operational taxonomic units were found in the natural metalliferous soil, compared to only 406 in the gamma-irradiated soil. *Lysobacter, Streptomyces, Agromyces, Nitrospira* and *Candidatus Chloracidobacterium* were present in 10- to 60-fold higher abundance in the untreated soil compared to the gamma-irradiated soil, and they were presumably involved in the higher metal uptake (Muehe et al., 2015). Despite no differences were detected in aboveground biomass of plants on native soils, the shift in rhizosphere microbial community structure, diversity and richness was associated with 100% more cadmium and 15% more zinc accumulation in the aerial parts of the plant, suggesting that rhizobacteria are strongly involved in metal mobilization and uptake. In another study, Luo et al. (2011) isolated a total of 30 Cd-resistant endophytic bacteria from the organs of *S. nigrum* belonging to four groups, Actinobacteria, Proteobacteria, Bacteroidetes and Firmicutes (Luo et al., 2011). Ma et al. (2015), isolated 42 endophytes from tissues of **Zn/Cd hyperaccumulator *S. plumbizincicola***, of which 5 were further studied in inoculation experiments due to their ability to promote plant growth and to utilize ACC as the sole nitrogen source. The five isolates were identified as *Bacillus pumilus* E2S2, *Bacillus* sp. E1S2, *Bacillus* sp. E4S1, *Achromobacter* sp. E4L5 and *Stenotrophomonas* sp. E1L (Ma et al., 2015). All of the isolates showed IAA production, siderophores and solubilization of phosphorus.

Phytoremediation of As-contaminated soil by **As hyperaccumulator *P. vittata*** has also received considerable interest. The rhizosphere of *P. vittata* grown in As-contaminated soil is assumed to be selectively enriched in microorganisms that are adapted to As stressed condition and to the utilization of specific plant compounds. A first metagenomics shotgun sequencing approach has been undertaken to investigate the plant-associated microbiota and functions in the rhizosphere of ***P. vittata***, the raw data are present on the NCBI Website but not published yet (https://www.ncbi.nlm.nih.gov/bioproject/?term=pteris%20vittata%20metagenome). Recently, Xu, Han, Chen, Zhu, and Ma (2016) isolated endophytic bacteria associated with all tissues of *P. vittata* when exposed to 200 mg arsenate per kg spiked soil for 60 days. The endophytic bacteria represented four phyla (Proteobacteria, Actinobacteria, Bacteroidetes and Firmicutes) and 17 genera, all having at least one plant growth enhancing ability (Xu et al., 2016). Similarly, another culture-dependent study isolated eight root endophytes from *P. vittata* plants growing on As-amended soil, belonging to Proteobacteria (*Enterobacter, Stenotrophomonas, Rhizobium*), Firmicutes (*Bacillus*) and Bacteroidetes (Tiwari, Sarangi, & Thul, 2016; Zhu, Guan, Luo, Rathinasabapathi, & Ma, 2014). Many were As tolerant with one showing tolerance up to 1000 mg/L concentration in minimal medium, and several strains were also IAA positive exhibited siderophore activity and phosphatase activity. Arsenic transformation experiments showed that some strains had dual activity in As(III) oxidation and As(V) reduction (Tiwari et al., 2016). The presence of arsenite oxidase genes (aox) and arsenite transporter (ArsB) were identified (Tiwari et al., 2016). Zhu et al. (2014) isolated 41 arsenic resistant endophytes from *P. vittata* and 33 from *Pteris multifida*. Similar to the previous studies, *Bacillus* sp. was a dominant isolate in both plant types, further *P. vittata* was colonized by *Paenibacillus* and *P. multifida* by *Lysinibacillus, Massilia, Micrococcus, Brevundimonas, Roseomonas, Staphylococcus, Sphingomonas* and *Microbacterium*, amongst others. Interestingly, *P. vittata* isolates were more tolerant to As(V) and *P. multifida* isolates were more tolerant to As(III).

Three hundred Mn-resistant endophytic bacteria were isolated from the **Mn hyperaccumulator *P. americana*,** grown at different levels of Mn (0, 1 and 10 mM) stress (Zhang, He, Wang, & Sheng, 2015). They were affiliated with 10 genera, with a dominance of *Bacillus* sp. (29%–67%). Moreover, a greater percentage of ACC-deaminase producing bacteria were found in Mn-treated *P. Americana* (21%–81%) versus 11% in nontreated soils, suggesting the importance of PGP-producing endophytic bacteria to alleviate the negative effects of Mn stress (Zhang, Chen, He, Wang, & Sheng, 2015).

Summarized, hyperaccumulator plants can be colonized by a high number of different genera and species of metal-tolerant bacteria. For the majority of the plants, Proteobacteria were frequently identified, as well as Actinobacteria and Firmicutes. However, the examples also highlight that we have not yet identified the functions of many yet-to-be-cultured isolates such as *Candidatus Chloracidobacterium* and many other Acidobacteria, as detected by Muehe et al. (2015) in the rhizosphere of *A. halleri*.

5.2 Mycorrhizal fungi

Different plant ecotypes with variable uptake abilities of trace metals have been reported around the world (Khan et al., 2014). Therefore, it is important to analyze hyperaccumulators also for their association with over 400-million-year-old endophytic AM fungi, as well as the biodiversity of the fungal community in a symbiotic relationship with them. Mycorrhizal fungi are known to have a close relationship with their host, sometimes closer than bacteria–plant relationships. Moreover endophytic fungi are taxonomically highly diverse and can shape through multiple mechanisms the morphological and physiological functions of the host plant, allowing it to withstand multiple stresses including exposure to trace metals. AM fungi have been shown to form symbiotic relationships with *P. vittata*, a common arsenic hyperaccumulator. This is due to the similar chemical properties of phosphorous and arsenic, and as AM fungi have been shown to aid in uptake of plant phosphorous, they could also assist in plant arsenic uptake (Al Agely, Sylvia, & Ma, 2005). It is therefore important to analyze the diversity of endophytic and EM fungi of many hyperaccumulators to determine which fungi are the best candidates to use for phytoextraction of soil contaminated with trace metals. Though we have only a limited understanding of the phylogenetic distribution of fungi in trace metal contaminated soils, let alone the rhizosphere or endosphere of hyperaccumulators.

In a recent study, metal concentrations (Zn and Cd) together with soil organic matter content were found to be the most important factors shaping the pioneer soil fungal communities of a single young Scots pine plantation on metal-polluted soil in the Campine region in Belgium (Op De Beeck et al., 2015). A restricted number of ectomycorrhizal fungal species in the vicinity of pine was found to dominate the collection, including *Suillus luteus* and *Sistotrema* sp. (Basidiomycota) and *Wilcoxina mikolae* and *Cadophora finlandica* (Ascomycota) (Op De Beeck et al., 2015). In another study investigating the AM fungi community structure of *Plantago major* on sites polluted with trace metals (As, Cd, Cu, Pb, Sn and Zn) in Montreal, Canada, it was found that ribotypes *Glomus mosseae* and *Glomus* spp. B9 and B13 were dominant in trace metal-polluted soils, suggesting the tolerance of this species to metal stress, as well as potential use for phytoremediation (Hassan Sel et al., 2011). In a recent study, Khan et al. (2016) isolated endophytic fungi associated with the Cd-hyperaccumulator ***S. nigrum*** and characterized them for Cd tolerance and host plant modulation under Cd contamination. Forty-two culturable endophytic fungal isolates were obtained representing various morphotypes and among which 14 different genera of the fungal phylum Ascomycota based on analyzes of the ITS region of the rRNA gene (Khan et al., 2016). Along the identified genera, *Colletotrichum* (18%) represented the dominant genus followed by *Alternaria* (14%) and *Fusarium* (14%). The majority of the endophytic isolates were sensitive to a lower concentration (0.5 mM) of Cd. However, *Glomerella truncata* PDL-1 and *Phomopsis fukushii* PDL-10 showed the highest tolerance to a concentration gradient of Cd (0.5–2 mM) (Khan et al., 2016). In another study, it was shown that the spores of fungi can be sensitive to the trace metals in the soil when plants are not present, which results in negative effects for the fungi when trace metal concentration is too high (Shalaby, 2003). In particular, it was shown that trace metals in the soil with the absence of plants terminate hyphal growth in all cases and that the resistance to the trace metals is not genetic, but rather phenotypic plasticity (Shalaby, 2003). With this is mind, contaminated sites could be inoculated with trace metal-resistant fungi and then introduce plants, rather than just introducing plants into the soil without a preestablished symbiosis (Gohre & Paszkowski, 2006).

Overall, with the lack of studies using high-throughput analyses to characterize mycorrhizal fungi associated with hyperaccumulators, the ecological function of many of these very abundant microorganisms is still unknown and requires further detailed analysis. Because it is apparent that symbiotic fungi play a key role in the rhizosphere of hyperaccumulators, having knowledge as to how the fungal community is established is essential in furthering our understanding.

5.3 Archaea

Not much is known about the rhizosphere and archaea, and much less is known about the rhizosphere of hyperaccumulators and the archaea they attract. All that can be stated with confidence is that archaea are present in the rhizosphere of hyperaccumulators. Various culture-independent methods, such as 16S rRNA gene amplification show that Crenarchaeota and Euryarchaeota exist in the rhizosphere, but only about 0.16% of the archaeal sequences were received. Further studies on archaea in the rhizosphere look at *Lycopersicon esculentum* as a hyperaccumulator for Fe, Cu and Ni (Ochsenreiter, Selezi, Quaiser, Bonch-Osmolovskaya, & Schleper, 2003). Using Fluorescent in situ hybridization, the results indicated that nonthermophylic Crenarchaeota extensively colonize the roots (Ochsenreiter et al., 2003).

6. FUNCTIONS OF THE HYPERACCUMULATOR RHIZOSPHERE MICROBIOME

Microbial community studies focussing only on the DNA-level do not give any information on the activity of cells, and even dead cells may be detected as the DNA has not been destroyed, thus more RNA-based analyses are needed. In addition, development of genetic tools such as in vivo expression technology (Barr, East, Leonard, Mauchline, & Poole, 2008), along with 'omic' technologies, (e.g., genomics, transcriptomics, proteomics, metabolomics) has provided opportunities to investigate global expression profiles of different microorganisms (bacteria and fungi) in response to plant signals (Bell et al., 2014; Cravatt & Kodadek, 2015) (Fig. 1). When expression profiles of all active (micro)organisms are studied, the term metatranscriptomics is used, which allows to determine whether particular classes of genes are actively expressed in rhizosphere microbial communities or specific populations (Turner et al., 2013).

Considerable attention has been paid to plant physiological mechanisms of metal transport, storage and tolerance of metals in hyperaccumulators, but less studies have been conducted to identify the role of total plant-associated microbial communities on plant metal tolerance and hyperaccumulation. However, there is strong evidence that plant-associated microbes have a large influence on plant tolerance and metal uptake (see chapter: Potential Role of Plant-Associated Bacteria in Plant Metal Uptake and Implications in Phytotechnologies by Kidd, Alvarez-Lopez, Becerra-Castro, Cabello-Conejo, & Prieto-Fernández, 2017). For example, in a study to test the influence of the rhizosphere microflora on metal accumulation, it was shown that *A. halleri* grown with their native rhizosphere microbiota accumulated double the amount of cadmium and 15% more zinc compared to plants grown on gamma-irradiated soil (Muehe et al., 2015). Another study by Farinati et al. (2011), showed that the diversity of rhizosphere soil microorganisms influenced the uptake of metals by *A. halleri*. In this experiment, *A. halleri* inoculated with an 'extracted' rhizosphere microbial community leads to the accumulation of more Cd and Zn compared to plants in the absence of the 'extracted' rhizosphere community in hydroponic cultures (Farinati

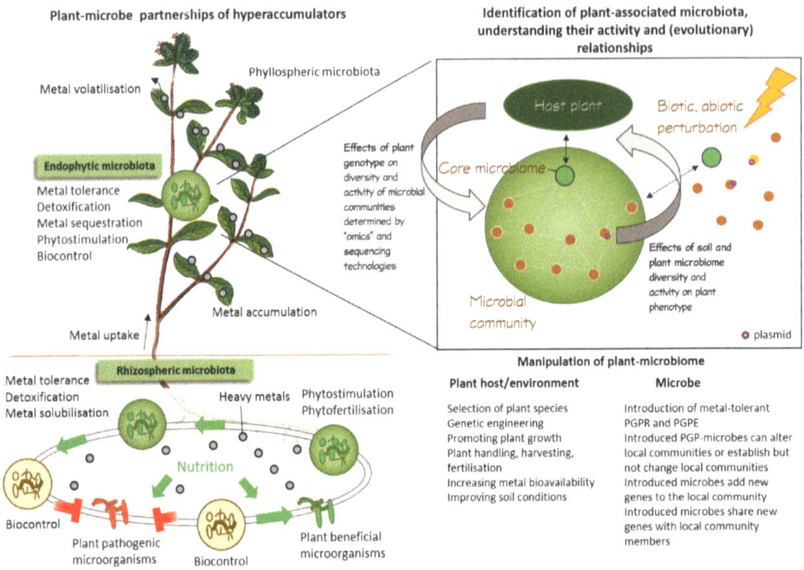

Figure 1 ***Identification, characterization and manipulation of the plant microbiome to enhance metal phytoextraction and phytomining.*** Trace metals in soil can be mobilized and transported into the roots through activity of metal-transporters in the plant root cells, and the activity of metal-tolerant microbiota. Metals can be transported out of the roots in the xylem and translocated into the shoots. Shoot endophytic bacteria can help to improve metal tolerance and increase plant growth through various mechanisms. By thoroughly characterizing the microbial diversity associated with hyperaccumulators, their function, and ecological relationship, metal phytoextraction may be enhanced through more targeted manipulation of the plant host, soil environment and microbial communities. Introduction of plant growth promoting (PGP)-microbes with appropriate metal-tolerant and detoxification systems on plasmids may be a promizing strategy to adapt the local microbial community for metal tolerance, and this may result in increased metal phytoextraction.

et al., 2011). As such the presence of microorganisms in the rhizosphere or endosphere seems to have a direct effect on metal uptake, and this can be directly related to the activity of the microbes. Also indirectly, symbiotic microorganisms (bacteria and fungi in particular) can shape morphological and physiological functions of the plant host to withstand multiple stresses including the exposure to trace metals (Dourado et al., 2013).

6.1 Metal Tolerance

There is a large body of evidence supporting that bacterial and fungal isolates colonizing the host plant growing in trace metal contaminated soil show high tolerance for trace metals, compared to isolates obtained from nonpolluted soils (Lodewyckx et al., 2002; Muehe et al., 2015). However, most studies are based on culturable isolates, and only a handful studies have used functional culture-independent analyses such as metagenome sequencing, functional gene-arrays or metatranscriptomics to investigate the distribution of metal-tolerance genes in the total microbial community of hyperaccumulators. The study of Xiong et al. (2010) used Geochip 3.0 to assess the functional potential of the rhizosphere microflora of the arsenic hyperaccumulator *P. vittata* under natural conditions. They reported a high occurrence of arsenic resistance genes, sulphur reduction, phosphate utilization and denitrification genes in *P. vittata* rhizosphere compared to nonrhizosphere soils, suggesting a strong link between the level of arsenic contamination in the soil and rhizosphere community and functional gene distribution (Xiong et al., 2010). Canonical correspondence analyses revealed that arsenic was the primary driver in reducing the functional gene diversity in the soil community, suggesting that rhizobacteria may play an important role in soil arsenic uptake and hyperaccumulation in *P. vittata* (Xiong et al., 2010).

Different transporters involved in nickel, copper, zinc and chromate resistance were detected by genomic analysis of three legume growth-promoting rhizobia, *Mesorhizobium amorphae* CCNWGS0123, *Sinorhizobium meliloti* CCNWSX0020 and *Agrobacterium tumefaciens* CCNWGS0286, that survive at regions with high levels of trace metals in China (Xie et al., 2015). *M. amorphae* contains metal transporters from P-type ATPase, Cation Diffusion Facilitator, HupE/UreJ and chromate ion transporter family involved in copper, zinc, nickel as well as chromate resistance and homoeostasis. The identified CopA/CueO system was expected to mediate copper resistance in *S. meliloti* while ZntA transporter, assisted with putative CzcD, determines zinc tolerance in *A. tumefaciens* (Xie et al., 2015). To develop more efficient systems for nickel phytomining, growing leguminous crops with metal-resistant rhizobacteria, side by side with hyperaccumulator plants has been proposed to enhance soil fertility (Saad, Kobaissi, Robin, Echevarria, & Benizri, 2016). In another study Tiwari et al. (2016) investigated genes for arsenic mobilization and uptake in the endophytic microbiome of the As hyperaccumulator *P. vittata*. They found that the presence of As transporters (ars) and As resistance gene (aoxB) in *P vittata* endophytes could be playing important roles in As-amended soils, and their presence might be enhancing As uptake and accumulation ability of this plant (Tiwari et al., 2016). Zhu et al. (2014) isolated several As-resistant endophytic bacteria from *P. vittata* (2 genera) and *P. multifida* (12 genera). In further wet-lab experiments, they had proven that the bacterial arsenic tolerance was related to their ability in As(III) oxidation and As(V) reduction, and their ability to retain As in the biomass to a varying extent, which has an important influence on As toxicity, tolerance and speciation in plants (Zhu et al., 2014).

Trace metal-tolerant fungal endophytes also have sophisticated molecular mechanisms comprized of extracellular or intracellular detoxification system to control trace metals uptake and concentration in various cellular organelles (see chapter: Mycorrhiza-Assisted Phytoremediation by Coninx et al., 2017; Zhao, Li, shen, Wang, & Zhao, 2015). For example, Zhao et al. (2015) cloned a trace metal transporter gene from Cd-resistant *Exophiala pisciphila*-dark septate endophytes (Zhao et al., 2015). The EpNramp gene (natural resistance associated macrophage protein) was identified as bivalent cation plasma membrane transporter, capable of iron uptake (Fe^{+2}), conferring high resistance to *E. pisciphila* against Cd toxicity. Further transcriptomic analyses of this fungal endophyte showed that many genes were differently expressed during Cd exposure (Wei et al., 2016), and 40% of them were associated in 10 important trace metal-tolerant pathways, including pathways for metal ion binding, organic acid production and their transportation, transcription factor, sulphate assimilation, DNA repair, cell wall maintenance, scavenging of reactive oxygen species and glutathione S-transferases enzymes. From the studies described above, it follows that presence and expression of metal-resistance and mobilizing genes is important for growth, survival and function of microbes in the rhizosphere, and consequently their effect on metal uptake and toxicity for the host plant.

6.2 Metal Mobilization

Low bioavailability of soil metals is one limiting factor for the success of phytoextraction in the field. We can enhance the uptake of metals by plants by the addition of chemical chelates, pH-ameliorators and stimulation of soil processes to promote the mobilization from minerals (Kidd et al., 2015). On the other hand, addition of organic amendments (Alvarez-Lopez et al., 2016) and composted municipal solid waste (Touceda-González et al., 2017) have been shown to decrease Cu availability and plant uptake. The use of expensive chelating compounds can increase the metal-leaching risk and impact negative effects on plant growth and soil structure. Plants themselves can mobilize metals by excreting low molecular weight chelators such as organic compounds, by lowering the pH in proximity of the roots to dissolve metals from minerals, or by acid-reduction-mediated processes to dissolute metals from poorly accessible organic and mineral forms, as reviewed (Kidd et al., 2009; Verbruggen et al., 2009). Though, besides the plant, plant–microbe interactions strongly determine the efficiency of metal phytoextraction (Cabello-Conejo et al., 2014; see chapter: Potential Role of Plant-Associated Bacteria in Plant Metal Uptake and Implications in Phytotechnologies by Kidd et al., 2017). Rhizosphere microorganisms can actively contribute to metal mobilization, metal tolerance and increasing metal uptake, thereby extending metal mobilization further than the direct proximity of the roots which is reached by the plant-derived chelators alone (Becerra-Castro et al., 2013). Bacteria and fungi may achieve this by the production of extracellular organic chelators, and ligands which can complex mobilized metals and thereby facilitate their transport to plant roots (Kolbas et al., 2015; Sessitsch et al., 2013). Microorganisms may also stimulate the production of metal transporters in plants (de Souza, Huang, Chee, & Terry, 1999). Some of the identified bacterial genera present in the rhizosphere of hyperaccumulators known to produce extracellular ligands, siderophores or catalyze iron reduction, include *Acidovorax*, *Bradyrhizobium*, *Chryseobacterium*, *Lysobacter*, *Massilia* and *Pseudomonas* (Sessitsch & Puschenreiter, 2008). In addition, mineral-dissolving microorganisms such as Fe(III) reducing bacteria, sulphur-oxidizing and phosphate solubilizing bacteria, fermenting bacteria and organic acid and proton releasing functional groups of microorganisms can increase metal bioavailability (Borch et al., 2010) (Fig. 1). Sulphur and phosphate for example are important nutrients for plant growth, and phosphate-associated trace metals could be mobilized during the transformation of these minerals.

Production of H+ and organic acids by rhizospheric bacteria appear to be the most significant mechanism for metal mobilization. Saravanan, Kalaiarasan, Madhaiyan, and Thangaraju (2007) reported the production of 5-ketogluconic acid that aides the solubilization of different Zn compounds by the endophyte *Gluconacetobacter diazotrophicus* under in vitro conditions. Majewska, Kurek, and Rogalski (2007) found that increases in microbially produced citric acid, acetic acid, catechol siderophores and Fe-chelators may have contributed to cadmium mobilization within soils, decreasing the pH from 6.5 to 5 after 48 h. The production of oxalic acid, tartaric acid, formic acid and acetic acid had a significant correlation with the concentrations of Zn and Cd mobilized from $CdCO_3$ and ZnO by rhizosphere bacteria associated with Cd/Zn hyperaccumulator *S. alfredii* (Li, Ye, & Wong, 2010). The uptake of trace metals such as Cd can also be enhanced significantly by its solubilization in soil through acidification or reaction with ligands, which allow the formation of mobile complexes (Sessitsch et al., 2013). Abou-Shanab et al. (2003) reported that the concentration of extractable Ni was increased from a high Ni soil of 2.2–2.6 mg/kg when the soil was inoculated with *Microbacterium* sp. which had the ability of producing acid and siderophores and solubilizing inorganic phosphate. All these studies indicate that the activity of soil bacteria would likely have a significant effect on increasing the bioavailability of metals in soils.

The study of Idris et al. (2004) showed that siderophore production is a frequently found trait among bacteria associated with trace metal accumulating plants. Trace metals have even been shown to stimulate the production of bacterial siderophores (Gilis et al., 1996; Lodewyckx et al., 2002). Siderophore compounds show high affinity for ferric iron but also form complexes with bivalent trace metal ions such as Al, Cd, Cu, Ga, In, Pb, Zn and radionuclides (Rajkumar et al., 2010) that can be assimilated by the plant. Hence, siderophores can directly assist in metal availability to plants (Rajkumar et al., 2010; Schalk, Hannauer, & Braud, 2011) and indirectly reduce trace metal toxicity by increasing the supply of iron to the plant (Qi & Zhao, 2013). The siderophore synthesized by *Pseudomonas fluorescens* improves Fe uptake in tomato, carnation, oats, vine and maize (Sharma & Johri, 2003). High levels of resistance to Cd (300 mg/L), Zn (730 mg/L) and Pb (1400 mg/L) were reported for a PGP strain of *Bacillus* sp. (SC2b), which was isolated from the rhizosphere of *S. plumbizincicola* grown in Pb/Zn mine soils (Ma et al., 2015). The siderophore-producing and arsenate-reducing *Pseudomonas* sp. bacterial strain plays a key role in the ability to convert arsenate to arsenite as well as promote plant growth and increase in the biomass of the fern *P. vittata* (Lampis, Santi, Ciurli, Andreolli, & Vallini, 2015). This suggests that the presence of rhizobacteria in soil can improve the efficiency of arsenic phytoextraction in hyperaccumulator plant species as well (Lampis et al., 2015). It has also been suggested that bacterial exopolymers may complex trace metals leading to reduced availability for plants (Kunito, Saeki, Nagaoka, Oyaizu, & Matsumoto, 2001). Exopolymers produced by bacteria can strongly bind trace metals, leading to the formation of difficult to degrade organic metal complexes (Kunito et al., 2001; Sessitsch et al., 2013). In addition, trace metal concentrations can induce the microbial production of exopolymers leading to enhanced metal resistance in plants (Bitton & Freihofer, 1977).

Fungi can also take part in the mechanism of enhancing trace metal solubilization by producing organic compounds such as citric acid and oxalic acid, which form complexes and enhance trace metal reclamation processes (see chapter: Mycorrhiza-Assisted Phytoremediation by Coninx et al., 2017; Gadd, 1999). All of these biogeochemical processes then contributes to an enhanced transport of dissolved metals into the roots by diffusion. Previously, it was reported that fungal inoculation of mine tailing soil has higher concentrations of available trace metals than noninoculated soil solutions (Babu, Shim, Bang, Shea, & Oh, 2014). In another study, it was shown that *P. fukushii* PDL-10 inoculation of *S. nigrum* increased the Cd concentration in plants, and thus PDL-10 was suggested to enhance the metal availability to plants (Khan et al., 2016). Similarly, the mycorrhizae associated with *Populus canescens* were reported to enhance the Cd uptake and lead to increased alleviation of oxidative stress than observed in nonmycorrhizal poplar plants (Ma et al., 2014). In general, a higher availability of trace metals in soil does not necessarily ensure higher level of translocation in different tissues of plants. For example, in the PDL-10 inoculated plants, the Cd contents were significantly higher in shoots of plants (Khan et al., 2016). This effect can be attributed to a certain capacity for metal transfer from stem to leaf, which is regulated to minimize further translocation once a threshold concentration is attained. A relatively higher accumulation capacity for Cd in roots of *S. nigrum* may be beneficial for the phytostabilization of Cd-contaminated soil, whereas, higher contents of Cd accumulated in the roots of PDL-10 inoculated plants better allow for phytoextraction (Khan et al., 2016).

6.3 Plant Growth Promotion

Another important aspect of phytoremediation is the employment of plants with high biomass and rapid growth rates. However, one limitation of using hyperaccumulator plants for phytoextraction is their small biomass and slow growth (Rascio & Navari-Izzo, 2011). Bacteria mainly plant growth—promoting rhizobacteria, and fungi, mainly AM fungi can significantly assist plant growth on contaminated soils by improving nutrient uptake by the plant or through the production of bioactive metabolites that promote host plant growth and protect against biotic and abiotic stresses (Li et al., 2010; Lugtenberg & Kamilova, 2009) (Fig. 1). Some PGPR that have been described to promote plant growth and used as pure cultures or cocultures for bioaugmentation and metal extraction with plants include, *Agrobacterium, Alcaligenes, Arthrobacter, Azospirillum, Bacillus, Burkholderia, Serratia, Pseudomonas and Rhizobium* (Glick, 2012).

Barzanti et al. (2007) reported that 83% of bacterial isolates recovered from *A. bertolonii* could produce siderophores and promote the plant growth under Ni stress. Xu et al. (2016) isolated endophytic bacteria from all tissues of the arsenic hyperaccumulator *P. vittata* when exposed to 200 mg/kg arsenate and found that many of the isolates showed at least one plant growth enhancing ability (Xu et al., 2016). Zhang, Chen, et al. (2015), Zhang, He, et al. (2015), and X. Zhang et al. (2015) showed that Mn-stress significantly affected the ratios of the IAA- and siderophore-producing endophytic bacteria, as well as ACC-deaminase producing bacteria, suggesting that Mn stress may be alleviated by the highly IAA- and siderophore-producing endophytic bacteria that existed in the plant tissues. Numerous other studies have shown that rhizospheric or endospheric bacteria of hyperaccumulator plants can promote plant growth and thereby indirectly increase trace metal accumulation of plants (Ma et al., 2011; Rajkumar, Sandhya, Prasad, & Freitas, 2012; Sheng et al., 2012; Zhang, He, et al., 2015).

Symbiotic fungi have also been reported to promote plant growth via different mechanisms including the production of phytohormones, enhancing nutrient availability and protecting against biotic and abiotic stresses (Waqas et al., 2012). Mycorrhiza can form an extensive hyphal net around the roots increasing the volume of soil that can be explored for nutrients by an order of magnitude. In exchange for plant organic carbon, mycorrhizal fungi make minerals in complex humus and lignin structures more available. *S. nigrum* plants inoculated with *G. truncata* PDL-1 and *P. fukushii* PDL-10 under Cd spiking of 0, 5, 15 and 25 mg/kg showed that PDL-10 inoculated plants had significantly higher Cd content in shoots as well as in roots than observed in the PDL-1 inoculated plants (Khan et al., 2016). Additionally, irrespective of Cd stress, PDL-1 and PDL-10 inoculation significantly improved plant growth attributes such as shoot and root length, chlorophyll content, and fresh and dry shoot/root weight as compared to those of noninoculated control plants (Khan et al., 2016). A possible explanation for the observed effects is that the fungi enhanced the host plant growth by inducing biochemical pathways of the plant to produce phytohormones or increasing the availability of nutritional elements for the host plant. Certain *Trichoderma* spp. can enhance host plant growth through the production of IAA, ACC deaminase, siderophores, acid phosphatase or phosphate solubilization under biotic and abiotic stresses (Su, Zeng, Bai, Jiang, & Li, 2010; Viterbo, Landau, Kim, Chernin, & Chet, 2010). In all, it can be concluded that endophytic fungal association with plants can alleviate the negative effects of trace metal stress, and this was confirmed for different plant species, *Festuca arundinacea* and *Festuca pratensis* (Soleimani et al., 2010), *Brassica napus* (Deng, Wang, Tan, & Cao, 2012), and *S. nigrum* (Khan et al., 2016).

6.4 Stress Alleviation

In addition to plant growth-promoting effects and direct effect on metal solubilization and uptake, microorganisms can also improve the stress tolerance of plants (Fig. 1). This may be achieved by the enzyme ACC deaminase leading to a reduction of stress-induced ethylene levels in the plant (Glick, Cheng, Czarny, & Duan, 2007). It has been found that Cd, like many other trace metals, significantly retards the growth of plants including root and shoot length, and dry weight, with increasing effects noticed at higher Cd concentrations. However, biomass and other growth parameters can be significantly improved after inoculation with PGP microorganisms that possess the ability to use ACC as sole nitrogen source. The enzyme ACC-deaminase (ACCd) has been isolated from several bacteria (Glick, 2010; Glick et al., 2007), it cleaves ACC an immediate precursor of ethylene in plants, and thus modulates ethylene levels, which can be particularly high under stress conditions. In this way, several PGPR can contribute to plant growth by lowering the level of ethylene in plants growing in the presence of trace metals. In a recent study, primers targeting the acdS gene were devolved to screen Ni-resistant bacteria having ACCd activity in *A. murale* rhizosphere (S. Lopez, pers. communication). They quantified ACCd activity, which is known to stimulate the growth of plants, directly from soil DNA extracted as a tool to predict phytomining efficiency.

7. BIOAUGMENTATION

Because the presence of the indigenous rhizosphere microflora may increase plant metal accumulation capacity and plant fitness (Farinati et al., 2009), there is a possibility to a certain extent to control metal accumulation in the shoots, by selecting the appropriate microbial consortium in the rhizosphere (Fig. 1). Until now, efforts to manipulate natural rhizospheric microbial communities by inoculating beneficial microbes or changing the environment and microbial nutrients have shown mixed successes. That is because the ecological functioning of most of occurring microbial taxa is yet incompletely understood. Though, the rhizospheric and endophytic microbes associated with hyperaccumulators have gained increased interest due to their extraordinary properties in enhancing or inhibiting trace metal uptake, and detoxification processes, in addition to their plant growth and developmental characteristics.

The introduction of nonendemic microbial species in the rhizosphere is difficult because they face tough competition from the native flora and may lose their activity (Raaijmakers, Paulitz, Steinberg, Alabouvette, & Moënne-Loccoz, 2008). To improve these uncertainties the time of inoculation and method of inoculation need to be optimized. However, much easier in this respect is the use of endophytic microbes, which are in close association with the plants and have the desired characteristics to improve plant growth in a synergistic way (Kolbas et al., 2015). Therefore, a lot of attention has been focused on manipulating the endophytes (bacterial/fungal) associated with hyperaccumulator plants to improve metal uptake and plant growth. Several successful approaches of plant inoculation with indigenous metal-resistant endophytes to increase metal extraction are known (Rajkumar et al., 2009). Ma et al. (2015) observed that the inoculation of *S. plumbizincicola* with Zn- and Cd-resistant bacterial endophytes increased the concentrations of water-extractable Cd and Zn in soil. *Bacillus pumilis* E2S2 significantly increased root (146%) and shoot (17%) length, fresh (37%) and dry biomass (32%) as well as plant Cd uptake (43%), whereas *Bacillus* sp. E1S2 significantly enhanced accumulation of Zn (18%) in plants compared to uninoculated controls (Ma et al., 2015). The possible explanation for enhanced plant growth and metal uptake might be the production of IAA, siderophore, phosphate solubilization and ACC-deaminase production. Another study showed that inoculation with strain 1Y31 was found to increase the root and aboveground tissue mass and total Mn-uptake of aboveground tissues (64%) compared to the control. Here, the inoculation with strain 1Y31 increased the ratio of IAA-producing bacteria in the rhizosphere and bulk soil of hybrid *Penisetum* grown in Mn-added soil (Zhang, Chen, et al., 2015). Inoculation with strains IV8L2 (*Pseudomonas veronii*), II8L4 (*P. fluorescens*), and IV8R2 (*P. fluorescence*) increased shoot Zn concentrations by 20%–52% and root Zn concentration by 64%–85% in *S. alfredii* grown on artificially $ZnCO_3$-contaminated soil (Long, Chen, Wong, Wei, & Wu, 2013). All strains had intrinsic ability of production of IAA, siderophore and solubilization of phosphate, increasing growth and P supply. *F. arundinacea* and *F. pratensis* inoculated with *Neotyphodium* spp. fungal endophytes showed higher production of biomass and Cd accumulation than plants not inoculated with endophytes in Cd-contaminated soils (Soleimani et al., 2010). Such trace metal-resistant endophytic fungi are capable of promoting host plant growth, biomass production and enhanced metal extraction. Furthermore, they alleviate the toxic effect of trace metals by regulating various biochemical processes inside the plant through the production of different metabolites and phytohormones that help the host plant avoid metal stress toxicity (Hoffman, Gunatilaka, Wijeratne, Gunatilaka, & Arnold, 2013; Singh, Gill, & Tuteja, 2011). From these studies and several others as reviewed (Rajkumar et al., 2009), it can be concluded that inoculating metal hyperaccumulator plants with their own selected functional endophytic bacterial strains, it should be possible to improve metal extraction efficiency of hyperaccumulator plants.

Variable results (positive, negative, no effect) have been reported for the effect of inoculating metal-resistant rhizospheric bacteria or fungi on metal uptake. Madhaiyan, Poonguzhali, and Sa (2007) showed that inoculation with *Methylobacterium oryzae* and *Burkholderia* sp. isolated from rice tissues reduced the Ni and Cd uptake in roots and shoots of tomatoes and also their availability in soil, whereas an increase in plant biomass was found. Hence protection against metal toxicity is in this case governed by reducing trace metal accumulations in plants, which may be a desired strategy for phytostabilization but not extraction (Madhaiyan et al., 2007). To better exploit rhizosphere microbiota to enhance metal phytoextraction, much more information needs to be obtained of the structure and function of metal-resistant microbes in the rhizosphere of hyperaccumulators (Fig. 1). The study of Bell et al. (2015) for example, used Ion Torrent sequencing and correlation networks to predict host–microbe interactions in the rhizosphere, which may be associated with Zn uptake in willows growing on a former landfill. Total Zn accumulation in the shoots was better explained by the rhizosphere microbiome composition 4 months postplanting, than 16 months postplanting, suggesting that initial recruitment of fungal endophytes by willow cultivars is important (Bell et al., 2015). Furthermore, they found a plant–fungus specificity that may be crucial for further optimization of metal accumulation (Bell et al., 2015). When this information becomes also available for hyperaccumulators, much more progress is expected in this field.

8. FUTURE PROSPECTS

Studies have demonstrated that plant-associated bacteria can greatly increase the accumulation of trace metal and the stress resistance of plants (Kidd et al., 2009; Muehe et al., 2015; Rajkumar et al., 2009; Sessitsch et al., 2013). But it can be hypothesized that the plant-associated microbial community structures and functions are shaped by the trace metal contamination and various complex interactions between microbiota and host plants (Kidd et al., 2009). Studies of plant-associated microorganisms of hyperaccumulators are thus important not only to better understand their role in trace metal hyperaccumulation but also exploring the possible uses of these microorganisms (bacteria/fungi) for metal phytoremediation.

In the 30 years, most studies have focussed on the culturable microorganisms of hyperaccumulators (Visioli et al., 2014), or research was defined to interactions of specific rhizosphere microbial members and in nonnatural substrates. Based on the results obtained by the study of Muehe et al. (2015), culture-independent tools combined with high-throughput sequencing should be more widely adopted to gain insights into the complex interplay of soil—plant microbe interactions that could impact metal uptake. Comparing the microbial communities of plant species or cultivars growing in the same soil, but with high metal-accumulation rates, and plants with lower accumulation rates for example, can allow the discrimination of supportive and beneficial microorganisms which can be further tested in the laboratory for their ability to mobilize metals. Also more studies should be carried out to characterize the potential in metal tolerance and transformation. Characterization of the diversity of hyperaccumulator-associated microorganisms and understanding the ecological relationship is the first step to better use them to enhance bioremediation of metal-contaminated sites. These strains will also be useful in further research on metal tolerance and transformation mechanisms involved in metal-detoxification.

In addition, many insights have come from single microbe—plant studies, and these were mainly restricted to hydroponics (Li, Ye, & Wong, 2007) rather than the study of the microorganisms in the natural environment where multiple plant—microbe interactions ranging from synergistic to antagonistic take place simultaneously (Fig. 1). In addition, available metal concentrations in hydroponics often exceed real-world concentrations, and metal mobility and transport processes are very different in hydroponics compared to soil. Therefore, the study of the microbiome of many more hyperaccumulators growing in the natural environment is necessary to reveal mineral—metal—plant microbe interactions. In addition, much more association studies are needed to elucidate how individual strains in microbial communities can promote metal hyperaccumulation by plants. Genome-wide association studies can reveal new strains in the rhizosphere of hyperaccumulators with poorly represented culturable isolates. For example, for many of the most abundant taxa found in the rhizosphere of *A. halleri* (Muehe et al., 2015) such as *Sediminibacterium* and *Candidatus Chloracidobacterium,* we do not know their potential effects on plant growth and metal uptake. Similarly, for *A. murale*, a Ni-hyperaccumulator commonly found on ultramafic soils around the Mediterranean, we should characterize the total rhizosphere microorganisms of these plants to find good indicators for the success of Ni extraction and to select interesting PGPR to boost the efficiency of phytomining. However, most studies have focused on analyzing soils with techniques that provide little detail about the phylogenetic structure of the bacterial communities.

To be able to model and manage the host-microbiome, a strong understanding of ecological principles is necessary (Fig. 1). In animals and humans, the assembly of host-associated microbiomes is known to be driven by ecologically important attributes, such as diet that covary with the host evolutionary history (Muegge et al., 2011). Also in insects like the attine ant, ecological modelling has allowed a comprehensive understanding of the drivers of the ant's cuticular microbiome. Here, vertically transmitted antibiotic producing bacteria that coexist with a fungus were found to protect the ant from parasites (Scheuring & Yu, 2012). A similar understanding of the drivers of plant microbiome assembly, let alone hyperaccumulator microbiomes, is largely lacking. That is because most knowledge comes from studies focussing on individual strains and individual host species. Different plant species and even genotypes have characteristic rhizosphere microbial communities (Agler et al., 2016; Lebeis, 2014). This variation in rhizospheric community structure are presumably due to phenotypic variation in ecologically important plant functional traits (van der Heijden & Hartmann, 2016), though the influence of the host functional traits on variation in rhizosphere community structure across diverse hyperaccumulator host species has not been largely quantified. It remains to be understood how e.g., metal hyperaccumulation traits in the plant genome can influence the microbial community in the rhizosphere. An ecological understanding of the plant–microbe interactions networks is necessary to better understand and predict community assembly, and the influence of distribution and functional biogeography of the host on its microbiome (Fig. 1). To address these knowledge gaps, future association studies should focus on characterizing the rhizosphere microbiome of multiple host hyperaccumulator species in diverse metal-contaminated soils, and combine these data with datasets on the attributes of the host, to be able to quantify the microbial diversity of hyperaccumulators, the host plant attributes that influence microbiome community assembly in the rhizosphere, and understand relationships between diversity and host plant traits such as metal uptake. This understanding can then be used to better manage hyperaccumulators for metal phytoremediation.

The future of molecular and physiological plant–microbe research is bright because of the great progress in detection and imaging technologies. For example, using transcriptomic and proteomics approaches (Cvjetko,

Zovko, & Balen, 2014), recent studies have attempted to unravel the mechanism of trace metal transport and accumulation in plants. Additionally, metabolomic analysis is being used to help to identify the metabolites associated with trace metal stresses (Kumar, Dubey, Tripathi, Chakrabarty, & Trivedi, 2015). In addition, suitable techniques for imaging can further our knowledge about trace metal metabolism in plants and microorganisms. Information of the complete genome sequence of important microbial members of hyperaccumulators is also expected to become more available in the near future, following the trends of model-plants such as *Arabidopsis thaliana* and agricultural crops. Genomic information together with insights from experiments with synthetic communities would allow us to better draw functional ecological networks to predict plant–microbe interactions and how it affects metal uptake, which in turn can be validated or rejected through functional experiments (Bodenhausen, Bortfeld-Miller, Ackermann, & Vorholt, 2014). Mass spectrometry-based proteomics can be used to study trace metal stresses in candidate microorganisms such as bacteria (Zakeri et al., 2012) and plants (Cvjetko et al., 2014). Furthermore, proteogenomics, the combination of proteomics and genomics, is being used to study the genomic and proteomic properties of microorganisms that tolerate high concentrations of contaminants and high levels of stress (Delmotte et al., 2009; Rubiano-Labrador et al., 2014).

Using the omics based discoveries, new strategies could be developed which aim to improve the tolerance, uptake and hyperaccumulation of trace metals using genomic and metabolic engineering approaches (Mosa, Saadoun, Kumar, Helmy, & Dhankher, 2016) (Fig. 1). Pathways in the host plant itself that control the uptake, detoxification, transport from root to shoot tissues and translocation and hyperaccumulation in the aboveground storage tissues can be engineered. Additionally, plant breeding programmes can be improved to not only enhance the biomass and growth habits of natural hyperaccumulators, but also to select for those traits that support a highly efficient microbiome in terms of stimulating metal phytoextraction, fast growth and plant health (Thijs, Sillen, Rineau, Weyens, & Vangronsveld, 2016). Additionally, efforts towards combining the phytoremediation approach with bioenergy through the dual use of plants for phytoremediation and biofuel production on contaminated lands should be supported. This approach would allow phytoremediation of the site and simultaneously produce renewable energy that can offset the costs of applying these type of methodologies. Collectively, all these efforts promise a new generation of tailored plant–microbe systems with higher bio/phytoremediation efficiencies, predictability and lower costs.

9. CONCLUSION

This chapter presented an overview of the current knowledge on microbial diversity and function of plant-associated microorganisms of some hyperaccumulator plants. From this literature review it can be concluded that the microbial community associated with hyperaccumulators is diverse despite the high concentrations of metals in the environment and contains a significant potential of bacteria and fungi with traits related to metal uptake, detoxification, plant stress reduction and growth promotion. The majority of hyperaccumulator plant species adapted to extreme ecosystems, and with unique traits for metal hyperaccumulation, have not yet been studied for their associated microbial communities. Hence a lot of work lays ahead of plant microbiome research to explore new aspects about the phylogenetic, and functional diversity, of plant-associated microbiota. Current next generation sequencing tools enable insights not only on the individual level but also community level and the ecology of entire metacommunities of microbes or metaorganisms. Functional microbial networks are needed to correlate microbial function with host effects, and for this important input from theoretical ecology is indispensable. Microbial diversity is an important trait with important effects on plant health, fitness and metal mobilization. Using the current technologies, we will be able to exert better control of metal accumulation in plants by selecting the most appropriate (engineered) host plant with the most supportive microbial consortia.

REFERENCES

Abou-Shanab, R. I., Delorme, T. A., Angle, J. S., Chaney, R. L., Ghanem, K., Moawad, H., & Ghozlan, H. A. (2003). Phenotypic characterization of microbes in the rhizosphere of *Alyssum murale*. *International Journal of Phytoremediation, 5*(4), 367–379.

Agler, M. T., Ruhe, J., Kroll, S., Morhenn, C., Kim, S. T., Weigel, D., & Kemen, E. M. (2016). Microbial hub taxa link host and abiotic factors to plant microbiome variation. *PLoS Biology, 14*(1), e1002352.

Al Agely, A., Sylvia, D. M., & Ma, L. Q. (2005). Mycorrhizae increase arsenic uptake by the hyperaccumulator Chinese brake fern (*Pteris vittata* L.). *Journal of Environmental Quality, 34*(6), 2181–2186.

Alford, É. R., Pilon-Smits, E. A., & Paschke, M. W. (2010). Metallophytes—A view from the rhizosphere. *Plant and Soil, 337*(1–2), 33–50.

Almås, Å. R., Bakken, L. R., & Mulder, J. (2004). Changes in tolerance of soil microbial communities in Zn and Cd contaminated soils. *Soil Biology and Biochemistry, 36*(5), 805–813.

Alvarez-Lopez, V., Prieto-Fernandez, A., Cabello-Conejo, M. I., & Kidd, P. S. (2016). Organic amendments for improving biomass production and metal yield of Ni-hyperaccumulating plants. *Science of the Total Environment, 548–549*, 370–379.
Arkhipova, T. N., Prinsen, E., Veselov, S. U., Martinenko, E. V., Melentiev, A. I., & Kudoyarova, G. R. (2007). Cytokinin producing bacteria enhance plant growth in drying soil. *Plant and Soil, 292*(1), 305–315.
Arshad, M., Saleem, M., & Hussain, S. (2007). Perspectives of bacterial ACC deaminase in phytoremediation. *Trends in Biotechnology, 25*(8), 356–362.
Arumugam, M., Raes, J., Pelletier, E., Le Paslier, D., Yamada, T., Mende, D. R., … Bork, P. (2011). Enterotypes of the human gut microbiome. *Nature, 473*(7346), 174–180.
Azarbad, H., Niklinska, M., van Gestel, C. A., van Straalen, N. M., Roling, W. F., & Laskowski, R. (2013). Microbial community structure and functioning along metal pollution gradients. *Environmental Toxicology and Chemistry, 32*(9), 1992–2002.
Azevedo, H., Azinheiro, S. G., Muñoz-Mérida, A., Castro, P. H., Huettel, B., Aarts, M. M., & Assunção, A. G. (2016). Transcriptomic profiling of *Arabidopsis* gene expression in response to varying micronutrient zinc supply. *Genomics Data, 7*, 256–258.
Babu, A. G., Shim, J., Bang, K. S., Shea, P. J., & Oh, B. T. (2014). *Trichoderma virens* PDR-28: A heavy metal-tolerant and plant growth-promoting fungus for remediation and bioenergy crop production on mine tailing soil. *Journal of Environmental Management, 132*, 129–134.
Baker, A., & Brooks, R. (1989). Terrestrial higher plants which hyperaccumulate metallic elements. A review of their distribution, ecology and phytochemistry. *Biorecovery, 1*(2), 81–126.
Baker, A., & Smith, A. (2000). Metal hyperaccumulator plants: Biological resources for exploitation in the phytoextraction of metal contaminated soils. In *Proceedings InterCost Workshop for bioremediation. Sorrento, Italia.*
Bakker, P. A., Pieterse, C. M., & van Loon, L. C. (2007). Induced systemic resistance by fluorescent *Pseudomonas* spp. *Phytopathology, 97*(2), 239–243.
Barr, M., East, A. K., Leonard, M., Mauchline, T. H., & Poole, P. S. (2008). In vivo expression technology (IVET) selection of genes of *Rhizobium leguminosarum* biovar viciae A34 expressed in the rhizosphere. *FEMS Microbiology Letters, 282*(2), 219–227.
Barzanti, R., Ozino, F., Bazzicalupo, M., Gabbrielli, R., Galardi, F., Gonnelli, C., & Mengoni, A. (2007). Isolation and characterization of endophytic bacteria from the nickel hyperaccumulator plant *Alyssum bertolonii*. *Microbial Ecology, 53*(2), 306–316.
Becerra-Castro, C., Kidd, P., Kuffner, M., Prieto-Fernandez, A., Hann, S., Monterroso, C., … Puschenreiter, M. (2013). Bacterially induced weathering of ultramafic rock and its implications for phytoextraction. *Applied and Environmental Microbiology, 79*(17), 5094–5103.
Bell, T. H., Cloutier-Hurteau, B., Al-Otaibi, F., Turmel, M. C., Yergeau, E., Courchesne, F., & St-Arnaud, M. (2015). Early rhizosphere microbiome composition is related to the growth and Zn uptake of willows introduced to a former landfill. *Environmental Microbiology, 17*(8), 3025–3038.
Bell, T. H., Joly, S., Pitre, F. E., & Yergeau, E. (2014). Increasing phytoremediation efficiency and reliability using novel omics approaches. *Trends in Biotechnology, 32*(5), 271–280.
Berendsen, R. L., Pieterse, C. M., & Bakker, P. A. (2012). The rhizosphere microbiome and plant health. *Trends in Plant Science, 17*(8), 478–486.
Berg, G., Krause, R., & Mendes, R. (2015). Cross-kingdom Similarities in microbiome ecology and biocontrol of pathogens. *Frontiers in Microbiology, 6*, 1311.
Bert, V., Bonnin, I., Saumitou-Laprade, P., De Laguérie, P., & Petit, D. (2002). Do *Arabidopsis halleri* from nonmetallicolous populations accumulate zinc and cadmium more effectively than those from metallicolous populations? *New Phytologist, 155*(1), 47–57.
Bitton, G., & Freihofer, V. (1977). Influence of extracellular polysaccharides on the toxicity of copper and cadmium toward *Klebsiella aerogenes*. *Microbial Ecology, 4*(2), 119–125.

Bodenhausen, N., Bortfeld-Miller, M., Ackermann, M., & Vorholt, J. A. (2014). A synthetic community approach reveals plant genotypes affecting the phyllosphere microbiota. *PLoS Genetics, 10*(4), e1004283.

Bonfante, P., & Anca, I. A. (2009). Plants, mycorrhizal fungi, and bacteria: A network of interactions. *Annual Review of Microbiology, 63*(1), 363–383.

Bonfante, P., & Genre, A. (2008). Plants and arbuscular mycorrhizal fungi: An evolutionary-developmental perspective. *Trends in Plant Science, 13*(9), 492–498.

Borch, T., Kretzschmar, R., Kappler, A., Cappellen, P. V., Ginder-Vogel, M., Voegelin, A., & Campbell, K. (2010). Biogeochemical redox processes and their impact on contaminant dynamics. *Environmental Science and Technology, 44*(1), 15–23.

Bottini, R., Cassán, F., & Piccoli, P. (2004). Gibberellin production by bacteria and its involvement in plant growth promotion and yield increase. *Applied Microbiology and Biotechnology, 65*(5), 497–503.

Brader, G., Compant, S., Mitter, B., Trognitz, F., & Sessitsch, A. (2014). Metabolic potential of endophytic bacteria. *Current Opinion in Biotechnology, 27*(100), 30–37.

Cabello-Conejo, M., Becerra-Castro, C., Prieto-Fernández, A., Monterroso, C., Saavedra-Ferro, A., Mench, M., & Kidd, P. (2014). Rhizobacterial inoculants can improve nickel phytoextraction by the hyperaccumulator *Alyssum pintodasilvae*. *Plant and Soil, 379*(1–2), 35–50.

Caporaso, J. G., Lauber, C. L., Walters, W. A., Berg-Lyons, D., Huntley, J., Fierer, N., … Knight, R. (2012). Ultra-high-throughput microbial community analysis on the Illumina HiSeq and MiSeq platforms. *ISME Journal, 6*(8), 1621–1624.

Chaney, R. L., Angle, J. S., Baker, A. J., & Li, Y.-M. (1998). *Method for phytomining of nickel, cobalt and other metals from soil* (Google Patents).

Chaney, R. L., Angle, J. S., Broadhurst, C. L., Peters, C. A., Tappero, R. V., & Sparks, D. L. (2007). Improved understanding of hyperaccumulation yields commercial phytoextraction and phytomining technologies. *Journal of Environmental Quality, 36*(5), 1429–1443.

Chen, Y. X., Lin, Q., Luo, Y. M., He, Y. F., Zhen, S. J., Yu, Y. L., … Wong, M. H. (2003). The role of citric acid on the phytoremediation of heavy metal contaminated soil. *Chemosphere, 50*(6), 807–811.

Compant, S., Clément, C., & Sessitsch, A. (2010). Plant growth-promoting bacteria in the rhizo- and endosphere of plants: Their role, colonization, mechanisms involved and prospects for utilization. *Soil Biology and Biochemistry, 42*(5), 669–678.

Compant, S., Duffy, B., Nowak, J., Clement, C., & Barka, E. A. (2005). Use of plant growth-promoting bacteria for biocontrol of plant diseases: Principles, mechanisms of action, and future prospects. *Applied and Environmental Microbiology, 71*(9), 4951–4959.

Coninx, L., Martinova, V., & Rineau, F. (2017). Mycorrhiza-assisted phytoremediation. In A. Cuypers, & J. Vangronsveld (Eds.), *Phytoremediation* (Vol. 83, pp. 127–188).

Cravatt, B. F., & Kodadek, T. (2015). Editorial overview: Omics: Methods to monitor and manipulate biological systems: Recent advances in 'omics'. *Current Opinion in Chemical Biology, 24*, v–vii.

Croes, S., Weyens, N., Janssen, J., Vercampt, H., Colpaert, J. V., Carleer, R., & Vangronsveld, J. (2013). Bacterial communities associated with *Brassica napus* L. grown on trace element-contaminated and non-contaminated fields: A genotypic and phenotypic comparison. *Microbial Biotechnology, 6*(4), 371–384 (Electronic Resource).

Cvjetko, P., Zovko, M., & Balen, B. (2014). Proteomics of heavy metal toxicity in plants. *Arhiv Za Higijenu Rada I Toksikologiju, 65*(1), 1–17.

Dahmani-Muller, H., Van Oort, F., Gelie, B., & Balabane, M. (2000). Strategies of heavy metal uptake by three plant species growing near a metal smelter. *Environmental Pollution, 109*(2), 231–238.

De Bruijn, F. (2015). *Biological nitrogen fixation. Principles of plant-microbe interactions. Microbes for sustainable agriculture.* B. Lugtenberg. Switzerland, Springer International Publishing.

Delmotte, N., Knief, C., Chaffron, S., Innerebner, G., Roschitzki, B., Schlapbach, R., … Vorholt, J. A. (2009). Community proteogenomics reveals insights

into the physiology of phyllosphere bacteria. *Proceedings of the National Academy of Sciences of the United States of America, 106*(38), 16428–16433.
Deng, Z., Wang, W., Tan, H., & Cao, L. (2012). Characterization of heavy metal-resistant endophytic yeast *Cryptococcus* sp. CBSB78 from rapes (*Brassica chinensis*) and its potential in promoting the growth of *Brassica* spp. in metal-contaminated soils. *Water, Air, and Soil Pollution, 223*(8), 5321–5329.
Dobbelaere, S., Vanderleyden, J., & Okon, Y. (2003). Plant growth-promoting effects of diazotrophs in the rhizosphere. *Critical Reviews in Plant Sciences, 22*(2), 107–149.
Dourado, M. N., Bogas, A. C., Pomini, A. M., Andreote, F. D., Quecine, M. C., Marsaioli, A. J., & Araújo, W. L. (2013). Methylobacterium-plant interaction genes regulated by plant exudate and quorum sensing molecules. *Brazilian Journal of Microbiology, 44*(4), 1331–1339.
Echevarria, G., Baker, A. J., Benizri, E., Houzelot, V., Laubie, B., Kidd, P. S., … Zhang, X. (2015). Agromining for nickel: A complete chain that optimizes ecosystem services rendered by ultramafic landscapes. In *13th International conference on the biogeochemistry of trace elements (ICOBTE), ICOBTE.*
van der Ent, A., Baker, A. J., Reeves, R. D., Pollard, A. J., & Schat, H. (2013). Hyperaccumulators of metal and metalloid trace elements: Facts and fiction. *Plant and Soil , 362*(1–2), 319–334.
Evangelou, M. W., Ebel, M., & Schaeffer, A. (2007). Chelate assisted phytoextraction of heavy metals from soil. Effect, mechanism, toxicity, and fate of chelating agents. *Chemosphere, 68*(6), 989–1003.
Farinati, S., DalCorso, G., Bona, E., Corbella, M., Lampis, S., Cecconi, D., … Furini, A. (2009). Proteomic analysis of *Arabidopsis halleri* shoots in response to the heavy metals cadmium and zinc and rhizosphere microorganisms. *Proteomics, 9*(21), 4837–4850.
Farinati, S., DalCorso, G., Panigati, M., & Furini, A. (2011). Interaction between selected bacterial strains and *Arabidopsis halleri* modulates shoot proteome and cadmium and zinc accumulation. *Journal of Experimental Botany, 62*(10), 3433–3447.
Freeman, J. L., Lindblom, S. D., Quinn, C. F., Fakra, S., Marcus, M. A., & Pilon-Smits, E. A. (2007). Selenium accumulation protects plants from herbivory by Orthoptera via toxicity and deterrence. *New Phytologist, 175*(3), 490–500.
Gadd, G. M. (1999). Fungal production of citric and oxalic acid: Importance in metal speciation, physiology and biogeochemical processes. *Advances in Microbial Physiology, 41*, 47–92.
Galardi, F., Mengoni, A., Pucci, S., Barletti, L., Massi, L., Barzanti, R., … Gonnelli, C. (2007). Intra-specific differences in mineral element composition in the Ni-hyperaccumulator *Alyssum bertolonii*: A survey of populations in nature. *Environmental and Experimental Botany, 60*(1), 50–56.
Gerhardt, K. E., Huang, X.-D., Glick, B. R., & Greenberg, B. M. (2009). Phytoremediation and rhizoremediation of organic soil contaminants: Potential and challenges. *Plant Science, 176*(1), 20–30.
Gilis, A., Khan, M. A., Cornelis, P., Meyer, J.-M., Mergeay, M., & Van Der Lelie, D. (1996). Siderophore-mediated iron uptake in *Alcaligenes eutrophus* CH34 and identification of aleB encoding the ferric iron-alcaligin E receptor. *Journal of Bacteriology, 178*(18), 5499–5507.
Glick, B. R. (2010). Using soil bacteria to facilitate phytoremediation. *Biotechnology Advances, 28*(3), 367–374.
Glick, B. R. (2012). Plant growth-promoting bacteria: Mechanisms and applications. *Scientifica (Cairo), 2012*, 963401.
Glick, B. R. (2014). Bacteria with ACC deaminase can promote plant growth and help to feed the world. *Microbiological Research, 169*(1), 30–39.
Glick, B. R., Cheng, Z., Czarny, J., & Duan, J. (2007). Promotion of plant growth by ACC deaminase-producing soil bacteria. *European Journal of Plant Pathology, 119*(3), 329–339.

Glick, B. R., Penrose, D. M., & Li, J. (1998). A model for the lowering of plant ethylene concentrations by plant growth-promoting bacteria. *Journal of Theoretical Biology, 190*(1), 63–68.

Gohre, V., & Paszkowski, U. (2006). Contribution of the arbuscular mycorrhizal symbiosis to heavy metal phytoremediation. *Planta, 223*(6), 1115–1122.

Goolsby, E. W., & Mason, C. M. (2015). Toward a more physiologically and evolutionarily relevant definition of metal hyperaccumulation in plants. *Frontiers in Plant Science, 6*, 33.

Gremion, F., Chatzinotas, A., & Harms, H. (2003). Comparative 16S rDNA and 16S rRNA sequence analysis indicates that Actinobacteria might be a dominant part of the metabolically active bacteria in heavy metal-contaminated bulk and rhizosphere soil. *Environmental Microbiology, 5*(10), 896–907.

Hardoim, P. R., van Overbeek, L. S., Berg, G., Pirttila, A. M., Compant, S., Campisano, A., … Sessitsch, A. (2015). The hidden world within plants: Ecological and evolutionary Considerations for defining functioning of microbial endophytes. *Microbiology and Molecular Biology Reviews: MMBR, 79*(3), 293–320.

Hardoim, P. R., van Overbeek, L. S., & Elsas, J. D. (2008). Properties of bacterial endophytes and their proposed role in plant growth. *Trends in Microbiology, 16*(10), 463–471.

Hassan Sel, D., Boon, E., St-Arnaud, M., & Hijri, M. (2011). Molecular biodiversity of arbuscular mycorrhizal fungi in trace metal-polluted soils. *Molecular Ecology, 20*(16), 3469–3483.

van der Heijden, M. G., & Hartmann, M. (2016). Networking in the plant microbiome. *PLoS Biology, 14*(2), e1002378.

Hengstler, J. G., Bolm-Audorff, U., Faldum, A., Janssen, K., Reifenrath, M., Götte, W., … Gebhard, S. (2003). Occupational exposure to heavy metals: DNA damage induction and DNA repair inhibition prove co-exposures to cadmium, cobalt and lead as more dangerous than hitherto expected. *Carcinogenesis, 24*(1), 63–73.

Hoffman, M. T., Gunatilaka, M. K., Wijeratne, K., Gunatilaka, L., & Arnold, A. E. (2013). Endohyphal bacterium enhances production of indole-3-acetic acid by a foliar fungal endophyte. *PLoS One, 8*(9), e73132.

Hogervorst, J., Plusquin, M., Vangronsveld, J., Nawrot, T., Cuypers, A., Van Hecke, E., … Staessen, J. A. (2007). House dust as possible route of environmental exposure to cadmium and lead in the adult general population. *Environmental Research, 103*(1), 30–37.

Hu, H., Jin, Q., & Kavan, P. (2014). A study of heavy metal pollution in China: Current status, pollution-control policies and countermeasures. *Sustainability, 6*(9), 5820–5838.

Idris, R., Kuffner, M., Bodrossy, L., Puschenreiter, M., Monchy, S., Wenzel, W. W., & Sessitsch, A. (2006). Characterization of Ni-tolerant methylobacteria associated with the hyperaccumulating plant *Thlaspi goesingense* and description of *Methylobacterium goesingense* sp. nov. *Systematic and Applied Microbiology, 29*(8), 634–644.

Idris, R., Trifonova, R., Puschenreiter, M., Wenzel, W. W., & Sessitsch, A. (2004). Bacterial communities associated with flowering plants of the Ni hyperaccumulator *Thlaspi goesingense*. *Applied and Environmental Microbiology, 70*(5), 2667–2677.

Jaishankar, M., Tseten, T., Anbalagan, N., Mathew, B. B., & Beeregowda, K. N. (2014). Toxicity, mechanism and health effects of some heavy metals. *Interdisciplinary Toxicology, 7*(2), 60–72.

Kandeler, E., Tscherko, D., Bruce, K., Stemmer, M., Hobbs, P. J., Bardgett, R. D., & Amelung, W. (2000). Structure and function of the soil microbial community in microhabitats of a heavy metal polluted soil. *Biology and Fertility of Soils, 32*(5), 390–400.

Khan, A. R., Ullah, I., Khan, A. L., Hong, S.-J., Waqas, M., Park, G.-S., … Park, M. (2014). Phytostabilization and physicochemical responses of Korean ecotype *Solanum nigrum* L. to cadmium contamination. *Water, Air, and Soil Pollution, 225*(10), 1–11.

Khan, A. R., Waqas, M., Ullah, I., Khan, A. L., Khan, M. A., Lee, I.-J., & Shin, J.-H. (2016). Culturable endophytic fungal diversity in the cadmium hyperaccumulator *Solanum nigrum* L. and their role in enhancing phytoremediation. *Environmental and Experimental Botany*. http://dx.doi.org/10.1016/j.envexpbot.2016.03.005.

Kidd, P., Alvarez-Lopez, V., Becerra-Castro, C., Cabello-Conejo, M. I., & Prieto-Fernández, Á. (2017). Potential role of plant-associated bacteria in plant metal uptake and implications in phytotechnologies. In A. Cuypers, & J. Vangronsveld (Eds.), *Phytoremediation* (Vol. 83, pp. 87–126).

Kidd, P., Barceló, J., Bernal, M. P., Navari-Izzo, F., Poschenrieder, C., Shilev, S., … Monterroso, C. (2009). Trace element behaviour at the root–soil interface: Implications in phytoremediation. *Environmental and Experimental Botany, 67*(1), 243–259.

Kidd, P., Mench, M., Álvarez-López, V., Bert, V., Dimitriou, I., Friesl-Hanl, W., … Puschenreiter, M. (2015). Agronomic practices for improving gentle remediation of trace element-contaminated soils. *International Journal of Phytoremediation, 17*(11), 1005–1037.

Kloepper, J. W., Ryu, C.-M., & Zhang, S. (2004). Induced systemic resistance and promotion of plant growth by *Bacillus* spp. *Phytopathology, 94*(11), 1259–1266.

Kolbas, A., Kidd, P., Guinberteau, J., Jaunatre, R., Herzig, R., & Mench, M. (2015). Endophytic bacteria take the challenge to improve Cu phytoextraction by sunflower. *Environmental Science and Pollution Research International, 22*(7), 5370–5382.

Kumar, S., Dubey, R. S., Tripathi, R. D., Chakrabarty, D., & Trivedi, P. K. (2015). Omics and biotechnology of arsenic stress and detoxification in plants: Current updates and prospective. *Environment International, 74*, 221–230.

Kunito, T., Saeki, K., Nagaoka, K., Oyaizu, H., & Matsumoto, S. (2001). Characterization of copper-resistant bacterial community in rhizosphere of highly copper-contaminated soil. *European Journal of Soil Biology, 37*(2), 95–102.

Kutschera, U. (2007). Plant-associated *Methylobacteria* as co-evolved phytosymbionts: A hypothesis. *Plant Signaling Behavior, 2*(2), 74–78.

Lampis, S., Santi, C., Ciurli, A., Andreolli, M., & Vallini, G. (2015). Promotion of arsenic phytoextraction efficiency in the fern *Pteris vittata* by the inoculation of As-resistant bacteria: A soil bioremediation perspective. *Frontiers in Plant Science, 6*, 80.

Lebeis, S. L. (2014). The potential for give and take in plant-microbiome relationships. *Frontiers in Plant Science, 5*, 287.

Leitenmaier, B., & Kupper, H. (2013). Compartmentation and complexation of metals in hyperaccumulator plants. *Frontiers in Plant Science, 4*, 374.

Leung, H.-M., Zhen-Wen, W., Zhi-Hong, Y. E., Kin-Lam, Y., Xiao-Ling, P., & Cheung, K.-C. (2013). Interactions between arbuscular mycorrhizae and plants in phytoremediation of metal-contaminated soils: A review. *Pedosphere, 23*(5), 549–563.

Li, W. C., Ye, Z. H., & Wong, M. H. (2007). Effects of bacteria on enhanced metal uptake of the Cd/Zn-hyperaccumulating plant, *Sedum alfredii*. *Journal of Experimental Botany, 58*(15–16), 4173–4182.

Li, W. C., Ye, Z. H., & Wong, M. H. (2010). Metal mobilization and production of short-chain organic acids by rhizosphere bacteria associated with a Cd/Zn hyperaccumulating plant, *Sedum alfredii*. *Plant and Soil, 326*(1), 453–467.

Lodewyckx, C., Mergeay, M., Vangronsveld, J., Clijsters, H., & Van Der Lelie, D. (2002). Isolation, characterization, and identification of bacteria associated with the zinc hyper-accumulator *Thlaspi caerulescens* subsp. *Calaminaria*. *International Journal of Phytoremediation, 4*(2), 101–115.

Long, X.-x., Chen, X.-m., Wong, J. W.-C., Wei, Z.-b., & Wu, Q.-t. (2013). Feasibility of enhanced phytoextraction of Zn contaminated soil with Zn mobilizing and plant growth promoting endophytic bacteria. *Transactions of Nonferrous Metals Society of China, 23*(8), 2389–2396.

Lopez-Guerrero, M. G., Ormeno-Orrillo, E., Rosenblueth, M., Martinez-Romero, J., & Martinez-Romero, E. (2013). Buffet hypothesis for microbial nutrition at the rhizosphere. *Frontiers in Plant Science, 4*, 188.

Lugtenberg, B., & Kamilova, F. (2009). Plant-growth-promoting rhizobacteria. *Annual Review of Microbiology, 63*, 541–556.

Luo, S. L., Chen, L., Chen, J. L., Xiao, X., Xu, T. Y., Wan, Y., ... Zeng, G. M. (2011). Analysis and characterization of cultivable heavy metal-resistant bacterial endophytes isolated from Cd-hyperaccumulator *Solanum nigrum* L. and their potential use for phytoremediation. *Chemosphere, 85*(7), 1130–1138.

Luo, Z.-B., Wu, C., Zhang, C., Li, H., Lipka, U., & Polle, A. (2014). The role of ectomycorrhizas in heavy metal stress tolerance of host plants. *Environmental and Experimental Botany, 108*, 47–62.

Ma, L. Q., Komar, K. M., Tu, C., Zhang, W., Cai, Y., & Kennelley, E. D. (2001). A fern that hyperaccumulates arsenic. *Nature, 409*(6820), 579.

Ma, Y., Prasad, M. N., Rajkumar, M., & Freitas, H. (2011). Plant growth promoting rhizobacteria and endophytes accelerate phytoremediation of metalliferous soils. *Biotechnology Advances, 29*(2), 248–258.

Ma, Y., He, J., Ma, C., Luo, J., Li, H., Liu, T., ... Luo, Z. B. (2014). Ectomycorrhizas with *Paxillus involutus* enhance cadmium uptake and tolerance in Populus × canescens. *Plant, Cell and Environment, 37*(3), 627–642.

Ma, Y., Oliveira, R. S., Nai, F., Rajkumar, M., Luo, Y., Rocha, I., & Freitas, H. (2015). The hyperaccumulator *Sedum plumbizincicola* harbors metal-resistant endophytic bacteria that improve its phytoextraction capacity in multi-metal contaminated soil. *Journal of Environmental Management, 156*, 62–69.

Madhaiyan, M., Poonguzhali, S., & Sa, T. (2007). Metal tolerating methylotrophic bacteria reduces nickel and cadmium toxicity and promotes plant growth of tomato (*Lycopersicon esculentum* L.). *Chemosphere, 69*(2), 220–228.

Mahmood, T., Malik, S. A., & Hussain, S. T. (2010). Biosorption and recovery of heavy metals from aqueous solutions by *Eichhornia crassipes* (water hyacinth) ash. *BioResources, 5*(2), 1244–1256.

Majewska, M., Kurek, E., & Rogalski, J. (2007). Microbially mediated cadmium sorption/desorption processes in soil amended with sewage sludge. *Chemosphere, 67*(4), 724–730.

Mastretta, C., Taghavi, S., van der Lelie, D., Mengoni, A., Galardi, F., Gonnelli, C., ... Vangronsveld, J. (2009). Endophytic bacteria from seeds of *Nicotiana tabacum* can reduce cadmium phytotoxicity. *International Journal of Phytoremediation, 11*(3), 251–267.

McGrath, S., Lombi, E., Gray, C., Caille, N., Dunham, S., & Zhao, F. (2006). Field evaluation of Cd and Zn phytoextraction potential by the hyperaccumulators *Thlaspi caerulescens* and *Arabidopsis halleri*. *Environmental Pollution, 141*(1), 115–125.

Mendes, R., Garbeva, P., & Raaijmakers, J. M. (2013). The rhizosphere microbiome: Significance of plant beneficial, plant pathogenic, and human pathogenic microorganisms. *FEMS Microbiology Reviews, 37*(5), 634–663.

Mengoni, A., Barzanti, R., Gonnelli, C., Gabbrielli, R., & Bazzicalupo, M. (2001). Characterization of nickel-resistant bacteria isolated from serpentine soil. *Environmental Microbiology, 3*(11), 691–698.

Mengoni, A., Grassi, E., Barzanti, R., Biondi, E. G., Gonnelli, C., Kim, C. K., & Bazzicalupo, M. (2004). Genetic diversity of bacterial communities of serpentine soil and of rhizosphere of the nickel-hyperaccumulator plant *Alyssum bertolonii*. *Microbial Ecology, 48*(2), 209–217.

Mengoni, A., Pini, F., Huang, L. N., Shu, W. S., & Bazzicalupo, M. (2009). Plant-by-plant variations of bacterial communities associated with leaves of the nickel hyperaccumulator *Alyssum bertolonii* Desv. *Microbial Ecology, 58*(3), 660–667.

Miethke, M., & Marahiel, M. A. (2007). Siderophore-based iron acquisition and pathogen control. *Microbiology and Molecular Biology Reviews: MMBR, 71*(3), 413–451.
Miransari, M. (2011a). Hyperaccumulators, arbuscular mycorrhizal fungi and stress of heavy metals. *Biotechnology Advances, 29*(6), 645–653.
Miransari, M. (2011b). Interactions between arbuscular mycorrhizal fungi and soil bacteria. *Applied Microbiology and Biotechnology, 89*(4), 917–930.
Mirete, S., de Figueras, C. G., & Gonzalez-Pastor, J. E. (2007). Novel nickel resistance genes from the rhizosphere metagenome of plants adapted to acid mine drainage. *Applied and Environmental Microbiology, 73*(19), 6001–6011.
Mohite, B. (2013). Isolation and characterization of indole acetic acid (IAA) producing bacteria from rhizospheric soil and its effect on plant growth. *Journal of Soil Science and Plant Nutrition, 13*(3), 638–649.
Mosa, K. A., Saadoun, I., Kumar, K., Helmy, M., & Dhankher, O. P. (2016). Potential biotechnological strategies for the cleanup of heavy metals and metalloids. *Frontiers in Plant Science,* 7, 303.
Muegge, B. D., Kuczynski, J., Knights, D., Clemente, J. C., Gonzalez, A., Fontana, L., … Gordon, J. I. (2011). Diet drives convergence in gut microbiome functions across mammalian phylogeny and within humans. *Science, 332*(6032), 970–974.
Muehe, E. M., Weigold, P., Adaktylou, I. J., Planer-Friedrich, B., Kraemer, U., Kappler, A., & Behrens, S. (2015). Rhizosphere microbial community composition affects cadmium and zinc uptake by the metal-hyperaccumulating plant *Arabidopsis halleri*. *Applied and Environmental Microbiology, 81*(6), 2173–2181.
O'Bryant, S. E., Edwards, M., Menon, C. V., Gong, G., & Barber, R. (2011). Long-term low-level arsenic exposure is associated with poorer neuropsychological functioning: A project FRONTIER study. *International Journal of Environmental Research and Public Health, 8*(3), 861–874.
Ochsenreiter, T., Selezi, D., Quaiser, A., Bonch-Osmolovskaya, L., & Schleper, C. (2003). Diversity and abundance of Crenarchaeota in terrestrial habitats studied by 16S RNA surveys and real time PCR. *Environmental Microbiology, 5*(9), 787–797.
Op De Beeck, M., Lievens, B., Busschaert, P., Rineau, F., Smits, M., Vangronsveld, J., & Colpaert, J. V. (2015). Impact of metal pollution on fungal diversity and community structures. *Environmental Microbiology, 17*(6), 2035–2047.
Pence, N. S., Larsen, P. B., Ebbs, S. D., Letham, D. L., Lasat, M. M., Garvin, D. F., … Kochian, L. V. (2000). The molecular physiology of heavy metal transport in the Zn/Cd hyperaccumulator *Thlaspi caerulescens*. *Proceedings of the National Academy of Sciences of the United States of America, 97*(9), 4956–4960.
Pham, V. H., & Kim, J. (2012). Cultivation of unculturable soil bacteria. *Trends in Biotechnology, 30*(9), 475–484.
Pilon-Smits, E. (2005). Phytoremediation. *Annual Review of Plant Biology, 56,* 15–39.
Ponraj, P., Shankar, M., Ilakkiam, D., & Gunasekaran, P. (2012). Influence of siderophore pyoverdine synthesis and iron-uptake on abiotic and biotic surface colonization of *Pseudomonas putida* S11. *Biometals, 25*(6), 1113–1128.
Prieto, P., Schiliro, E., Maldonado-Gonzalez, M. M., Valderrama, R., Barroso-Albarracin, J. B., & Mercado-Blanco, J. (2011). Root hairs play a key role in the endophytic colonization of olive roots by *Pseudomonas* spp. with biocontrol activity. *Microbial Ecology, 62*(2), 435–445.
Puschenreiter, M., Horak, O., Friesl, W., & Hartl, W. (2005). Low-cost agricultural measures to reduce heavy metal transfer into the food chain—a review. *Plant, Soil and Environment, 51*(1), 1–11.
Qi, W., & Zhao, L. (2013). Study of the siderophore-producing *Trichoderma asperellum* Q1 on cucumber growth promotion under salt stress. *Journal of Basic Microbiology, 53*(4), 355–364.

Qin, R., Hirano, Y., & Brunner, I. (2007). Exudation of organic acid anions from poplar roots after exposure to Al, Cu and Zn. *Tree Physiology, 27*(2), 313–320.
Raaijmakers, J. M., Paulitz, T. C., Steinberg, C., Alabouvette, C., & Moënne-Loccoz, Y. (2008). The rhizosphere: A playground and battlefield for soilborne pathogens and beneficial microorganisms. *Plant and Soil, 321*(1–2), 341–361.
Rajapaksha, R. M. C. P., Tobor-Kaplon, M. A., & Baath, E. (2004). Metal toxicity affects fungal and bacterial activities in soil differently. *Applied and Environmental Microbiology, 70*(5), 2966–2973.
Rajkumar, M., Ae, N., & Freitas, H. (2009). Endophytic bacteria and their potential to enhance heavy metal phytoextraction. *Chemosphere, 77*(2), 153–160.
Rajkumar, M., Ae, N., Prasad, M. N., & Freitas, H. (2010). Potential of siderophore-producing bacteria for improving heavy metal phytoextraction. *Trends in Biotechnology, 28*(3), 142–149.
Rajkumar, M., Nagendran, R., Lee, K. J., Lee, W. H., & Kim, S. Z. (2006). Influence of plant growth promoting bacteria and Cr6+ on the growth of Indian mustard. *Chemosphere, 62*(5), 741–748.
Rajkumar, M., Sandhya, S., Prasad, M., & Freitas, H. (2012). Perspectives of plant-associated microbes in heavy metal phytoremediation. *Biotechnology Advances, 30*(6), 1562–1574.
Rascio, N., & Navari-Izzo, F. (2011). Heavy metal hyperaccumulating plants: How and why do they do it? And what makes them so interesting? *Plant Science, 180*(2), 169–181.
Reeves, R. D. (2003). Tropical hyperaccumulators of metals and their potential for phytoextraction. *Plant and Soil, 249*(1), 57–65.
Reeves, R. D., & Baker, A. J. (2000). Metal-accumulating plants. In *Phytoremediation of toxic metals: Using plants to clean up the environment* (pp. 193–229). New York: Wiley.
Rhoads, A., & Au, K. F. (2015). PacBio sequencing and its applications. *Genomics Proteomics Bioinformatics, 13*(5), 278–289.
Rooks, M. G., Veiga, P., Wardwell-Scott, L. H., Tickle, T., Segata, N., Michaud, M., ... Garrett, W. S. (2014). Gut microbiome composition and function in experimental colitis during active disease and treatment-induced remission. *ISME Journal, 8*(7), 1403–1417.
Rosenblueth, M., & Martinez-Romero, E. (2006). Bacterial endophytes and their interactions with hosts. *Molecular Plant-Microbe Interactions: MPMI, 19*(8), 827–837.
Rubiano-Labrador, C., Bland, C., Miotello, G., Guérin, P., Pible, O., Baena, S., & Armengaud, J. (2014). Proteogenomic insights into salt tolerance by a halotolerant alpha-proteobacterium isolated from an Andean saline spring. *Journal of Proteomics, 97*, 36–47.
Saad, R., Kobaissi, A., Robin, C., Echevarria, G., & Benizri, E. (2016). Nitrogen fixation and growth of Lens culinaris as affected by nickel availability: A pre-requisite for optimization of agromining. *Environmental and Experimental Botany, 131*, 1–9.
Saleem, M., Arshad, M., Hussain, S., & Bhatti, A. S. (2007). Perspective of plant growth promoting rhizobacteria (PGPR) containing ACC deaminase in stress agriculture. *Journal of Industrial Microbiology and Biotechnology, 34*(10), 635–648.
Saravanan, V. S., Kalaiarasan, P., Madhaiyan, M., & Thangaraju, M. (2007). Solubilization of insoluble zinc compounds by *Gluconacetobacter diazotrophicus* and the detrimental action of zinc ion (Zn2+) and zinc chelates on root knot nematode *Meloidogyne incognita. Letters in Applied Microbiology, 44*(3), 235–241.
Schalk, I. J., Hannauer, M., & Braud, A. (2011). New roles for bacterial siderophores in metal transport and tolerance. *Environmental Microbiology, 13*(11), 2844–2854.
Scheuring, I., & Yu, D. W. (2012). How to assemble a beneficial microbiome in three easy steps. *Ecology Letters, 15*(11), 1300–1307.
Schulz, B., & Boyle, C. (2006). What are endophytes. In B. Schulz, C. Boyle, & T. N. Sieber (Eds.), *Microbial root endophytes* (Vol. 9). Berlin Heidelberg, Springer.

Segura, A., Rodríguez-Conde, S., Ramos, C., & Ramos, J. L. (2009). Bacterial responses and interactions with plants during rhizoremediation. *Microbial Biotechnology, 2*(4), 452–464 (Electronic Resource).

Sentchilo, V., Mayer, A. P., Guy, L., Miyazaki, R., Green Tringe, S., Barry, K., ... van der Meer, J. R. (2013). Community-wide plasmid gene mobilization and selection. *ISME Journal, 7*(6), 1173–1186.

Sessitsch, A., Hardoim, P., Doring, J., Weilharter, A., Krause, A., Woyke, T., ... Reinhold-Hurek, B. (2012). Functional characteristics of an endophyte community colonizing rice roots as revealed by metagenomic analysis. *Molecular Plant-Microbe Interactions: MPMI, 25*(1), 28–36.

Sessitsch, A., Kuffner, M., Kidd, P., Vangronsveld, J., Wenzel, W. W., Fallmann, K., & Puschenreiter, M. (2013). The role of plant-associated bacteria in the mobilization and phytoextraction of trace elements in contaminated soils. *Soil Biology and Biochemistry, 60*(100), 182–194.

Sessitsch, A., & Puschenreiter, M. (2008). Endophytes and rhizosphere bacteria of plants growing in heavy metal-containing soils. In P. Dion, & C. Nautiyal (Eds.), *13. Microbiology of extreme soils* (pp. 317–332). Springer Berlin Heidelberg.

Shalaby, A. M. (2003). Responses of arbuscular mycorrhizal fungal spores isolated from heavy metal-polluted and unpolluted soil to Zn, Cd, Pb and their interactions in vitro. *Pakistan Journal of Biological Sciences, 6*(16), 1416–1422.

Sharma, A., & Johri, B. (2003). Growth promoting influence of siderophore-producing *Pseudomonas* strains GRP3A and PRS 9 in maize (*Zea mays* L.) under iron limiting conditions. *Microbiological Research, 158*(3), 243–248.

Sheng, X., Sun, L., Huang, Z., He, L., Zhang, W., & Chen, Z. (2012). Promotion of growth and Cu accumulation of bio-energy crop (*Zea mays*) by bacteria: Implications for energy plant biomass production and phytoremediation. *Journal of Environmental Management, 103*, 58–64.

Singh, A. P., Goel, R. K., & Kaur, T. (2011). Mechanisms pertaining to arsenic toxicity. *Toxicology International, 18*(2), 87.

Singh, L. P., Gill, S. S., & Tuteja, N. (2011). Unraveling the role of fungal symbionts in plant abiotic stress tolerance. *Plant Signaling and Behavior, 6*(2), 175–191.

Slonecker, T., Haack, B., & Price, S. (2009). Spectroscopic analysis of arsenic uptake in *Pteris ferns*. *Remote Sensing, 1*(4), 644–675.

Soleimani, M., Afyuni, M., Hajabbasi, M. A., Nourbakhsh, F., Sabzalian, M. R., & Christensen, J. H. (2010). Phytoremediation of an aged petroleum contaminated soil using endophyte infected and non-infected grasses. *Chemosphere, 81*(9), 1084–1090.

de Souza, M. P., Huang, C. P. A., Chee, N., & Terry, N. (1999). Rhizosphere bacteria enhance the accumulation of selenium and mercury in wetland plants. *Planta, 209*(2), 259–263.

Spaepen, S., & Vanderleyden, J. (2011). Auxin and plant-microbe interactions. *Cold Spring Harbor Perspectives in Biology, 3*(4), a001438 (Electronic Resource).

Spaepen, S., Vanderleyden, J., & Remans, R. (2007). Indole-3-acetic acid in microbial and microorganism-plant signaling. *FEMS Microbiology Reviews, 31*(4), 425–448.

Stefanowicz, A. M., Niklinska, M., & Laskowski, R. (2008). Metals affect soil bacterial and fungal functional diversity differently. *Environmental Toxicology and Chemistry, 27*(3), 591–598.

Su, S., Zeng, X., Bai, L., Jiang, X., & Li, L. (2010). Bioaccumulation and biovolatilisation of pentavalent arsenic by *Penicillin janthinellum*, *Fusarium oxysporum* and *Trichoderma asperellum* under laboratory conditions. *Current Microbiology, 61*(4), 261–266.

Thijs, S., Sillen, W., Rineau, F., Weyens, N., & Vangronsveld, J. (2016). Towards an enhanced understanding of plant-microbiome interactions to improve phytoremediation: Engineering the metaorganism. *Frontiers in Microbiology, 7*, 341.

Tiwari, S., Sarangi, B. K., & Thul, S. T. (2016). Identification of arsenic resistant endophytic bacteria from *Pteris vittata* roots and characterization for arsenic remediation application. *Journal of Environmental Management, 180*, 359–365.
Toth, G., Hermann, T., Da Silva, M. R., & Montanarella, L. (2016). Heavy metals in agricultural soils of the European Union with implications for food safety. *Environment International, 88*, 299–309.
Touceda-González, M., Álvarez-López, V., Prieto-Fernández, Á., Rodríguez-Garrido, B., Trasar-Cepeda, C., Mench, M., … Kidd, P. S. (2017). Aided phytostabilisation reduces metal toxicity, improves soil fertility and enhances microbial activity in Cu-rich mine tailings. *Journal of Environmental Management, 186*(Pt 2), 301–313.
Truyens, S., Beckers, B., Thijs, S., Weyens, N., Cuypers, A., & Vangronsveld, J. (2015). Cadmium-induced and transgenerational changes in the cultivable and total seed endophytic community of *Arabidopsis thaliana*. *Plant Biology, 18*(3), 376–381.
Truyens, S., Weyens, N., Cuypers, A., & Vangronsveld, J. (2015). Bacterial seed endophytes: Genera, vertical transmission and interaction with plants. *Environmental Microbiology Reports,* 7(1), 40–50 (Electronic Resource).
Turner, T. R., Ramakrishnan, K., Walshaw, J., Heavens, D., Alston, M., Swarbreck, D., … Poole, P. S. (2013). Comparative metatranscriptomics reveals kingdom level changes in the rhizosphere microbiome of plants. *ISME Journal,* 7(12), 2248–2258.
Vangronsveld, J., Herzig, R., Weyens, N., Boulet, J., Adriaensen, K., Ruttens, A., … Mench, M. (2009). Phytoremediation of contaminated soils and groundwater: Lessons from the field. *Environmental Science and Pollution Research International, 16*(7), 765–794.
Verbruggen, N., Hermans, C., & Schat, H. (2009). Molecular mechanisms of metal hyperaccumulation in plants. *New Phytologist, 181*(4), 759–776.
Visioli, G., D'Egidio, S., & Sanangelantoni, A. M. (2014). The bacterial rhizobiome of hyperaccumulators: Future perspectives based on omics analysis and advanced microscopy. *Frontiers in Plant Science, 5*, 752.
Viterbo, A., Landau, U., Kim, S., Chernin, L., & Chet, I. (2010). Characterization of ACC deaminase from the biocontrol and plant growth-promoting agent *Trichoderma asperellum* T203. *FEMS Microbiology Letters, 305*(1), 42–48.
Waqas, M., Khan, A. L., Kamran, M., Hamayun, M., Kang, S.-M., Kim, Y.-H., & Lee, I.-J. (2012). Endophytic fungi produce gibberellins and indoleacetic acid and promotes host-plant growth during stress. *Molecules: A Journal of Synthetic Chemistry and Natural Product Chemistry, 17*(9), 10754–10773.
Wei, Y., Hou, H., Li, J., ShangGuan, Y., Xu, Y., Zhang, J., … Wang, W. (2014). Molecular diversity of arbuscular mycorrhizal fungi associated with an Mn hyperaccumulator—*Phytolacca americana*, in Mn mining area. *Applied Soil Ecology, 82*, 11–17.
Wei, Y.-F., Li, T., Li, L.-F., Wang, J.-L., Cao, G.-H., & Zhao, Z.-W. (2016). Functional and transcript analysis of a novel metal transporter gene EpNramp from a dark septate endophyte (*Exophiala pisciphila*). *Ecotoxicology and Environmental Safety, 124*, 363–368.
Wei, S., Zhou, Q., & Koval, P. V. (2006). Flowering stage characteristics of cadmium hyperaccumulator *Solanum nigrum* L. and their significance to phytoremediation. *Science of the Total Environment, 369*(1), 441–446.
Weyens, N., van der Lelie, D., Taghavi, S., Newman, L., & Vangronsveld, J. (2009). Exploiting plant-microbe partnerships to improve biomass production and remediation. *Trends in Biotechnology, 27*(10), 591–598.
Weyens, N., van der Lelie, D., Taghavi, S., & Vangronsveld, J. (2009). Phytoremediation: Plant-endophyte partnerships take the challenge. *Current Opinion in Biotechnology, 20*(2), 248–254.
Whipps, J. M. (2001). Microbial interactions and biocontrol in the rhizosphere. *Journal of Experimental Botany, 52*, 487–511.

White, C., Sayer, J. A., & Gadd, G. M. (1997). Microbial solubilization and immobilization of toxic metals: Key biogeochemical processes for treatment of contamination. *FEMS Microbiology Reviews, 20*(3–4), 503–516.

Wu, L., Li, N., & Luo, Y. (2008). Phytoextraction of heavy metal contaminated soil by *Sedum plumbizincicola* under different agronomic strategies. In *Proc 5th Int Phytotech Conf, Nanjing, China.*

Xie, P., Hao, X., Herzberg, M., Luo, Y., Nies, D. H., & Wei, G. (2015). Genomic analyses of metal resistance genes in three plant growth promoting bacteria of legume plants in Northwest mine tailings, China. *Journal of Environmental Sciences, 27*, 179–187.

Xiong, J., Wu, L., Tu, S., Van Nostrand, J. D., He, Z., Zhou, J., & Wang, G. (2010). Microbial communities and functional genes associated with soil arsenic contamination and the rhizosphere of the arsenic-hyperaccumulating plant *Pteris vittata* L. *Applied and Environmental Microbiology, 76*(21), 7277–7284.

Xu, J. Y., Han, Y. H., Chen, Y., Zhu, L. J., & Ma, L. Q. (2016). Arsenic transformation and plant growth promotion characteristics of As-resistant endophytic bacteria from As-hyperaccumulator *Pteris vittata. Chemosphere, 144*, 1233–1240.

Yang, X., Long, X., Ye, H., He, Z., Calvert, D., & Stoffella, P. (2004). Cadmium tolerance and hyperaccumulation in a new Zn-hyperaccumulating plant species (*Sedum alfredii* Hance). *Plant and Soil, 259*(1–2), 181–189.

Zakeri, B., Fierer, J. O., Celik, E., Chittock, E. C., Schwarz-Linek, U., Moy, V. T., & Howarth, M. (2012). Peptide tag forming a rapid covalent bond to a protein, through engineering a bacterial adhesin. *Proceedings of the National Academy of Sciences of the United States of America, 109*(12), E690–E697.

Zhang, W. H., Chen, W., He, L. Y., Wang, Q., & Sheng, X. F. (2015). Characterization of Mn-resistant endophytic bacteria from Mn-hyperaccumulator *Phytolacca americana* and their impact on Mn accumulation of hybrid *Penisetum. Ecotoxicology and Environmental Safety, 120*, 369–376.

Zhang, W. H., He, L. Y., Wang, Q., & Sheng, X. F. (2015). Inoculation with endophytic *Bacillus megaterium* 1Y31 increases Mn accumulation and induces the growth and energy metabolism-related differentially-expressed proteome in Mn hyperaccumulator hybrid *Pennisetum. Journal of Hazardous Materials, 300*, 513–521.

Zhang, X., Zhong, T., Liu, L., & Ouyang, X. (2015). Impact of soil heavy metal pollution on food safety in China. *PLoS One, 10*(8), e0135182.

Zhao, D., Li, T., shen, M., Wang, J., & Zhao, Z. (2015). Diverse strategies conferring extreme cadmium (Cd) tolerance in the dark septate endophyte (DSE), *Exophiala pisciphila*: Evidence from RNA-seq data. *Microbiological Research, 170*, 27–35.

Zhou, N., Liu, P., Wang, Z. Y., & Xu, G. D. (2011). The effects of rapeseed root exudates on the forms of aluminum in aluminum stressed rhizosphere soil. *Crop Protection, 30*(6), 631–636.

Zhu, L. J., Guan, D. X., Luo, J., Rathinasabapathi, B., & Ma, L. Q. (2014). Characterization of arsenic-resistant endophytic bacteria from hyperaccumulators *Pteris vittata* and *Pteris multifida. Chemosphere, 113*, 9–16.

Zhu, Y.-G., Pilon-Smits, E. A., Zhao, F.-J., Williams, P. N., & Meharg, A. A. (2009). Selenium in higher plants: Understanding mechanisms for biofortification and phytoremediation. *Trends in Plant Science, 14*(8), 436–442.

CHAPTER THREE

Potential Role of Plant-Associated Bacteria in Plant Metal Uptake and Implications in Phytotechnologies

Petra S. Kidd[1], Vanessa Álvarez-López, Cristina Becerra-Castro[†], Maribel Cabello-Conejo and Ángeles Prieto-Fernández
Consejo Superior de Investigaciones Científicas (CSIC), Santiago de Compostela, Spain
[1]Corresponding author: E-mail: pkidd@iiag.csic.es

Contents

Abstract

Plants interact closely with microbes and these can enhance plant growth and health by increasing nutrient uptake and improving plant resistance to pathogens and stress. Plant-associated microorganisms are commonly used as 'biofertilisers' in agriculture but their incorporation into phytoremediation systems to improve plant establishment and growth, and modify plant metal accumulation, is more recent. This review focuses on the recent advances in the use of plant—bacterial associations to enhance phytoextraction (phytomining) processes in trace element (TE)—contaminated or —enriched sites. Experimental evidence shows that plant-associated bacteria play an important role in plant TE bioaccumulation, and bench level studies suggest bacterial inoculants could enhance phytoextraction efficiency. However, the performance of these bacterial inoculants under natural conditions will have to be investigated under a field scale.

[†] deceased.

Advances in Botanical Research, Volume 83
ISSN 0065-2296
http://dx.doi.org/10.1016/bs.abr.2016.12.004

1. PHYTOMANAGEMENT OF TRACE ELEMENT–ENRICHED SOILS

Worldwide environmental pollution by trace elements (TE), both metals and metalloids (with common concentrations of <100 mg/kg dry weight (DW) in living organisms), along with organic pollutants, has led to severe and diffuse contamination of soils with significant environmental effects and health risks, e.g., soil erosion, loss of biodiversity, water pollution, food safety risks, etc. (Bhargava, Carmona, Bhargava, & Srivastava, 2012). Metal(loid)s are amongst the most frequently occurring soil contaminants at polluted sites across Europe (Panagos, Van Liedekerke, Yigini, & Montanarella, 2013), and their presence in elevated concentrations has been identified by the European Commission as one of the eight major threats to European soils (COM, 2002). Conventional remediation techniques are generally based on civil engineering techniques which can be highly destructive and expensive. Over the last few decades there has been growing interest in the use of *Gentle soil Remediation Options* (GRO) which include in situ contaminant stabilisation ('inactivation') and plant-based ('phytoremediation') options. GRO are mainly based on the combined use of plants and their associated microorganisms, partly assisted by the use of organic and inorganic amendments, and soil management practices which decrease the labile pool and/or total TE content. These techniques are considered to be less invasive and more cost-effective than techniques such as encapsulation, vitrification or soil washing, and more sustainable than 'dig and dump' strategies (Mench et al., 2009, 2010; Vangronsveld et al., 2009). In addition to potentially restoring soil structure, functions and quality, and positively influencing other ecosystem services (e.g., increased biodiversity, improved surface and ground water quality, C-storage, soil erosion, temperature regulation, etc.) these methods may also provide valuable sources of renewable biomass for the bio-based economy (e.g., bioenergy, biocatalysis and platform molecules for green chemicals, and ecomaterials).

Several GRO have been developed for targeting distinct contaminant scenarios: for TE-contaminated soils (TECS), the objective is to decrease the labile ('bioavailable') pool and/or total contents of TE in the soil, or to reduce any related pollutant linkages (e.g., leaching from the root zone, soil erosion and water runoff, etc.). Phytoextraction aims to remove TE from contaminated soils through their uptake and accumulation in plant parts that can be harvested. Harvested plant biomass can be burned to produce metal(loid)-enriched ash or 'bio-ore' (the process is then known as phytomining). The energy gained from biomass combustion can support the profitability of the technology. Bio-ores can also be used for TE recovery (smelting, pytometallurgy); although phytomining has, so far, only been proven economically viable for Ni (Chaney et al., 2014; chapter: Phytoremediation and Phytomining: Status and Promise by Chaney & Baklanov, 2017). In contrast, phytostabilisation aims to establish a vegetation cover and progressively promote in situ inactivation of metal(loid)s by combining the use of TE-excluding plants and soil amendments (Mench, Renella, Gelsomino, Landi, & Nannipieri, 2006; Vangronsveld et al., 2009; Vangronsveld, Sterckx, Van Assche, Clijsters, 1995). Although this technology does not lead to a clean-up of the soil, by altering TE speciation and mobility it moderates their potential negative environmental impacts and pollutant linkages.

The success of GRO is highly dependent on the solubility and speciation of TE in soil, as well as their tolerance and uptake by plants. Adequate plant selection for each technique is vital: in phytoextraction-related technologies a high uptake of specific TE into the shoots is desirable, while successful phytostabilisation requires that TE be excluded from the aerial portions. TE-hyperaccumulators which are able to accumulate extreme concentrations of metal(loid)s (e.g., Cd, Ni, Zn, Se and As) in their above-ground biomass (often endemic to metal-enriched substrates, such as ultramafic or calamine soils) have been proposed for phytoextraction (van der Ent et al., 2015; chapter: Metallophytes of Serpentine and Calamine Soils – Their Unique Ecophysiology and Potential for Phytoremediation by Wojcik et al., 2017). However, the efficiency of phytoextraction can be limited due to slow growth, low biomass or/and shallow root systems of TE-hyperaccumulating plants (except some Ni-hyperaccumulators such as *Berkheya coddii*), leading to a high number of cropping cycles required for clean-up (if the objective is to reduce total TE concentrations in soils) (Robinson, Bañuelos, Conesa, Evangelou, & Schulin, 2009). High-biomass crops (annuals or perennials) and woody plants are recognised as viable alternatives to hyperaccumulators for phytoextraction of TE (particularly Cd, Se and Zn) if they also show relevant shoot TE removals (i.e., moderate-high bioconcentration factor and high shoot yield) (Kidd et al., 2015). The success of GRO can also be limited by TE mobility and availability and the development of methods for modifying soil TE bioavailability could have a positive impact on phytoremediation efficiency. Increasing contaminant bioavailability would improve processes such as phytoextraction, whereas the use of amendments which reduce contaminant bioavailability could substantially improve phytostabilisation (Wenzel, 2009). During the last decade, agronomic and biotechnological methods have been developed to optimise the effectiveness and success of GRO. Plant metal(loid) tolerance, establishment and growth in TECS can be improved using plant growth-promoting bacteria (PGPB) and fungi (including mycorrhizae and endophytic fungi). Moreover, such plant-associated bacteria can modify TE mobility and hence plant uptake and bioaccumulation (Becerra-Castro et al., 2009; Sessitsch et al., 2013; chapter: The Bacterial and Fungal Microbiota of Hyperaccumulator Plants: Small Organisms, Large Influence by Thijs, Langill, & Vangronsveld, 2017; Weyens, van der Lelie, Taghavi, Newman, & Vangronsveld, 2009).

The main focus of this chapter will be on the beneficial effects of plant-associated bacteria on phytoremediation processes in TECS and, in particular, on improving phytoextraction or phytomining techniques. In the early 2000s several studies suggested that the accumulation of metals by (hyper)accumulating plants was influenced by their rhizospheric microbiota (Abou-Shanab et al., 2003; Delorme, Gagliardi, Angle, & Chaney, 2001; Lodewyckx, Mergeay, Vangronsveld, Clijsters, & Van der Lelie, 2002; Mengoni, Barzanti, Gonnelli, Gabbrielli, & Bazzicalupo, 2001; Whiting, Leake, McGrath, & Baker, 2001a, 2001b). Since then a large number of studies can be found indicating that plant-associated microorganisms are indeed essential players during metal phytoextraction or phytomining (Becerra-Castro et al., 2013; Glick, 2010, 2014; Kidd et al., 2009; Lebeau, Braud, & Jezequel, 2008; Muehe et al., 2015; Sessitsch et al., 2013).

2. PLANT-ASSOCIATED MICROORGANISMS

Plants interact closely with microbes, and these can enhance plant growth and health by increasing nutrient uptake and improving plant resistance to pathogens and stress (Göhre & Paszkowski, 2006; Lebeau et al., 2008; Lugtenberg & Kamilova, 2009). Moreover, they can also help plants adapt to the presence of phytotoxic concentrations of metals in soils (Filion, St-Arnaud, & Fortin, 1999; Gadd, 2010; Wehner, Antunes, Powell, Mazukatow, & Rillig, 2010; Weissenhorn, Leyval, Belgy, & Berthelin, 1995). Fig. 1 presents an overview of the plant–microbial–soil interactions involved in phytoremediation processes.

Endophytic fungi and bacteria colonise the inner tissues of living plants; these establish a harmonious relationship with the host plants and cause no negative effects on plant health (Compant, Clément, & Sessitsch, 2010; Zheng et al., 2016). They have been isolated from a wide range of plant species, from woody tree species to herbaceous crop plants, suggesting a

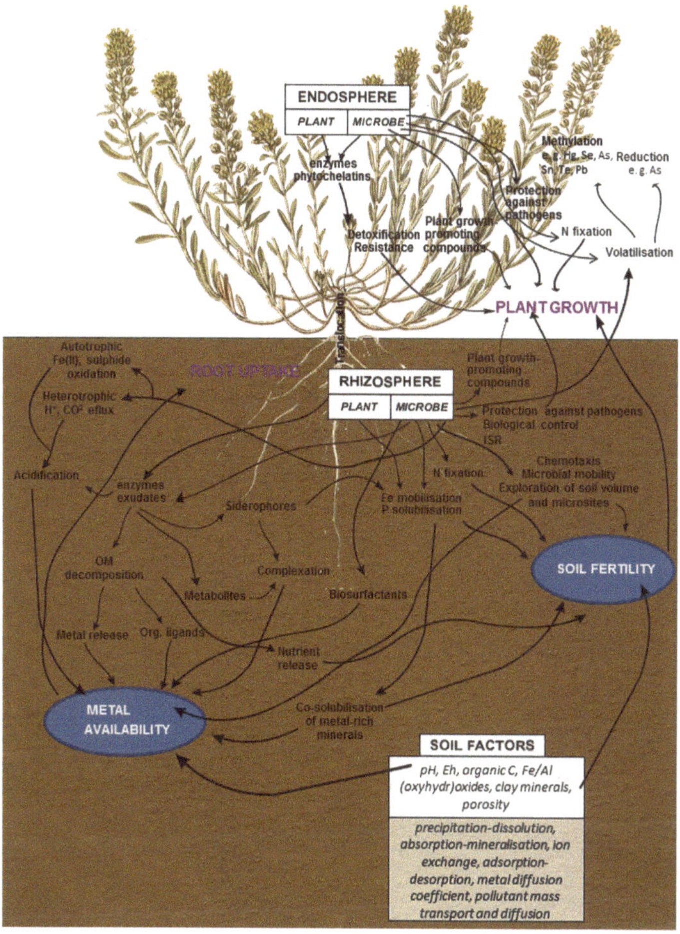

Figure 1 Overview of plant–microbe–soil interactions involved in phytoextraction processes. *Modified from Wenzel, W., Lombi, E., & Adriano, D. (1999) Biogeochemical processes in the rhizosphere: Role in phytoremediation of metal-polluted soils. In* Heavy metal stress in plants *(pp. 273–303). Springer; Pilon-Smits, E. (2005) Phytoremediation.* Annual Review of Plant Biology, 56, *15–39, http://dx.doi.org/10.1146/annurev.arplant.56.032604.144214.*

ubiquitous existence in nearly all higher plants (Luo et al., 2011). Colonising microbial communities are often specific to the different plant compartments or organs, including roots, stem, leaves, flowers as well as fruits and seeds (Compant et al., 2010). Phyllospheric bacteria inhabit the external surfaces of plant parts; and rhizobacteria are present in the rhizosphere soil influenced by plant root activity (Sessitsch & Puschenreiter, 2008; Weyens et al., 2009). The root–soil interface or rhizosphere is defined as the volume of soil surrounding living plant roots that is influenced by root activity (Hiltner, 1904), and its volume can vary both spatially and temporally according to the plant species and the soil physico-chemical properties (Hinsinger, Gobran, Gregory, & Wenzel, 2005). The rhizosphere represents both a unique and dynamic zone of plant–microbial–soil interactions (Fig. 1), where intense biological activity and root exudation create gradients in soil physico-chemical and biological parameters. Plants can promote the abundance of beneficial microorganisms in the rhizosphere (Thijs, Sillen, Rineau, Weyens, & Vangronsveld, 2016; Vessey, 2003) (Fig. 1). Rhizosphere organisms that have received much attention because of their beneficial effects on plant growth and health are the plant growth-promoting rhizobacteria (PGPR) including nitrogen-fixing bacteria or biocontrol microorganisms, mycorrhizal fungi, mycoparasitic fungi and protozoa (Mendes, Garbeva, & Raaijmakers, 2013; chapter: Mycorrhiza-Assisted Phytoremediation by Coninx, Martinova, & Rineau, 2017). All these plant-microbial symbioses are mutualistic: the plant host provides exudates and creates habitats for the microorganisms, and, in return, the PGP bacteria and fungi promote plant growth and health (Mendes et al., 2013; Raaijmakers, Vlami, & de Souza, 2002).

The use of microorganisms influencing the availability of plant nutrients (such as N, Fe or P) in agriculture as 'biofertilisers' to enhance plant nutrient uptake and alleviate nutrient deficiencies is well established (Vessey, 2003). However, the idea of incorporating plant-associated microbes into phytoremediation systems is more recent. The use of bacteria, which are able to fix atmospheric N_2 or to solubilise poorly available P and Fe, have been identified as good candidates for sustainable biomass production in TE-contaminated lands (Weyens et al., 2009). Higher plants present associations with diazotrophic bacteria, occurring in the form of nodules (symbionts) or as free living cells, which can reduce N_2 to the NH_4^+ form (that is then available to plants for uptake). Inorganic P-solubilising bacteria are able to mobilise insoluble phosphate and make it accessible for plant uptake (Rodriguez & Fraga, 1999; Vessey, 2003). Siderophore compounds are Fe(III)-specific chelating agents; these compounds can be produced by bacteria and can bind insoluble Fe(III) making it available for plants (Crowley & Kraemer, 2007). Bacteria can also produce phytohormones, such as auxins, cytokinins and gibberellins, which stimulate plant growth and development (Taghavi et al., 2009; Tanimoto, 2005). Phytohormones are involved in root growth and root hair proliferation, the stimulation of cell division and tissue expansion, and in stomatal opening or modifying plant morphology and tissue extension (Hare, Cress, & van Staden, 1997; Taghavi et al., 2009; Vessey, 2003). Plant-associated bacteria can also reduce plant stress through the synthesis of 1-aminocyclopropane-1-carboxylic acid (ACC) deaminase which reduces the high levels of ethylene by consuming its immediate precursor, the ACC (Glick, 2014). Other well-known mechanisms of PGPBs include increasing water uptake, alteration of root morphology, production of antibiotics and the induction of plant defence mechanisms (Kidd et al., 2009; Lin, Okon, & Hardy, 1983; van Loon & Bakker, 2003). Bacteria that inhibit or reduce plant diseases are often referred to as biocontrol agents and protect plants indirectly by competing for nutrients and space with pathogenic bacteria (niche exclusion), producing antimicrobial compounds or through the induction of plant defence mechanisms (Compant et al., 2010; Lemanceau, Maurhofer, & Défago, 2007; Podile & Kishore, 2006). The production of siderophores by PGP bacteria in the rhizosphere are thought to deprive pathogens of Fe (Compant, Duffy, Nowak, Clément, & Barka, 2005). Antibiosis is an important mode of action of many biocontrol agents, and compounds with biocontrol capacity include antimicrobial compounds (antibiotics), biosurfactants and chitinolytic enzymes (Lemanceau et al., 2007). A wide range of antimicrobial compounds (most of them with a broad-spectrum activity) are produced by bacteria: 2,4-diacetylphloroglucinol (DAPG), hydrogen cyanide (HCN), kanosamine, phenazines, oomycin A, pyrrolnitrin, viscosinamide, pyoluteorin, butyrolactones, pantocin A and B, xanthobaccins and zwittermycin A (Raaijmakers et al., 2002). Antimicrobial compounds act, mainly, on four different targets: cell wall synthesis, protein synthesis, nucleic acid replication or cellular membranes (Raaijmakers et al., 2002). PGP bacteria also have the capacity to produce cell wall hydrolases, such as quitinases or glucanases, which attack the cell wall of phytopathogenic fungi (Compant et al., 2005; Podile & Kishore, 2006). Finally, some bacterial strains can activate plant defence mechanisms without causing visible symptoms of stress on the host plant, a phenomenon referred to as induced systemic resistance [ISR; van Loon, Bakker, and Pieterse (1998)]. Bacterial determinants of ISR include lipopolysaccharides (LPS), siderophores, salicylic acid (SA) and other macromolecules (van Loon et al., 1998). Of all these PGP traits, indoleacetic acid (IAA) is often considered of major relevance since it promotes plant growth directly by stimulating plant cell elongation or by affecting cell division (Duca, Lorv, Patten, Rose, & Glick, 2014), and bacterial-induced stimulation of plant growth in metal-contaminated soils has often been associated with this characteristic

(Dell'Amico, Cavalca, & Andreoni, 2008; Luo et al., 2011; Pereira & Castro, 2014; Sheng, Xia, Jiang, He, & Qian, 2008; Shilev, Fernández, Benlloch, & Sancho, 2006).

Plant–microbial interactions greatly influence soil TE mobility and availability to plants for uptake (Anderson, Guthrie, & Walton, 1993; Gadd, 2004; Haferburg & Kothe, 2007; Marschner, Crowley, & Yang, 2004; Sessitsch et al., 2013; Weyens et al., 2009). Three specific mechanisms of microbial-induced mobilisation of TE and increased plant metal(loid) accumulation are as follows: (1) increasing root surface area and hair production, (2) increasing element solubility and (3) increasing soluble element transfer from the rhizosphere to the plant (Whiting et al., 2001b). An increase in soil metal(loid) mobility and availability can result in their increased uptake and therefore the overall phytoextracted TE (Kidd et al., 2009; Sessitsch et al., 2013; Weyens et al., 2009). In contrast, microorganisms can immobilise metal(loid)s through sorption to cell components or exopolymers, transport and intracellular sequestration, release of metal(loid) binding compounds or precipitation as insoluble organic or inorganic molecules (Gadd, 2004). Mineralisation of colloidal metal(loid)-organic complexes may be another cause of microbial-mediated immobilisation (Gadd, 2004). In contrast, precipitated metal(loid)s can be solubilised by acidification, chelation and ligand-induced dissolution (Gadd, 2004). Oxalate, malate and citrate are some of the most important organic acids identified in root and microbial exudates (Ehrlich, 1998; Jones, 1998).

3. BACTERIA ASSOCIATED WITH PLANT METALLOPHYTES AND (HYPER)ACCUMULATORS

Natural metal(loid)-enriched areas (such as serpentine soils) or TE-contaminated sites are not only a source of interesting plant species for application in phytoremediation but also of microorganisms (Batty, 2005; Mengoni, Schat, & Vangronsveld, 2010; Schippers, Hallmann, Wentzien, & Sand, 1995). Negative effects of elevated metal concentrations in soils have been found on microbial communities, with high metal contents being correlated with decreases in total microbial biomass (Liao & Xie, 2007) and reduction of microbial activity (Smolders, Buekers, Oliver, & McLaughlin, 2004). Trace metal contamination also exerts selective pressure on soil microbial communities, leading to shifts in community structure (Kelly, Häggblom, & Tate, 2003) and to decreases in functional diversity (Liao & Xie, 2007). Despite this, the bacterial communities present in these soils often exhibit high levels of tolerance to several metals and can be a valuable source of plant-associated bacterial strains for phytoremediation purposes. Over the last decade numerous studies have focused on the isolation and characterisation of culturable microorganisms from this type of substrate in the search for bioinoculants (Abou-Shanab et al., 2003; Barzanti et al., 2007; Grandlic, Mendez, Chorover, Machado, & Maier, 2008; Idris, Trifonova, Puschenreiter, Wenzel, & Sessitsch, 2004; Lodewyckx et al., 2002; Pal, Dutta, Mukherjee, & Paul, 2005; Weyens et al., 2009). To be useful in phytoremediation processes the selected bacterial inoculants must be able to tolerate the potentially toxic concentrations of metal(loid)s present in the substrate but should also be adapted to the specific conditions of the site (such as nutrient deficiency).

3.1 Plant-Associated Bacterial Communities in Naturally Metal-Rich Soils

Serpentine soils are formed through the weathering of ultramafic rocks which are comprised of at least 70% ferromagnesian (or mafic) minerals (particularly within the olivine and pyroxene groups) and less than 45% of silica (SiO_2) (Kruckeberg, 2002). These soils are ubiquitous, but patchily distributed around the world (occupying approximately 1% of the earth's crust (Proctor, 1999)), and although some variations can occur between sites, they are typically characterised by a low Ca:Mg ratio and elevated concentrations of TE such as Ni, Cr and Co. Serpentine soils are also often deficient in essential nutrients such as N, K or P (Kruckeberg, 2002). The particular plant communities which develop in these areas (including metal-hyperaccumulating plant species) have been well studied for their morphological and physiological adaptations to these conditions (known as the serpentine syndrome) (Proctor, 1971; chapter: Metallophytes of Serpentine and Calamine Soils – Their Unique Ecophysiology and Potential for Phytoremediation by Wojcik et al., 2017). However, the attention of microbiologists towards their associated microbiome, and its potential influence on plant metal uptake and accumulation, is more recent. For Ni phytomining, serpentine soils represent a source of potentially beneficial plant-growth promoting and Ni-tolerant microorganisms (Álvarez-López, Prieto-Fernández, Becerra-Castro, Monterroso, & Kidd, 2016; Cabello-Conejo et al., 2014; Durand et al., 2016).

In serpentine areas, nutrient deficiency and metal phytotoxicity have often been suggested to be the cause of the low number of microorganisms in these soils (Acea & Carballas, 1986; Lipman, 1926; Pal et al., 2005). Nonetheless, bacterial communities in serpentine soils present a high genetic diversity, and highly TE-tolerant bacterial strains have been isolated from such sites

(Abou-Shanab et al., 2003; Amir & Pineau, 2003; Mengoni et al., 2001; Pal et al., 2005; chapter: The Bacterial and Fungal Microbiota of Hyperaccumulator Plants: Small Organisms, Large Influence by Thijs et al., 2017). Schlegel, Cosson, and Baker (1991) found bacterial strains isolated from serpentine soils tolerated up to 10–20 mM Ni (in the culture medium), while strains from other soil types tolerated only 1 mM Ni. Turgay, Gormez, and Bilen (2012) found bacterial strains, isolated from Turkish serpentine soils, tolerated up to 34 mM Ni in the growth medium. Furthermore, the rhizosphere bacterial communities associated with metal-hyperaccumulating plants (such as the Ni-hyperaccumulators *Alyssum bertolonii*, *Noccaea goesingensis* and *Sebertia acuminata*) have been shown to differ from those of nonaccumulating plants growing at the same site or of nonvegetated soil (Abou-Shanab et al., 2003; Idris et al., 2004; Lodewyckx et al., 2002; Mengoni et al., 2001, 2004, 2010; Schlegel et al., 1991; chapter: The Bacterial and Fungal Microbiota of Hyperaccumulator Plants: Small Organisms, Large Influence by Thijs et al., 2017). A higher number of Ni-tolerant bacteria have been found in the rhizosphere of Ni-hyperaccumulators, and this selective enrichment has been correlated with an increase in soil Ni availability (Becerra-Castro et al., 2009; chapter: The Bacterial and Fungal Microbiota of Hyperaccumulator Plants: Small Organisms, Large Influence by Thijs et al., 2017). Abou-Shanab et al. (2003) found the majority of rhizobacteria isolated from serpentine soils in Oregon (United States) able to grow in the presence of 8 mM Ni, while Mengoni et al. (2001) found rhizobacterial isolates from serpentine soils in Tuscany (Italy) resisted between 7 and 10 mM Ni in their growth medium. Becerra-Castro et al. (2009) also showed a higher proportion of Ni-tolerant bacteria in the rhizosphere of *Alyssum serpyllifolium* ssp. *lusitanicum* than in surrounding soil. Álvarez-López, Prieto-Fernández, Becerra-Castro, et al. (2016) confirmed higher densities of Ni-tolerant bacteria associated with the Ni-hyperaccumulators *A. serpyllifolium* ssp. *lusitanicum* and *Alyssum serpyllifolium* ssp. *malacitanum* but observed significant differences in this selective enrichment amongst different plant populations of these subspecies across the Iberian Peninsula. Likewise, Zn tolerance has often been described in bacterial communities exposed to this metal: Delorme et al. (2001) found higher densities of Zn-tolerant bacteria in the rhizosphere of the Zn-hyperaccumulator *Noccaea caerulescens* compared to the nonaccumulator *Trifolium pratense* or to nonvegetated soil.

Amongst the culturable Ni-tolerant bacterial strains isolated from serpentine soils and/or associated with Ni-hyperaccumulating plants members of the *Actinobacteria*, *Acidobacteria*, *Chlorobi*, *Firmicutes*, *Verrucomicrobia* and *Proteobacteria* have been described (Mengoni et al., 2001; Oline, 2006;

Turgay et al., 2012). Culturable rhizobacteria associated with the Ni-hyperaccumulator *N. goesingensis* were mainly represented by *Methylobacterium* spp., an alphaproteobacterial genus, as well as *Rhodococcus* spp. and *Okibacterium* spp., belonging to the *Actinobacteria* (Idris et al., 2004). Bacteria of the phyla *Proteobacteria* (particularly *α-Proteobacteria*) and *Actinobacteria* were the most numerous within the rhizosphere of the Ni-hyperaccumulator *Alyssum murale* (Abou-Shanab et al., 2010). In the rhizosphere of *A. bertolonii* the culturable bacterial population was dominated by Ni-resistant *Pseudomonas* strains (*γ-Proteobacteria*) (Mengoni et al., 2001). Gremion, Chatzinotas, and Harms (2003) constructed clone libraries based on the 16S rRNA and 16S rDNA and found *Actinobacteria* to be a dominant part of the metabolically active bacteria in the rhizosphere of the Cd/Zn-hyperaccumulator *N. caerulescens*. Similarly, Álvarez-López, Prieto-Fernández, Becerra-Castro, et al. (2016) characterised the culturable bacterial community associated with different populations of the two Ni-hyperaccumulating subspecies of *Alyssum serpyllifolium* (subsp. *lusitanicum* and subsp. *malacitanum*) of the Iberian Peninsula. In general, the culturable rhizobacterial community of these Ni-hyperaccumulators was dominated by the phyla *Proteobacteria* and, in particular, *Actinobacteria*. Isolates were affiliated with genera, such as *Arthrobacter*, *Streptomyces*, *Rhodococcus* or *Microbacterium*, which have been frequently described amongst soil bacteria. However, these authors showed population-specific differences in the composition of the rhizosphere bacterial community (Fig. 2). Although *Actinobacteria* were the most diverse taxon, the proportion of isolates belonging to this phylum differed between plant populations, being more dominant in the rhizosphere of *A. serpyllifolium* subsp. *malacitanum* from the serpentine outcrops in S Spain (Sierra Bermeja, Málaga) compared to *A. serpyllifolium* subsp. *lusitanicum* in NW Spain (Barazón, Galicia). The phylum *Firmicutes* was primarily associated with the *Alyssum* subspecies from NE Portugal (representing 11.4% of isolates in this population), while isolates belonging to the *Bacteroidetes* were only found in the population from NW Spain (representing 5.7% of isolates) (Fig. 2).

The predominance of actinobacterial strains in serpentine soils has been related to the high adaptability of such Gram-positive bacteria to toxic concentrations of trace metals (DeGrood, Claassen, & Scow, 2005). Metal toxicity or nutrient deficiency has been shown to lower the frequency of r-strategists (bacteria capable of rapid growth and utilization of resources), such as *Bacillus* or *Pseudomonas*, as these are more sensitive to toxic substances (Kozdrój, 1995; Kunito, Saeki, Nagaoka, Oyaizu, & Matsumoto, 2001). This could partly explain the lower frequency of r-strategists in the

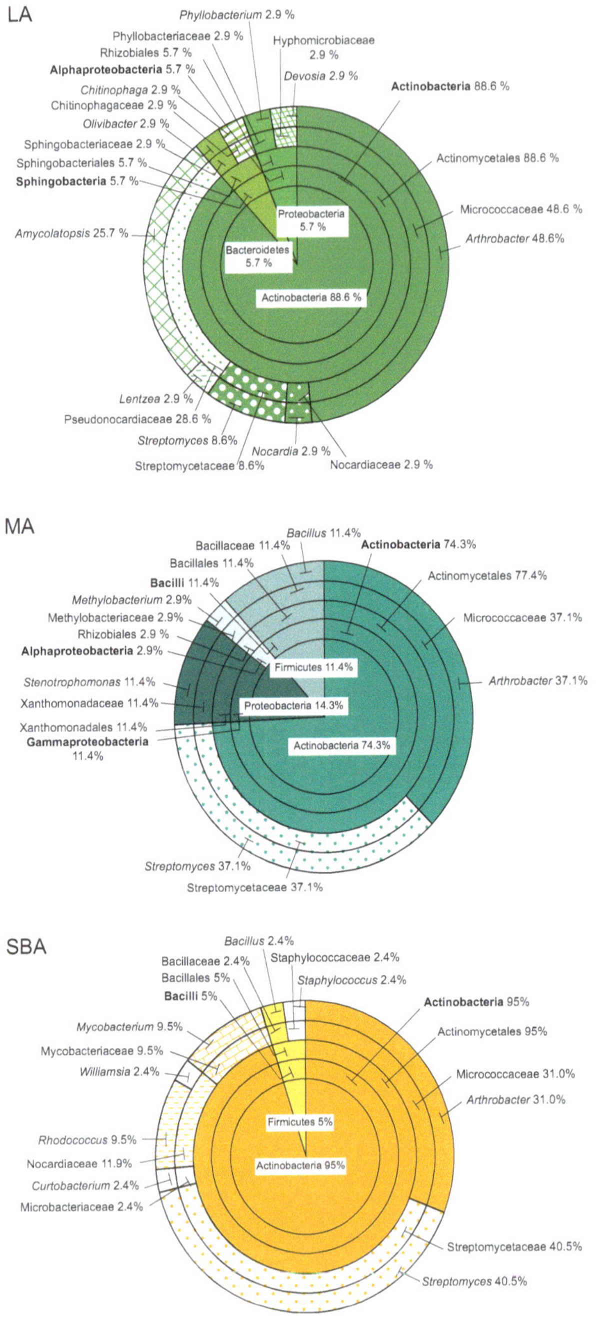

Figure 2 Taxonomic breakdown of 16s rDNA sequences of the rhizosphere culturable community associated with *Alyssum serpyllifolium* subsp. *lusitanicum* (Barazón

serpentine soils. On the other hand, despite a reduced growth and metabolic activity, k-strategists (such as the *Actinomycetales*) present stable populations over a long time period (making them recommended candidates for bioaugmentation purposes) (Lebeau et al., 2008).

Álvarez-López, Prieto-Fernández, Becerra-Castro, et al. (2016) compared the rhizosphere bacterial communities associated with hyperaccumulating *A. serpyllifolium* subspecies with those of nonhyperaccumulating plant species (the Ni-excluder *Dactylis glomerata* and the facultative serpentinophyte *Santolina semidentata*) growing at the same serpentine sites. Hyperaccumulating *Alyssum* subspecies hosted higher densities of bacteria compared to nonhyperaccumulators and showed a strong rhizosphere effect. Bacterial densities are expected to be higher in the rhizosphere compared to nonvegetated or bulk soil as a result of plant root exudates, secretions and lysates which contain labile C sources and growth factors and are well known to stimulate microbial growth and metabolic activity (Delorme et al., 2001; Grayston, Wang, Campbell, & Edwards, 1998). However, the higher bacterial densities associated with the *Alyssum* subspecies compared to the nonhyperaccumulators could be due to plant species—specific differences in root exudate composition and their influence on the composition of their associated bacterial community. Phosphorus-solubilisers and IAA producers were mostly associated with the hyperaccumulators, while siderophore-producers were mostly isolated from *D. glomerata*. Most P-solubilisers were *Actinobacteria* and identified as members of the genera *Arthrobacter*, *Streptomyces* or *Rhodococcus*. One P-solubiliser isolated from the Morais population of *Alyssum pintodasilvae* in NE Portugal was identified as *Methylobacterium* (*α-Proteobacteria*). The appearance of a specific phenotype linked to a specific plant species suggests a plant-driving effect in microorganism selection. Moreover, the maximum tolerable concentration (MTC) for isolates associated with the hyperaccumulators was higher than that of the Ni-excluder *D. glomerata*, which could indicate that the activity of the hyperaccumulator plant leads to an increase in labile Ni and hence enrichment in Ni-tolerant bacteria or that this plant group selects for Ni-tolerant bacteria which in turn modify soil Ni availability.

population (LA), NW Spain), *A. serpyllifolium* subsp. *lusitanicum* (Morais population (MA), NE Portugal), and *Alyssum serpyllifolium* subsp. *malacitanum* (S. Bermeja population (SBA), S Spain). The central pie shows percentages by phyla; each outer ring progressively breaks these down to finer taxonomic levels (phyla, class, family, genera).

Several studies have demonstrated that hyperaccumulator plants harbour endophytic bacterial populations with high genetic diversity, as well as a high level of resistance to heavy metals, representing a potential advantage to plant growth and metal uptake (Barzanti et al., 2007; Idris et al., 2004; chapter: The Bacterial and Fungal Microbiota of Hyperaccumulator Plants: Small Organisms, Large Influence by Thijs et al., 2017). The community structure of shoot endophytic bacteria associated with *N. goesingensis* was highly different from that observed in rhizosphere soil associated with this species. The majority of clones affiliated with the *Proteobacteria*, but bacteria belonging to the Cytophaga/Flexibacter/Bacteroides division, the Holophaga/Acidobacterium division and the low-GC Gram-positive bacteria, were also frequently found. A high number of 16S rRNA gene sequences closely related to *Sphingomonas* were detected. Strains belonging to this genus were also numerous among endophytic isolates. Endophytic bacteria isolated from roots, stems and leaves of *A. bertolonii* belonged to the genera *Bacillus*, *Paenibacillus*, *Leifsonia*, *Curtobacterium*, *Microbacterium*, *Micrococcus* and *Staphyloccoccus* (Barzanti et al., 2007).

3.2 Plant-Associated Bacterial Communities in Trace Element–Contaminated Soils

Numerous metal(loid)-tolerant bacterial strains belonging to diverse taxonomic groups have been isolated from mining sites, as well as domestic and industrial wastes (Batty, 2005; Schippers et al., 1995; Sprocati et al., 2006; Stoppel & Schlegel, 1995). Mine spoils and tailings generally present hostile environments for plant growth, due to low nutrient availability, low organic matter content, high acidity and often elevated trace metal content. Like plants, microorganisms have adapted to these extreme conditions and can aid the establishment and proliferation of colonising plant species (Grandlic et al., 2008; Hanbo et al., 2004). Numerous studies have shown a reduction in the density, metabolic activity and diversity of microbial communities after long-term exposure to trace metals (Giller, Witter, & McGrath, 1998; Kozdrój & van Elsas, 2000; Lorenz et al., 2006). Toxic concentrations of metal(loid)s at these sites typically induce a shift in species composition and the selection of metal-tolerant microorganisms. Becerra-Castro et al. (2012) characterised the rhizosphere bacterial community associated with three pseudometallophytes (*Betula celtiberica*, *Cytisus scoparius* and *Festuca rubra*) colonising highly contaminated Pb/Zn mine tailings. Total Cd, Pb and Zn in nonvegetated soils was up to 50, 3000 and 20,000 mg/kg DW, respectively. The density of culturable bacteria observed in the nonvegetated spots at this sampling site was extremely low [mean microbial density of 3.2 ± 0.1 log colony forming units (CFUs)/g soil] and, as expected, the rhizosphere always harboured higher bacterial densities (by up to one or two logarithmic units). This rhizosphere effect was attributed to the higher concentration of dissolved organic carbon in rhizosphere soil compared to nonvegetated soil. Moreover, all three plant species hosted higher metal-tolerant populations: some rhizobacterial isolates resisted up to 6 mM Cd and 25 mM Zn, concentrations which are similar to those tolerated by rhizobacteria and endophytes associated with *Salix caprea* trees from a Zn/Pb mining site in Austria (Kuffner et al., 2010). The culturable rhizosphere bacterial diversity was also low, and rhizobacterial strains were primarily affiliated with the phylum *Actinobacteria*. Although there are contradictions, several other studies have also found Gram-positive bacteria, and in particular the *Actinobacteria*, to dominate culturable bacterial collections from trace metal–contaminated soils (Pereira & Castro, 2014). At the genus level, the majority belonged to the genera *Streptomyces* (61%), *Tsukamurella* (18%) or *Pseudomonas* (18%). Isolates belonging to the *Pseudomonas* genera presented a lower metal tolerance (MTC of <0.5–2 mM for Cd and <0.5–2.5 mM for Zn in solid culture media), while all of the isolates belonging to the *Tsukamurella* genera presented a higher MTC (of up to 4–5 mM for Cd and 2.5–25 mM for Zn). Zhang, Huang, He, and Sheng (2012) and He et al. (2010) found both *Actinobacteria* and *γ-Proteobacteria* to be well represented in the rhizosphere of *Chenopodium ambrosioides* grown in Pb mine tailings and in the rhizosphere of Cu-tolerant plants grown in copper mine wasteland. Hanbo et al. (2004) found over 90% of the studied culturable bacteria from Pb–Zn mine soils were affiliated with the *Arthrobacter* genus (*Actinobacteria*).

Pereira et al. (2015) studied the rhizosphere bacterial communities associated with wetland species (*Phragmites australis* and *Juncus effusus*) affected by the discharge of solid residues and wastewaters from a large industrial chemical complex and contaminated by both metals (Zn, Pb, Hg) and metalloids (As). Phylogenetic analysis indicated that the bacterial strains isolated from the most contaminated site were mainly related to classes *Actinobacteria* (36%), *Bacilli* (24%), *β-Proteobacteria* (20%) and *γ-Proteobacteria* (12%), with a small fraction associated with classes *α-Proteobacteria* (6%) and *Sphingobacteria* (6%). Gram-positive isolates belonged to genera (in order of importance) *Bacillus*, *Microbacterium*, *Rhodococcus*, *Arthrobacter* and *Oerskovia*; while Gram-negatives were affiliated to genera *Achromobacter*, *Burkholderia*, *Cupriavidus*, *Variovorax*, *Pseudomonas* and *Mycoplana*. These authors suggested that both *Actinobacteria* and *γ-Proteobacteria* may be a dominant part in the metabolically active rhizosphere population of trace metal accumulators.

Pereira and Castro (2014) evaluated the phylogenetic diversity of culturable bacterial endophytes of *Zea mays* growing in an agricultural soil contaminated with Zn and Cd. Densities of endophytes was plant tissue–specific, being higher in roots than in shoots. Phylogenetic analysis showed that endophytes belonged to three major groups: *α-Proteobacteria* (31%), *γ-Proteobacteria* (26%) and *Actinobacteria* (26%). *Pseudomonas*, *Agrobacterium*, *Variovorax* and *Curtobacterium* were among the most represented genera. Montalbán et al. (2016) studied the culturable bacterial community associated with *Brassica napus* growing on a Zn-contaminated agricultural area affected by smelter emissions. The diversity of the isolated bacterial populations was similar in rhizosphere and roots, but lower in soil and stem compartments. Members of the genera *Burkoholderia*, *Alcaligenes*, *Agrococcus*, *Polaromonas*, *Stenotrophomonas*, *Serratia*, *Microbacterium* and *Caulobacter* were found as root endophytes exclusively. At the same Zn-contaminated site, Janssen et al. (2015) isolated bacterial strains associated with two willow clones, *Salix viminalis* and *Salix alba x alba*. Densities of culturable bacteria decreased from the rhizosphere to roots to twigs in both clones, and many genera isolated from the roots of both clones were also present in their rhizospheres. However, strains belonging to the genera *Curtobacterium* were only identified as root or twig endophytes.

4. APPLICATION OF BIOINOCULANTS OBTAINED FROM TRACE ELEMENT–ENRICHED SOILS IN PHYTOREMEDIATION

The potential application of TE-tolerant plant-associated bacteria in phytoremediation processes is based on their ability to improve plant growth and biomass production and/or modify soil metal availability and plant uptake. Plant growth promotion plays a major role in the extraction and removal of TEs since a simple improvement in biomass results in an increase in the overall metal yield (phytoextracted TE). Bacterial strains for bioaugmentation trials are frequently selected on the presence of PGP traits (such as the production of IAA, ACC deaminase or siderophore production) or their ability to release compounds which could potentially modify metal bioavailability (such as the production of biosurfactants, siderophores or organic acids) (Abou-Shanab et al., 2003; Cabello-Conejo et al., 2014; Idris et al., 2004; chapter: The Bacterial and Fungal Microbiota of Hyperaccumulator Plants: Small Organisms,

Large Influence by Thijs et al., 2017). At a bench level, examples of bacterial-induced plant growth promotion and metal accumulation in a phytoextraction context can be found in a wide array of plant species, including crop plants, hyperaccumulators and woody tree species. Moreover, these inoculants have been tested in various types of contaminated soils or natural metal-enriched soils. However, to date there are a limited number of field studies testing the potential benefits of these plant-associated microbial inoculants on the performance of GRO in TE-contaminated sites.

Screening methods using inert growth substrates have allowed for a rapid selection of interesting strains (Becerra-Castro et al., 2012; Cabello-Conejo et al., 2014). These strains can later be tested in more complex systems (e.g., soil system) where bacterial mechanisms of nutrient acquisition are more likely to be induced (Becerra-Castro et al., 2012). Fourteen bacterial strains which were isolated from the rhizosphere soil of pseudometallophytes in Pb/Zn mine tailings were evaluated for promoting plant growth of two species: *S. viminalis* and *Festuca pratensis* (Becerra-Castro et al., 2012). Thirteen inoculants enhanced growth of *F. pratensis*, while only three enhanced growth of *S. viminalis*. Two strains were capable of producing IAA (*Rhodococcus erythropolis* P30 and *Massilia niastensis* P87) and both enhanced biomass production in *F. pratensis* and *S. viminalis*. However, growth enhancement could not always be related to isolate PGP traits. For example, both isolates P56 and P64 are siderophore-producers, and both belong to the genus *Streptomyces*, but they caused contrasting effects in *S. viminalis* growth. Similarly, isolates P41 and P42 identified as *Pseudomonas lurida* are P-solubilisers, but again induced contradicting effects on the growth of this species. It is evident that mechanisms other than those normally studied must be involved in this growth enhancement. The environmental conditions to which an inoculant is exposed will also influence whether or not certain PGP traits are activated. Similarly, Cabello-Conejo et al. (2014) screened 15 bacterial isolates for their PGP capacities by growing *A. pintodasilvae* in a simple perlite:sand mixture watered with Ni-rich nutrient solution. This screening method successfully identified beneficial bacteria for Ni extraction using *A. pintodasilvae* which later also proved to be beneficial under more realistic soil conditions. These authors suggested that this type of screening method was more helpful in identifying potentially useful strains than basing selection on the in vitro phenotypical characterisation of the strains since this is not sufficient to predict the in vivo behaviour of inoculants in interaction with their host plants.

Most studies have found bacterial inoculants to be more successful in promoting plant growth and biomass production (hence increasing the metal yield and metal removal from the soil), rather than increasing the metal concentration in shoots. This was also indicated in a metaanalysis of phytoremediation-orientated inoculation studies carried out by Sessitsch et al. (2013). This analysis was based on the results of more than 70 publications and analysed 738 individual cases or treatments, to identify the most frequent effects of plant inoculation on shoot biomass production, trace metal concentration and yield in shoots. In 30% of the cases studied an increase in shoot biomass was observed (while the shoot metal concentration was unchanged), in contrast only 11% of treatments were found to increase shoot metal concentration, and 19% of treatments increased both shoot metal concentration and shoot biomass production.

Since metal-enriched soils are characteristically deficient in essential nutrients, the identification of bacterial inoculants which could potentially improve the nutritive state of the phytoextracting plant is of great interest. Phosphate-solubilising bacteria are effective in promoting plant growth and biomass because they release P from inorganic and organic P pools through solubilisation and mineralisation processes (Rodriguez & Fraga, 1999). Numerous studies have found a bacterial-induced stimulation in the growth of agricultural crops in TECS. Jeong, Moon, Shin, and Nam (2013) inoculated *Brassica juncea* with strains of phosphate-solubilising *Bacillus* sp. According to these authors the release of organic acids and the drop in soil pH led to a mobilisation of Cd and consequent increase in Cd uptake as well as an enhanced plant biomass. An increase in plant growth and P uptake were reported by Ma, Rajkumar, Vicente, and Freitas (2010) after inoculating the energy crops *Ricinus communis* and *Helianthus annuus* with the rhizobacterial strain *Psychrobacter* sp. SRS8. Inoculation of *B. juncea* seeds with a Cu-resistant PGPR strain, *Achromobacter xylosoxidans* Ax10, isolated from a Cu mine soil increased the root and shoot biomass of plants grown in a sterilised, Cu-spiked soil and improved their Cu uptake (Ma, Rajkumar, & Freitas, 2009). The incorporation of *Pseudomonas aspleni* into the soil also facilitated Cu uptake in *B. napus* by increasing its biomass (Reed & Glick, 2005). Seed inoculation with *Proteus vulgaris* increased germination, biomass and chlorophyll content and decreased root and shoot Cu accumulation of *Cajanus cajan* (Rani, Shouche, & Goel, 2008). Zaidi, Usmani, Singh, and Musarrat (2006) reported that an IAA-producing *Bacillus subtilis* strain was able to promote the growth of *B. juncea* and thereby increased Ni extraction. Beneficial effects of bacterial inoculants on the growth of metal-exposed plants have often been attributed to the production of this phytohormone (Dell'Amico et al., 2008).

Bacterial inoculants have also been successfully applied to metal-hyperaccumulating plant species to improve both their growth and metal yields. Enhanced phytoextraction efficiency after inoculation with either rhizosphere or endophytic bacterial strains was observed in *A. serpyllifolium* subsp. *lusitanicum* (Cabello-Conejo et al., 2014), *N. caerulescens* (Aboudrar, Schwartz, Morel, & Boularbah, 2013; Visioli, Vamerali, Mattarozzi, Dramis, & Sanangelantoni, 2015), *A. murale* (Abou-Shanab, Angle, & Chaney, 2006; Durand et al., 2016) or *Sedum alfredii* (Guo et al., 2011). The P-solubilising strain *Arthrobacter nicotinovorans* SA40 was shown to significantly enhance the biomass production of *A. serpyllifolium* subsp. *lusitanicum* when grown in contrasting Ni-rich soils (serpentine soils and sewage sludge–amended agricultural soils) (Cabello-Conejo et al., 2014). Strain SA40 was also characterised as a siderophore- and IAA-producer and not only promoted plant growth but also significantly enhanced shoot Ni concentrations in this species when grown in the serpentine soil. The authors attributed this to the result of an enhanced Ni phytoavailability and hence plant uptake. Three other rhizosphere isolates were found to stimulate plant growth but not shoot Ni accumulation when the same hyperaccumulator was grown in serpentine soil: *Arthrobacter nitroguajacolicus* strain LA44, which shows intense IAA-production, is an organic acid producer and highly Ni-resistant; *Microbacterium* sp. strain SA5b, which is an organic acid producer and shows intermediate Ni resistance; and finally, *Microbacterium hydrocarbonoxydans* strain SA17, which presents intermediate Ni resistance, produces organic acids and siderophores.

Janssen et al. (2015) found an increase in metal extraction potential of *S. viminalis* after inoculating with a *Rahnella* sp. strain in Cd–Zn–Pb contaminated soil due to a significantly increased in twig biomass. Kuffner et al. (2008) revealed a positive effect of rhizosphere bacteria on accumulated TE contents in willows (*S. caprea*). Moreover, microbial inoculants increased Cd and Zn translocation factors from roots to leaves of *S. caprea* grown in a moderately contaminated soil (De Maria et al., 2011).

PGP endophytic bacterial strains *Ochrobactrum haematophilum* ZR 3–5, *Acidovorax oryzae* ZS 1–7, *Frigoribacterium faeni* ZS 3–5 and *Pantoea allii* ZS 3–6 were shown to increase root elongation and biomass of maize seedlings grown in soil contaminated with Cd and Zn (Pereira & Castro, 2014). Endophytic bacteria may be of particular interest as bioinoculants since they have the advantage of proliferating within the plant tissue thus facing less competition for nutrients and being protected from high-stress environment—polluted soils (Sturz, Christie, & Nowak, 2000). Montalbán et al. (2016) characterised the PGP effects of rhizosphere and endophytic bacterial strains on the growth of *B. napus* in square vertical petri plates with metal-enriched Murashige and Skoog (MS) medium. The inoculation of seeds with *Pseudomonas* sp. strains 228 and 256, and *Serratia* sp. strain 246 facilitated the root development of *B. napus* at 1000 μM Zn. *Arthrobacter* sp. strain 222, *Serratia* sp. strain 246, and *Pseudomonas* sp. 228 and 262 increased the root length at 300 μM Cd. The inoculated bacterial strains were able to produce siderophores, IAA and exhibited ACC deaminase activity. Inoculating another high-biomass annual crop, *H. annuus*, with crude seed extract containing seed endophytes led to a promotion in root and shoot growth and reduced Cu phytotoxicity (Kolbas et al., 2015). Babu, Shea, Sudhakar, Jung, and Oh (2015) found that inoculation with the root endophyte (*Pseudomonas koreensis* AGB-1) increased plant growth of *Miscanthus sinensis* growing in mine tailings at the same time as an increase in plant metal(loid) uptake. In the same study a decrease in antioxidant enzyme activities (catalase or superoxide dismutase) which are induced under metal(loid) stress was also found. Extracellular sequestration of metal(loid)s by this bacterial strain was revealed using Transmission Electron Microscopy-micrographs, which were attributed to the formation of complexes between metal(loid)s and exopolymers produced by the strain to reduce metal(loid) toxicity.

Organic acid production is considered one of the principal mechanisms involved in metal mobilisation by bacterial strains (Sessitsch et al., 2013). Abou-Shanab et al. (2003) found that inoculating ultramafic soils with the actinobacterial *Microbacterium arabinogalactanolyticum* AY509224 increased soil Ni extractability and uptake by *A. murale*. Becerra-Castro et al. (2011) showed that by using the cell-free culture of 13 bacterial strains associated with hyperaccumulating *Alyssum* sp. the extractable Ni concentration from serpentine soils was increased. Likewise, TE-mobilising metabolites from various *Actinobacteria* mobilised soil Zn and/or Cd (Kuffner et al., 2010). *Pseudomonas jessenii* increased biomass of *R. communis* and was efficient at solubilising Cu (Rajkumar, Ae, & Freitas, 2009). A bacterial strain isolated from the rhizosphere of *Elsholtzia splendens* growing on Tonglu Mountain Cu mines increased soil water-soluble Cu, as well as root and shoot Cu accumulation (Chen, Wang, Lin, & Luo, 2005). However, in all of these studies, no clear relation between the capacity for TE mobilisation and the phenotypic traits of the strains was found, and the mechanisms involved in this process were not fully elucidated. In a more detailed study, Becerra-Castro et al. (2013) evaluated the capacity of two rhizobacterial strains isolated from the Ni-hyperaccumulators *A. serpyllifolium* subsp. *lusitanicum* and subsp. *malacitanum* to mobilise Ni from ultramafic rocks under in vitro conditions. Minimal culture medium containing ground ultramafic rock was inoculated with either of two *Arthrobacter* strains: LA44 (IAA-producer) or SBA82 (siderophore- and IAA-producer, PO_4 solubiliser). TE and organic compounds were determined at different time intervals after inoculation, and trace metal fractionation was carried out on the remaining rock at the end of the experiment (Fig. 3). Both strains presented a capacity to mobilise Ni, however, the results indicated that they preferentially acted upon different mineral phases. Strain LA44, identified as *A. nitroguajacolicus*, was a strong mobiliser of Ni (Fig. 3) and primarily solubilised Ni associated with Mn oxides. This was shown to be a result of intense oxalate exudation, compared to other organic acids (such as succinate) which were exuded at similar rates by both bacterial strains (Fig. 3). Strain SBA82 also led to release of Ni but to a much lower extent, and in this case, the concurrent mobilisation of Fe and Si into the growth medium indicated preferential weathering of Fe oxides and serpentine minerals, possibly related to the siderophore production capacity of the strain (Fig. 3). The same bacterial strains were tested in a serpentine soil—plant system: the Ni-hyperaccumulator *A. serpyllifolium* subsp. *malacitanum* was grown in serpentine soil in a rhizobox system and inoculated with each bacterial strain. Biomass production and shoot Ni concentrations were higher in plants from inoculated pots than from noninoculated pots, and Ni yield was significantly enhanced in plants inoculated with strain LA44. The authors proposed this Ni-mobilising strain as a potential inoculant for improving Ni uptake by hyperaccumulator plants in phytomining applications.

Most studies have evaluated the effects of re-inoculating host plants with their associated isolated strains (Abou-Shanab et al., 2003, 2006; Ghosh et al., 2011; Cabello-Conejo et al., 2014). However, the specificity of these plant-bacterial combinations is unclear, and some inoculants have been shown to have beneficial effects on a wide range of plant hosts (Grandlic et al., 2008; Ma et al., 2010; Becerra-Castro et al., 2012). The effects of the same bacterial inoculum on plant growth and metal bioaccumulation have also been found to be soil-specific (Cabello-Conejo et al., 2014; Grandlic, Palmer, & Maier, 2009). A prerequisite for successful bioaugmentation is the selection of microorganisms able to adapt and survive

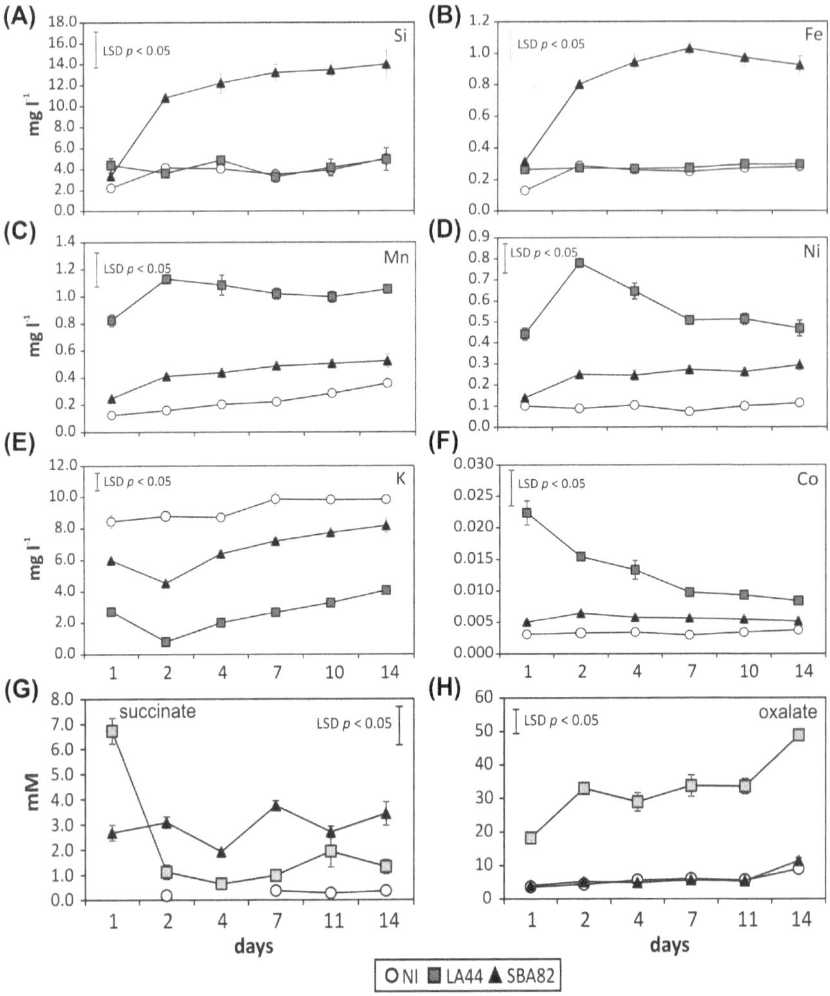

Figure 3 Evolution in Si (A), Fe (B), Mn (C), Ni (D), K (E), Co (F), succinate (G) and oxalate (H) concentrations over time in minimal culture media with ultramafic rock and inoculated with two *Arthrobacter* rhizobacterial strains (LA44 and SBA82) associated with the Ni-hyperaccumulating *Alyssum serpyllifolium* subspecies. Cultures were incubated during 14 days and periodically sampled. The error bars indicate standard errors. *LSD*, least significant difference. *From Becerra-Castro, C., Kidd, P., Kuffner, M., Prieto-Fernández, A., Hann, S., Monterroso, C.,…, Puschenreiter, M. (2013). Bacterially induced weathering of ultramafic rock and its implications for phytoextraction.* Applied and Environmental Microbiology, 79, *5094–5103.*

in the soil conditions to which they are exposed as well as an ability to colonise the rhizosphere. Cabello-Conejo et al. (2014) found the P-solubilising, siderophore- and IAA-producer *A. nicotinovorans* SA40 isolated from serpentine soils in association with *A. pintodasilvae* was able to promote the growth of this plant when grown in both its natural serpentine soil and also in a sewage sludge—amended agricultural soil with elevated concentrations of both Ni and Cd. The identification of potential bacterial strains which are able to respond under a range of soil conditions is interesting when transferring bioaugmentation strategies to real-life field scenarios.

Factors such as the host plant species, soil type and properties, and nature and type of contamination (mono- or poly-metallic contamination, co-contamination with organic and inorganic pollutants, pollutant concentration, anthropogenic, geogenic, spiked contamination) will ultimately determine whether or not a beneficial bacterial-induced effect is observed (Becerra-Castro et al., 2012; Cabello-Conejo et al., 2014; Sessitsch et al., 2013; Weyens et al., 2010). Bioaugmentation can also be challenged by the strong competition encountered in the soil when a selected microbe is introduced. The selected host plant species may not necessarily be compatible with the inoculated bacterial or fungal strain, since these are not naturally selected for by the host (Thijs et al., 2016). Johnson, Wilson, Bowker, Wilson, and Miller (2010) showed that nonnative plant species form less beneficial associations with soil mycorrhizal fungi than native plant species. On the other hand, endophytes do not have to compete with the large abundance and diversity of soil microorganisms which may enhance their chances of establishing a stable and active population. In addition, and on the basis of available literature, it is evident that most researchers choose their own method of inoculation and there has been little standardisation in this field (Table 1). However, the choice of an appropriate inoculation method is likely to be of crucial importance in the inoculation process. Table 1 summarises the different experimental conditions and inoculation methods used in some recent phytoextraction-orientated inoculation studies. The most common methods are seed and soil inoculation, while plant inoculation and experiments including more than one inoculation event are rare. Bacterial densities of the inoculum generally fluctuate between 10^7—10^8 CFUs/mL, however, the total density of inoculated bacteria per pot (CFUs per pot) is highly variable due to the different amounts of soil used. According to Malusá, Sas-Paszt, and Ciesielska (2012), the commercial use of microbial inoculants has been limited due to the inconsistency of the results obtained. There is therefore

Table 1 Bacterial Inoculation Methods Used in Recent Phytoextraction-Orientated Studies

Bacterial Inoculant	Plant Host	Soil/Substrate Type	Amount of Soil/ Substrate (g)	Experiment Duration (d)	Inoculation Method		Re-Inoculation (weeks)	Cellular Density	References
					Plant Seed	Substrate			
Pseudomonas brassicacearum subsp. *brassicacearum* and *Rhizobium leguminosarum* bv. *trifolii*	*Brassica juncea* and *Vicia sativa* subsp. *sativa*	Standard soil for pot experiments spiked using Zn sulphate solution	500	42	X			7.5×10^8 CFUs/mL	Adediran, Ngwenya, Mosselmans, and Heal (2016)
Mixed inoculum: *Bacillus pumilus* and *Micrococcus* sp	*Noccaea caerulescens*	Serpentine Ni-rich soil	100	32	X				Aboudrar et al. (2013)
Acidovorax avenae, *Clavibacter xyli*, *Microbacterium* sp., *Rhizobium galegae*	*Alyssum murale*	**1** low Ni soil **2** moderate Ni-serpentine soil **3** Ni-rich serpentine soil	500	60	X			20′ shaking in 7.4×10^8 CFUs/mL	Abou-Shanab et al. (2006)
Pseudomonas putida	*Eruca sativa*	Ni-contaminated agricultural soil			X			10^8 cells/mL	Kamran et al. (2016)
14 strains isolated from native plants	*Festuca pratensis* *Salix viminalis*	Perlite:quartz sand mixture with Hoagland solution suplemented with Cd and Zn.		49		X		9 mL 1×10^7 CFUs/mL	Becerra-Castro et al. (2012)

15 strains isolated from rhizosphere of *Alyssum serpyllifolium* *Microbacterium* sp., *Arthrobacter* sp., *Streptomyces* sp	*Alyssum pintodasilvae*	Perlite:quartz sand mixture with Hoagland solution suplemented with Ni. **1** Serpentinitic region of Trás-os-Montes **2** Contaminated Ni soil		70 150			X	9 mL 1 × 10^8 CFUs/mL 2 mL 1 × 10^8 CFUs/mL	Cabello-Conejo et al. (2014)
Pseudomonas sp, *Alcaligenes* sp, *Mycobacterium* sp	*Brassica napus*	Commercial soil artificially contaminated with $CdCl_2$	320	20		X		Incubated in 40 mL 1 × 10^8 CFUs/mL	Dell'Amico et al. (2008)
Streptomyces (AR17) and *Agromyces* (AR33)	*Salix caprea*, clone Boku 04 CZ-024	Heavy metal(loid) contaminated soil	800	84			X	10 mL 1 × 10^8 CFUs/mL	De Maria et al. (2011)
Variovorax paradoxus	Intercropping of *Bornmuellera tymphaea* – *Noccaea tymphaea* and *B. tymphaea* – *A. murale*	Natural forest ultramafic soil	1717	180			X	9.3 10^5 CFUs/g dry soil	Durand et al. (2016)
3 *Gordonia alkanivoran*, 1 *Cupriavidus necator* and 1 *Sporosarcina luteol*	*B. juncea* and *Lupinus albus*	As- and Hg-contaminated industrial soil	400	30			X	10^8 cells g/soil	Franchi et al. (2016)
Rahnella sp., *Sphingobacterium* sp., *Caulobacter* sp., *Curtobacterium* sp., *Pseudomonas* sp.	*S. viminalis* and *Salix alba x alba*	Former maize field close to a Zn smelter, Cd, Zn and Pb contaminated	4500	90	X			10^8 CFUs/mL	Janssen et al. (2015)

(*Continued*)

Table 1 Bacterial Inoculation Methods Used in Recent Phytoextraction-Orientated Studies—cont'd

Bacterial Inoculant	Plant Host	Soil/Substrate Type	Amount of Soil/ Substrate (g)	Experiment Duration (d)	Inoculation Method		Re-Inoculation (weeks)	Cellular Density	References
					Plant Seed	Substrate			
Mixture of *Pseudomonas* sp., *Pseudomonas fluorescens*, *Pabrys* sp.	*Helianthus annus*	Cu-contaminated soil from a former wood preservation site		30	X			10^7 cells/mL	Kolbas et al. (2015)
Rhizosphere isolate RX232 and the endophyte EX72	*S. caprea*, clone Boku 04 CZ-024	Moderately contaminated soil	800	84		X		10 mL 1×10^8 CFUs/mL	Kuffner et al. (2010)
Consortium 1 (German soil): 4 *Streptomyces* sp, 4 *Enterobacter* sp, 1 *Citrobacter* sp, 1 *Psychrobacillu* sp Consortium 2 (Swedish soil): 1 *Acinetobacter* sp, 1 *Rhodococcus* sp, 2 *Arthrobacter* sp, 1 *Pseudomonas* sp, 1 *Lysinibacillus*, 1 *Paenibacillus*	*Agrostis capillaris, Deschampsia flexuosa, Festuca rubra, Helianthus annuus*	**1** One former mining site from Germany, multi-metal(loid)-contaminated **2** One former mining site from Sweden, multi-metal(loid)-contaminated amended with compost		84		X		1×10^9 CFUs/pot with 10 mL	Langella et al. (2014)
Pseudomonas sp., *Deltia lacustris*, *Bacillus* sp., *Variovorax boronicumulans*, *Pseudoxanthomonas mexicana*	*Pteris vitatta*	As-contaminated landfill site	3000	120		X	Applied initially and after 2 months	10^8 CFUs/g soil	Lampis, Santi, Ciurli, Andreolli, and Vallini (2015)
8 *Pseudomonas* sp.	*Zea mays*	Artificially Cu contaminated soil using $CuCl_2$ solution	200	30	X	X	Applied together	7.5×10^8 CFUs/mL	Li and Ramakrishna (2011)

Achromobacter piechaudii	*Sedum plumbizinccola*	Cd- Zn- and Pb contaminated agricultural soil	750	75	X					Ma, Zhang, Oliveira, Freitas, and Luo (2016)
Ralstonia eutropha, Chryseobacterium humi	*H. annuus*	Agricultural soil artificially contaminated with Zn or Cd	300	140			X		10 mL 1×10^8 CFUs/mL	Marques, Moreira, Franco, Rangel, and Castro (2013)
Rhodococcus erythropoli; Achromobacter sp.; *Microbacterium* sp. and *Arthrobacter* sp.	*Trifolium repens*	Cd- and Zn-spiked soil	100	70			X		10^8 CFUs/mL	Pereira, Barbosa, et al. (2015)
4 *P. fluorescens*	*Mirabilis jalapa*	Contaminated soil with Cd, Cr,Cu, Ni, Pb and Zn by a smelter dump		75		X			$1–2 \times 10^9$ CFUs/mL	Petriccione et al. (2013)
Streptomyces acidiscabies and *Streptomyces tenda*	*Sorghum bicolor*	Multi-metal contaminated soil material from a former uranium mine	2500	90–180			X			Phieler, Merten, Roth, Büchel, and Kothe (2015)
Micrococcus sp.	*Zea mays*. cv. CPDK 888	Garden soil Cd-spiked	3000	112			X	Repeated at 2, 4 and 6 weeks	10^6 CFUs/g soil	Sangthong, Setkit, and Prapagdee (2016)
Arthrobacter sp., *Kocuria rhizophila Bacillus* sp. *Microbacterium oxydans, Bacillus amyloliquefaciens*	*N. caerulescens, Arabidopsis thaliana*	Ni-rich serpentine soil	750	60		X			10^8 CFUs/mL (5×10^8 cells/mL)	Visioli et al. (2015)
Bacillus subtilis	*B. juncea* var. Pusa Bold.	Soil amended with increasing amounts of $NiCl_2$		140		X (coated seeds)			2×10^5 CFUs/seed	Zaidi et al. (2006)

a need to optimise inoculation methods for plants growing in TECS, which is an aspect that has rarely been addressed. Álvarez-López, Prieto-Fernández, Janssen, et al. (2016) tested the effects of plant inoculation mode using the IAA-producing bacterial strain *R. erythropolis* P30 together with tobacco plants. A single soil inoculation event was found to give the best results together with a lower bacterial density (10^6 CFUs/mL). Both the inoculation mode (seed, plant or soil inoculation) and the cellular density of the inoculum were shown to be important factors to take into account when designing inoculation strategies since they can modify plant performance and soil metal removal. A similar conclusion was reached by Pereira, Barbosa, and Castro (2015).

Other strategies to improve the performance of bioaugmentation in phytoextraction techniques are to create formulations of microbial inoculants, or to combine inoculants with other biostimulants, e.g., plant hormones or extracts (humic acids, strigolactones, nod metabolites, etc.) or soil amendments (Thijs et al., 2016). Promising results have been found when using mixtures or consortia of different PGP strains with complementary actions (Rylott, 2014; Thijs et al., 2016). Visioli et al. (2015) found that co-inoculating the Ni-hyperaccumulator *N. caerulescens* when growing in serpentine soil with two root endophytes belonging to the *Arthrobacter* and *Microbacterium* genera had a more positive effect on plant growth, soil Ni removal, and Ni translocation, than when inoculated individually. Both strains were strong IAA-producers and presented ACC deaminase activity. Moreira, Pereira, Marques, Rangel, and Castro (2016) showed the benefits of combined inoculation of arbuscular mycorrhizal fungi and PGPR for the growth of energy maize in metal-contaminated soils and their potential application in phytomanagement strategies. From a collection of endophytic bacterial strains, obtained from seeds of tobacco plants grown on Cd/Zn-contaminated soils in Northern Europe, a Cd-resistant *Sanguibacter* sp., a *Pseudomonas* sp. and a consortium of Cd-resistant endophytes were found to increase threefold Cd accumulation in *Nicotiana tabacum* (Mastretta et al., 2009). Petrisor et al. (2004) found that co-inoculating the N_2-fixing *Azotobacter chroococcum* with the P-solubiliser *Bacillus megaterium* led to a better establishment and growth of indigenous herbaceous plants in phophogypsum mine tailings compared to inorganic fertilisation. Although in this study the objective of bacterial inoculation was to decrease the uptake of Cd, Zn and Mn by plants. Inoculation with consortia often results in more pronounced beneficial effects on plant biomass production as compared with inoculation with single strains, suggesting synergistic effects of the consortia members.

Figure 4 Plant biomass of *Nicotiana tabacum* and phytoextracted Zn [means ±standard errors (SE)] obtained when combining inoculation with four rhizobacterial stains and compost amendment in Cd–Pb–Zn mine soils.

Additionally, the success of bacterial inoculants may be limited by a low proliferation and survival in the inoculated medium. Plants and microbes compete for resources, including nutrients and water, and in soils with low nutrient availability (as is commonly the case at contaminated sites) resource competition may become a limiting factor of bacterial growth and establishment (Joner, Leyval, & Colpaert, 2006; Kaye & Hart, 1997; Moorhead, Westerfield, & Zak, 1998; Unterbrunner et al., 2007). Addition of soil amendments to overcome this competition for nutrients may be required to ensure bacterial survival. Few studies have evaluated the effects of combining soil amendments together with bacterial inoculants. Álvarez-López, Prieto-Fernández, Rodríguez-Garrido, et al. (2017) evaluated the metal phytoextraction efficiency of two high-yielding crops: *S. caprea* and *N. tabacum* and the potential improvement with soil organic amendment and/or bioaugmentation. The combined effects of soil inoculants (strains *M. niastensis* P87, *Pseudomonas costantinii* P29, *R. erythropolis* P30, *Tsukamurella strandjordii* P75) together with compost amendments on biomass production of tobacco were more beneficial than when either of these methods was used on their own (Fig. 4). Bacterial inoculants effectively improved plant growth, modified soil metal availability and increased plant metal accumulation. However, the effects of these rhizobacterial strains on plant growth or metal accumulation were plant species—specific.

It seems clear that plant-associated bacteria play an important role in plant TE bioaccumulation and experimental evidence suggests that bacterial inoculants can further enhance phytoextraction and phytomining efficiency. However, the effectivity of plant-associated bioinoculants is dependent on a complex array of interacting factors and the underlying mechanisms operating in the plant—microbial—soil system are not wholly understood. Finally, the performance of these bacterial inoculants under natural conditions will have to be investigated under a field scale.

ACKNOWLEDGEMENTS

This article is dedicated to the work of Cristina Becerra-Castro, our colleague and friend, who was taken from us too soon and will be greatly missed.

We acknowledge the financial support of the Spanish Ministerio de Economía y Competitividad (CTM2015-66439-R) and FEDER.

REFERENCES

Abou-Shanab, R., Angle, J., & Chaney, R. (2006). Bacterial inoculants affecting nickel uptake by *Alyssum murale* from low, moderate and high Ni soils. *Soil Biology and Biochemistry, 38*, 2882–2889. http://dx.doi.org/10.1016/j.soilbio.2006.04.045.

Abou-Shanab, R. A. I., van Berkum, P., Angle, J. S., Delorme, T. A., Chaney, R. L., Ghozlan, H. A., … Moawad, H. (2010). Characterization of Ni-resistant bacteria in the rhizosphere of the hyperaccumulator *Alyssum murale* by 16S rRNA gene sequence analysis. *World Journal of Microbiology and Biotechnology, 26*, 101–108. http://dx.doi.org/10.1007/s11274-009-0148-6.

Abou-Shanab, R. I., Delorme, T. A., Angle, J. S., Chaney, R. L., Ghanem, K., Moawad, H., & Ghozlan, H. A. (2003). Phenotypic characterization of microbes in the rhizosphere of *Alyssum murale*. *International Journal of Phytoremediation, 5*, 367–379.

Aboudrar, W., Schwartz, C., Morel, J. L., & Boularbah, A. (2013). Effect of nickel-resistant rhizosphere bacteria on the uptake of nickel by the hyperaccumulator *Noccaea caerulescens* under controlled conditions. *Journal of Soils and Sediments, 13*, 501–507.

Acea, M., & Carballas, T. (1986). Estudio de la población microbiana de diversos tipos de suelos de zona húmeda (N.O. de España). *An Edafol Agrobiology, 45*, 381–398.

Adediran, G. A., Ngwenya, B. T., Mosselmans, J. F. W., & Heal, K. V. (2016). Bacteria-zinc co-localization implicates enhanced synthesis of cysteine-rich peptides in zinc detoxification when *Brassica juncea* is inoculated with *Rhizobium leguminosarum*. *New Phytologist, 209*, 280–293. http://dx.doi.org/10.1111/nph.13588.

Álvarez-López, V., Prieto-Fernández, Á., Becerra-Castro, C., Monterroso, C., & Kidd, P. S. (2016). Rhizobacterial communities associated with the flora of three serpentine outcrops of the Iberian Peninsula. *Plant and Soil, 403*, 233–252. http://dx.doi.org/10.1007/s11104-015-2632-0.

Álvarez-López, V., Prieto-Fernández, A., Janssen, J., Herzig, R., Vangronsveld, J., & Kidd, P. S. (2016). Inoculation methods using *Rhodococcus erythropolis* strain P30 affects bacterial assisted phytoextraction capacity of *Nicotiana tabacum*. *International Journal of Phytoremediation, 18*, 406–415. http://dx.doi.org/10.1080/15226514.2015.1109600.

Álvarez-López, V., Prieto-Fernández, Á., Rodríguez-Garrido, B., Herzig, R., Puschenreiter, M., & Kidd, P. S. (2017). Evaluating phytoextraction efficiency of two high-biomass crops after soil amendment and inoculation with rhizobacterial strains. *Environmental Science Pollution Research*. http://dx.doi.org/10.1007/s11356-017-8366-1.

Amir, H., & Pineau, R. (2003). Release of Ni and Co by microbial activity in New Caledonian ultramafic soils. *Canadian Journal of Microbiology, 49*, 288–293. http://dx.doi.org/10.1139/w03-039.

Anderson, T. A., Guthrie, E. A., & Walton, B. T. (1993). Bioremediation in the rhizosphere. *Environmental Science & Technology, 27*, 2630–2636. http://dx.doi.org/10.1021/es00049a001.

Babu, A. G., Shea, P. J., Sudhakar, D., Jung, I.-B., & Oh, B.-T. (2015). Potential use of *Pseudomonas koreensis* AGB-1 in association with *Miscanthus sinensis* to remediate heavy metal(loid)-contaminated mining site soil. *Journal of Environmental Management, 151*, 160–166. http://dx.doi.org/10.1016/j.jenvman.2014.12.045.

Barzanti, R., Ozino, F., Bazzicalupo, M., Gabbrielli, R., Galardi, F., Gonnelli, C., & Mengoni, A. (2007). Isolation and characterization of endophytic bacteria from the nickel hyperaccumulator plant *Alyssum bertolonii*. *Microbial Ecology, 53*, 306–316. http://dx.doi.org/10.1007/s00248-006-9164-3.

Batty, L. C. (2005). The potential importance of mine sites for biodiversity. *Mine Water and the Environment, 24*, 101–103. http://dx.doi.org/10.1007/s10230-005-0076-0.

Becerra-Castro, C., Kidd, P., Kuffner, M., Prieto-Fernández, A., Hann, S., Monterroso, C., … Puschenreiter, M. (2013). Bacterially induced weathering of ultramafic rock and its implications for phytoextraction. *Applied and Environmental Microbiology, 79*, 5094–5103.

Becerra-Castro, C., Monterroso, C., García-Lestón, M., Prieto-Fernández, A., Acea, M. J., & Kidd, P. S. (2009). Rhizosphere microbial densities and trace metal tolerance of the nickel hyperaccumulator *Alyssum serpyllifolium* subsp. *lusitanicum*. *International Journal of Phytoremediation, 11*, 525–541. http://dx.doi.org/10.1080/15226510902717549.

Becerra-Castro, C., Monterroso, C., Prieto-Fernández Á, Rodríguez-Lamas, L., Loureiro-Viñas, M., Acea, M. J., & Kidd, P. S. (2012). Pseudometallophytes colonising Pb/Zn mine tailings: a description of the plant-microorganism-rhizosphere soil system and isolation of metal-tolerant bacteria. *Journal of hazardous materials, 217–218*, 350–359. http://dx.doi.org/10.1016/j.jhazmat.2012.03.039.

Becerra-Castro, C., Prieto-Fernández, Á., Álvarez-López, V., Monterroso, C., Cabello-Conejo, M. I., Acea, M. J., & Kidd, P. S. (2011). Nickel solubilizing capacity and characterization of rhizobacteria isolated from hyperaccumulating and non-hyperaccumulating subspecies of *Alyssum serpyllifolium*. *International Journal of Phytoremediation, 13*(Suppl. 1), 229–244. http://dx.doi.org/10.1080/15226514.2011.568545.

Bhargava, A., Carmona, F. F., Bhargava, M., & Srivastava, S. (2012). Approaches for enhanced phytoextraction of heavy metals. *Journal of Environmental Management, 105*, 103–120. http://dx.doi.org/10.1016/j.jenvman.2012.04.002.

Cabello-Conejo, M. I., Becerra-Castro, C., Prieto-Fernández, Á., Monterroso, C., Saavedra-Ferro, A., Mench, M., & Kidd, P. S. (2014). Rhizobacterial inoculants can improve nickel phytoextraction by the hyperaccumulator *Alyssum pintodasilvae*. *Plant and Soil, 379*, 35–50.

Chaney, R., & Baklanov, I. (2017). Phytoremediation and phytomining: Status and promise. In A. Cuypers, & J. Vangronsveld (Eds.), *Phytoremediation* (Vol. 83, pp. 189–222).

Chaney, R., Reeves, R. D., Baklanov, L. A., Centofanti, T., Broadhurst, C. L., Baker, A. J. M., … Roseberg, R. J. (2014). Phytoremediation and phytomining: Using plants to remediate contaminated or mineralized environments. In N. Rajakaruna, et al. (Eds.), *Plant ecology and evolution in harsh environments* (pp. 365–391). New York, NY: Nova Science Publishers.

Chen, Y. X., Wang, Y. P., Lin, Q., & Luo, Y. M. (2005). Effect of copper-tolerant rhizosphere bacteria on mobility of copper in soil and copper accumulation by *Elsholtzia splendens*. *Environment International, 31*, 861–866. http://dx.doi.org/10.1016/j.envint.2005.05.044.

COM. (2002). *Communication from the commission to the council and the European Parliament, the European Economic and Social Committee and the Committee of the Regions: Towards a thematic strategy for soil protection*. COM (2002) 179- final.

Compant, S., Clément, C., & Sessitsch, A. (2010). Plant growth-promoting bacteria in the rhizo- and endosphere of plants: Their role, colonization, mechanisms involved and prospects for utilization. *Soil Biology and Biochemistry, 42*, 669–678. http://dx.doi.org/10.1016/j.soilbio.2009.11.024.

Compant, S., Duffy, B., Nowak, J., Clément, C., & Barka, E. A. (2005). Use of plant growth-promoting bacteria for biocontrol of plant diseases: principles, mechanisms of action, and future prospects. *Applied and Environmental Microbiology, 71*, 4951–4959. http://dx.doi.org/10.1128/aem.71.9.4951-4959.2005.

Coninx, L., Martinova, V., & Rineau, F. (2017). Mycorrhiza-assisted phytoremediation. In A. Cuypers, & J. Vangronsveld (Eds.), *Phytoremediation* (Vol. 83, pp. 127–188).

Crowley, D., & Kraemer, S. M. (2007). Function of Siderophores in the plant rhizosphere. In R. Pinton, Z. Varanini, & P. Nannipieri (Eds.), *The Rhizosphere: Biochemistry and organic substances at the soil-plant interface* (p. 616). CRC Press.

DeGrood, S. H., Claassen, V. P., & Scow, K. M. (2005). Microbial community composition on native and drastically disturbed serpentine soils. *Soil Biology and Biochemistry, 37*, 1427–1435. http://dx.doi.org/10.1016/j.soilbio.2004.12.013.

Dell'Amico, E., Cavalca, L., & Andreoni, V. (2008). Improvement of *Brassica napus* growth under cadmium stress by cadmium-resistant rhizobacteria. *Soil Biology and Biochemistry, 40*, 74–84. http://dx.doi.org/10.1016/j.soilbio.2007.06.024.

Delorme, T. A., Gagliardi, J. V., Angle, J. S., & Chaney, R. L. (2001). Influence of the zinc hyperaccumulator *Thlaspi caerulescens* J. & C. Presl. and the non-metal accumulator *Trifolium pratense* L. on soil microbial populations. *Canadian Journal of Microbiology, 47*, 773–776. http://dx.doi.org/10.1139/cjm-47-8-773.

De Maria, S., Rivelli, A. R., Kuffner, M., Sessitsch, A., Wenzel, W. W., Gorfer, M., … Puschenreiter, M. (2011). Interactions between accumulation of trace elements and macronutrients in *Salix caprea* after inoculation with rhizosphere microorganisms. *Chemosphere, 84*, 1256–1261. http://dx.doi.org/10.1016/j.chemosphere.2011.05.002.

Duca, D., Lorv, J., Patten, C. L., Rose, D., & Glick, B. R. (2014). Indole-3-acetic acid in plant-microbe interactions. Antonie van Leeuwenhoek. *International Journal of General and Molecular Microbiology, 106*, 85–125.

Durand, A., Piutti, S., Rue, M., Morel, J. L., Echevarria, G., & Benizri, E. (2016). Improving nickel phytoextraction by co-cropping hyperaccumulator plants inoculated by plant growth promoting rhizobacteria. *Plant and Soil, 399*, 179–192. http://dx.doi.org/10.1007/s11104-015-2691-2.

Ehrlich, H. L. (1998). Geomicrobiology: its significance for geology. *Earth-Science Reviews, 45*, 45–60. http://dx.doi.org/10.1016/S0012-8252(98)00034-8.

van der Ent, A., Baker, A. J., Reeves, R. D., Chaney, R. L., Anderson, C. W., Meech, J. A., … Mulligan, D. R. (2015). Agromining: Farming for metals in the future? *Environmental Science & Technology, 49*, 4773–4780. http://dx.doi.org/10.1021/es506031u.

Filion, M., St-Arnaud, M., & Fortin, J. A. (1999). Direct interaction between the arbuscular mycorrhizal fungus *Glomus intraradices* and different rhizosphere microorganisms. *New Phytologist, 141*, 525–533. http://dx.doi.org/10.1046/j.1469-8137.1999.00366.x.

Franchi, E., Rolli, E., Marasco, R., Agazzi, G., Borin, S., Cosmina, P., … Petruzzelli, G. (2016). Phytoremediation of a multi contaminated soil: mercury and arsenic phytoextraction assisted by mobilizing agent and plant growth promoting bacteria. *Journal of Soils and Sediments*, 1–13. http://dx.doi.org/10.1007/s11368-015-1346-5.

Gadd, G. M. (2004). Microbial influence on metal mobility and application for bioremediation. *Geoderma, 122*, 109–119.

Gadd, G. M. (2010). Metals, minerals and microbes: geomicrobiology and bioremediation. *Microbiology, 156*, 609–643. http://dx.doi.org/10.1099/mic.0.037143-0.

Giller, K. E., Witter, E., & McGrath, S. P. (1998). Toxicity of heavy metals to microorganisms and microbial processes in agricultural soils: a review. *Soil Biology and Biochemistry, 30*, 1389–1414. http://dx.doi.org/10.1016/S0038-0717(97)00270-8.

Glick, B. R. (2010). Using soil bacteria to facilitate phytoremediation. *Biotechnology Advances, 28*, 367–374. http://dx.doi.org/10.1016/j.biotechadv.2010.02.001.

Glick, B. R. (2014). Bacteria with ACC deaminase can promote plant growth and help to feed the world. *Microbiological Research, 169*, 30–39.

Göhre, V., & Paszkowski, U. (2006). Contribution of the arbuscular mycorrhizal symbiosis to heavy metal phytoremediation. *Planta, 223*, 1115–1122. http://dx.doi.org/10.1007/s00425-006-0225-0.

Grandlic, C. J., Mendez, M. O., Chorover, J., Machado, B., & Maier, R. M. (2008). Plant growth-promoting bacteria for phytostabilization of mine tailings. *Environmental Science and Technology, 42*, 2079–2084.

Grandlic, C. J., Palmer, M. W., & Maier, R. M. (2009). Optimization of plant growth-promoting bacteria-assisted phytostabilization of mine tailings. *Soil biology & biochemistry, 41*, 1734–1740. http://dx.doi.org/10.1016/j.soilbio.2009.05.017.

Grayston, S. J., Wang, S., Campbell, C. D., & Edwards, A. C. (1998). Selective influence of plant species on microbial diversity in the rhizosphere. *Soil Biology and Biochemistry, 30*, 369–378. http://dx.doi.org/10.1016/S0038-0717(97)00124-7.

Gremion, F., Chatzinotas, A., & Harms, H. (2003). Comparative 16S rDNA and 16S rRNA sequence analysis indicates that Actinobacteria might be a dominant part of the metabolically active bacteria in heavy metal-contaminated bulk and rhizosphere soil. *Environmental Microbiology, 5*, 896–907. http://dx.doi.org/10.1046/j.1462-2920.2003.00484.x.

Guo, J., Tang, S., Ju, X., Ding, Y., Liao, S., & Song, N. (2011). Effects of inoculation of a plant growth promoting rhizobacterium *Burkholderia* sp. D54 on plant growth and metal uptake by a hyperaccumulator *Sedum alfredii* Hance grown on multiple metal contaminated soil. *World Journal Microbiology Biotechnology, 27*, 2835–2844.

Haferburg, G., & Kothe, E. (2007). Microbes and metals: interactions in the environment. *Journal of Basic Microbiology, 47*, 453–467. http://dx.doi.org/10.1002/jobm.200700275.

Hanbo, Z., Changqun, D., Qiyong, S., Weimin, R., Tao, S., Lizhong, C., … Bin, H. (2004). Genetic and physiological diversity of phylogenetically and geographically distinct groups of *Arthrobacter* isolated from lead–zinc mine tailings. *FEMS Microbiology Ecology, 49*, 333–341.

Hare, P. D., Cress, W. A., & van Staden, J. (1997). The involvement of cytokinins in plant responses to environmental stress. *Plant Growth Regulation, 23*, 79–103. http://dx.doi.org/10.1023/a:1005954525087.

He, L. Y., Zhang, Y. F., Ma, H. Y., Su, L. N., Chen, Z. J., Wang, Q. Y., … Sheng, X. F. (2010). Characterization of copper-resistant bacteria and assessment of bacterial communities in rhizosphere soils of copper-tolerant plants. *Applied Soil Ecology, 44*, 49–55. http://dx.doi.org/10.1016/j.apsoil.2009.09.004.

Hiltner, L. (1904). Über neuere erfahrungen und probleme auf dem geviet der bodenbakteriolgie unter besonderer berüksichtigung der gründüngung und brache. *Arb Dtsch Landwirt Ges, 98*, 59–78.

Hinsinger, P., Gobran, G. R., Gregory, P. J., & Wenzel, W. W. (2005). Rhizosphere geometry and heterogeneity arising from root-mediated physical and chemical processes. *New Phytologist, 168*, 293–303. http://dx.doi.org/10.2307/3694538.

Idris, R., Trifonova, R., Puschenreiter, M., Wenzel, W. W., & Sessitsch, A. (2004). Bacterial communities associated with flowering plants of the Ni hyperaccumulator *Thlaspi goesingense*. *Applied and Environmental Microbiology, 70*, 2667–2677. http://dx.doi.org/10.1128/aem.70.5.2667-2677.2004.

Janssen, J., Weyens, N., Croes, S., Beckers, B., Meiresonne, L., Van Peteghem, P., … Vangronsveld, J. (2015). Phytoremediation of metal contaminated soil using willow: Exploiting plant-associated bacteria to improve biomass production and metal uptake. *International Journal of Phytoremediation, 17*, 1123–1136. http://dx.doi.org/10.1080/15226514.2015.1045129.

Jeong, S., Moon, H. S., Shin, D., & Nam, K. (2013). Survival of introduced phosphate-solubilizing bacteria (PSB) and their impact on microbial community structure during the phytoextraction of Cd-contaminated soil. *Journal of Hazardous Materials, 263*(Part 2), 441–449. http://dx.doi.org/10.1016/j.jhazmat.2013.09.062.

Johnson, N. C., Wilson, G. W. T., Bowker, M. A., Wilson, J. A., & Miller, R. M. (2010). Resource limitation is a driver of local adaptation in mycorrhizal symbioses. *Proceedings of the National Academy of Sciences of the United States of America, 107*, 2093–2098.

Joner, E. J., Leyval, C., & Colpaert, J. V. (2006). Ectomycorrhizas impede phytoremediation of polycyclic aromatic hydrocarbons (PAHs) both within and beyond the rhizosphere. *Environmental Pollution, 142*, 34–38. http://dx.doi.org/10.1016/j.envpol.2005.09.007.

Jones, D. L. (1998). Organic acids in the rhizosphere – a critical review. *Plant and Soil, 205*, 25–44. http://dx.doi.org/10.1023/a:1004356007312.

Kamran, M. A., Kamran, M. A., Eqani, S. A., Bibi, S., Xu, R. K., Amna, … Chaudhary, H. J. (2016). Bioaccumulation of nickel by *E. sativa* and role of plant growth promoting rhizobacteria (PGPRs) under nickel stress. *Ecotoxicology and Environmental Safety, 126*, 256–263. http://dx.doi.org/10.1016/j.ecoenv.2016.01.002.

Kaye, J. P., & Hart, S. C. (1997). Competition for nitrogen between plants and soil microorganisms. *Trends in Ecology & Evolution, 12*, 139–143. http://dx.doi.org/10.1016/S0169-5347(97)01001-X.

Kelly, J. J., Häggblom, M. M., & Tate, R. L. (2003). Effects of heavy metal contamination and remediation on soil microbial communities in the vicinity of a zinc smelter as indicated by analysis of microbial community phospholipid fatty acid profiles. *Biology and Fertility of Soils, 38*, 65–71. http://dx.doi.org/10.1007/s00374-003-0642-1.

Kidd, P., Barcelo, J., Bernal, M. P., Navari-Izzo, F., Poschenrieder, C., Shilev, S., … Monterroso, C. (2009). Trace element behaviour at the root–soil interface: Implications in phytoremediation. *Environmental and Experimental Botany, 67*, 243–259. http://dx.doi.org/10.1016/j.envexpbot.2009.06.013.

Kidd, P., Mench, M., Álvarez-López, V., Bert, V., Dimitriou, I., Friesl-Hanl, W., … Puschenreiter, M. (2015). Agronomic practices for improving gentle remediation of trace element-contaminated soils. *International Journal of Phytoremediation, 17*, 1005–1037. http://dx.doi.org/10.1080/15226514.2014.1003788.

Kolbas, A., Kidd, P., Guinberteau, J., Jaunatre, R., Herzig, R., & Mench, M. (2015). Endophytic bacteria take the challenge to improve Cu phytoextraction by sunflower. *Environmental Science and Pollution Research, 22*, 5370–5382. http://dx.doi.org/10.1007/s11356-014-4006-1.

Kozdrój, J. (1995). Microbial responses to single or successive soil contamination with Cd or Cu. *Soil Biology and Biochemistry, 27*, 1459–1465. http://dx.doi.org/10.1016/0038-0717(95)00070-U.

Kozdrój, J., & van Elsas, J. D. (2000). Response of the bacterial community to root exudates in soil polluted with heavy metals assessed by molecular and cultural approaches. *Soil Biology and Biochemistry, 32*, 1405–1417. http://dx.doi.org/10.1016/S0038-0717(00)00058-4.

Kruckeberg, A. R. (2002). The influences of lithology on plant life. In *Geology and plant life: The effects of landforms and rock type on plants* (pp. 160–181). Seattle, London: Univ Wash Press.

Kuffner, M., De Maria, S., Puschenreiter, M., Fallmann, K., Wieshammer, G., Gorfer, M., … Sessitsch, A. (2010). Culturable bacteria from Zn- and Cd-accumulating *Salix caprea* with differential effects on plant growth and heavy metal availability. *Journal of Applied Microbiology, 108*, 1471–1484. http://dx.doi.org/10.1111/j.1365-2672.2010.04670.x.

Kunito, T., Saeki, K., Nagaoka, K., Oyaizu, H., & Matsumoto, S. (2001). Characterization of copper-resistant bacterial community in rhizosphere of highly copper-contaminated soil. *European Journal of Soil Biology, 37*, 95–102. http://dx.doi.org/10.1016/S1164-5563(01)01070-6.

Lampis, S., Santi, C., Ciurli, A., Andreolli, M., & Vallini, G. (2015). Promotion of arsenic phytoextraction efficiency in the fern *Pteris vittata* by the inoculation of As-resistant bacteria: a soil bioremediation perspective. *Frontiers in Plant Science, 6*, 80. http://dx.doi.org/10.3389/fpls.2015.00080.

Langella, F., Grawunder, A., Stark, R., Weist, A., Merten, D., Haferburg, G., … Kothe, E. (2014). Microbially assisted phytoremediation approaches for two multi-element contaminated sites. *Environmental Science and Pollution Research, 21*, 6845–6858.

Lebeau, T., Braud, A., & Jezequel, K. (2008). Performance of bioaugmentation-assisted phytoextraction applied to metal contaminated soils: a review. *Environmental Pollution, 153*, 497–522. http://dx.doi.org/10.1016/j.envpol.2007.09.015.

Lemanceau, P., Maurhofer, M., & Défago, G. (2007). Contribution of studies on suppressive soils to the identification of bacterial biocontrol agents and to the knowledge of their modes of action. In *Plant-associated bacteria* (pp. 231–267). Springer.

Li, K., & Ramakrishna, W. (2011). Effect of multiple metal resistant bacteria from contaminated lake sediments on metal accumulation and plant growth. *Journal of Hazardous Materials, 189*, 531–539. http://dx.doi.org/10.1016/j.jhazmat.2011.02.075.

Liao, M., & Xie, X. M. (2007). Effect of heavy metals on substrate utilization pattern, biomass, and activity of microbial communities in a reclaimed mining wasteland of red soil area. *Ecotoxicology and Environmental Safety, 66*, 217–223. http://dx.doi.org/10.1016/j.ecoenv.2005.12.013.

Lin, W., Okon, Y., & Hardy, R. W. F. (1983). Enhanced mineral uptake by *Zea mays* and *Sorghum bicolor* roots inoculated with *Azospirillum brasilense*. *Applied and Environmental Microbiology, 45*, 1775–1779.

Lipman, C. B. (1926). The bacterial flora of serpentine soils. *Journal of Bacteriology, 12*, 315–318.

Lodewyckx, C., Mergeay, M., Vangronsveld, J., Clijsters, H., & Van der Lelie, D. (2002). Isolation, characterization, and identification of bacteria associated with the zinc hyperaccumulator *Thlaspi caerulescens* subsp. *calaminaria*. *International Journal of Phytoremediation, 4*, 101–115. http://dx.doi.org/10.1080/15226510208500076.

van Loon, L. C., & Bakker, P. A. H. M. (2003). Signalling in rhizobacteria-plant interactions. In H. Kroon, & E. J. W. Visser (Eds.), *Root Ecology* (pp. 297–330). Berlin, Heidelberg: Springer Berlin Heidelberg. http://dx.doi.org/10.1007/978-3-662-09784-7_12.

van Loon, L. C., Bakker, P. A., & Pieterse, C. M. (1998). Systemic resistance induced by rhizosphere bacteria. *Annual Review of Phytopathology, 36*, 453–483. http://dx.doi.org/10.1146/annurev.phyto.36.1.453.

Lorenz, N., Hintemann, T., Kramarewa, T., Katayama, A., Yasuta, T., Marschner, P., & Kandeler, E. (2006). Response of microbial activity and microbial community composition in soils to long-term arsenic and cadmium exposure. *Soil Biology and Biochemistry, 38*, 1430–1437. http://dx.doi.org/10.1016/j.soilbio.2005.10.020.

Lugtenberg, B., & Kamilova, F. (2009). Plant-growth-promoting rhizobacteria. *Annual Review of Microbiology, 63*, 541–556. http://dx.doi.org/10.1146/annurev.micro.62.081307.162918.

Luo, S. L., Chen, L., Chen, J. L., Xiao, X., Xu, T. Y., Wan, Y., … Zeng, G. M. (2011). Analysis and characterization of cultivable heavy metal-resistant bacterial endophytes isolated from Cd-hyperaccumulator *Solanum nigrum* L. and their potential use for phytoremediation. *Chemosphere, 85*, 1130–1138. http://dx.doi.org/10.1016/j.chemosphere.2011.07.053.

Malusá, E., Sas-Paszt, L., & Ciesielska, J. (2012). Technologies for beneficial microorganisms inocula used as biofertilizers. *The Scientific World Journal, 2012*, 1–12. http://dx.doi.org/10.1100/2012/491206.

Ma, Y., Rajkumar, M., & Freitas, H. (2009). Inoculation of plant growth promoting bacterium *Achromobacter xylosoxidans* strain Ax10 for the improvement of copper phytoextraction by *Brassica juncea*. *Journal of Environmental Management, 90*, 831–837. http://dx.doi.org/10.1016/j.jenvman.2008.01.014.

Ma, Y., Rajkumar, M., Vicente, J. A. F., & Freitas, H. (2010). Inoculation of Ni-resistant plant growth promoting bacterium *Psychrobacter* sp. strain SRS8 for the improvement of nickel phytoextraction by energy crops. *International Journal of Phytoremediation, 13*, 126–139. http://dx.doi.org/10.1080/15226511003671403.

Ma, Y., Zhang, C., Oliveira, R. S., Freitas, H., & Luo, Y. (2016). Bioaugmentation with endophytic bacterium E6S homologous to *Achromobacter piechaudii* enhances metal rhizoaccumulation in host *Sedum plumbizincicola*. *Frontiers in Plant Science, 7*. http://dx.doi.org/10.3389/fpls.2016.00075.

Marques, A. P., Moreira, H., Franco, A. R., Rangel, A. O., & Castro, P. M. (2013). Inoculating *Helianthus annuus* (sunflower) grown in zinc and cadmium contaminated soils with plant growth promoting bacteria – Effects on phytoremediation strategies. *Chemosphere, 92*, 74–83. http://dx.doi.org/10.1016/j.chemosphere.2013.02.055.

Marschner, P., Crowley, D., & Yang, C. (2004). Development of specific rhizosphere bacterial communities in relation to plant species, nutrition and soil type. *Plant and Soil, 261*, 199–208. http://dx.doi.org/10.1023/B: PLSO.0000035569.80747.c5.

Mastretta, C., Taghavi, S., van der Lelie, D., Mengoni, A., Galardi, F., Gonnelli, C., … Vangronsveld, J. (2009). Endophytic bacteria from seeds of *Nicotiana*

tabacum can reduce cadmium phytotoxicity. *International Journal of Phytoremediation, 11*, 251–267.
Mench, M., Lepp, N., Bert, V., Schwitzguébel, J.-P., Gawronski, S., Schröder, P., & Vangronsveld, J. (2010). Successes and limitations of phytotechnologies at field scale: outcomes, assessment and outlook from COST Action 859. *Journal of Soils and Sediments, 10*, 1039–1070. http://dx.doi.org/10.1007/s11368-010-0190-x.
Mench, M., Renella, G., Gelsomino, A., Landi, L., & Nannipieri, P. (2006). Biochemical parameters and bacterial species richness in soils contaminated by sludge-borne metals and remediated with inorganic soil amendments. *Environmental Pollution, 144*, 24–31. http://dx.doi.org/10.1016/j.envpol.2006.01.014.
Mench, M., Schwitzguébel, J.-P., Schroeder, P., Bert, V., Gawronski, S., & Gupta, S. (2009). Assessment of successful experiments and limitations of phytotechnologies: contaminant uptake, detoxification and sequestration, and consequences for food safety. *Environmental Science and Pollution Research, 16*, 876–900. http://dx.doi.org/10.1007/s11356-009-0252-z.
Mendes, R., Garbeva, P., & Raaijmakers, J. M. (2013). The rhizosphere microbiome: significance of plant beneficial, plant pathogenic, and human pathogenic microorganisms. *FEMS Microbiology Reviews, 37*, 634–663. http://dx.doi.org/10.1111/1574-6976.12028.
Mengoni, A., Barzanti, R., Gonnelli, C., Gabbrielli, R., & Bazzicalupo, M. (2001). Characterization of nickel-resistant bacteria isolated from serpentine soil. *Environmental Microbiology, 3*, 691–698. http://dx.doi.org/10.1046/j.1462-2920.2001.00243.x.
Mengoni, A., Grassi, E., Barzanti, R., Biondi, E. G., Gonnelli, C., Kim, C. K., & Bazzicalupo, M. (2004). Genetic diversity of bacterial communities of serpentine soil and of rhizosphere of the nickel-hyperaccumulator plant *Alyssum bertolonii*. *Microbial Ecology, 48*, 209–217. http://dx.doi.org/10.1007/s00248-003-0149-1.
Mengoni, A., Schat, H., & Vangronsveld, J. (2010). Plants as extreme environments? Ni-resistant bacteria and Ni-hyperaccumulators of serpentine flora. *Plant and Soil, 331*, 5–16.
Montalbán, B., Croes, S., Weyens, N., Lobo, M. C., Pérez-Sanz, A., & Vangronsveld, J. (2016). Characterization of bacterial communities associated with *Brassica napus* L. growing on a Zn-contaminated soil and their effects on root growth. *International Journal of Phytoremediation, 18*, 985–993. http://dx.doi.org/10.1080/15226514.2016.1183566.
Moorhead, L. D., Westerfield, M. M., & Zak, C. J. (1998). Plants retard litter decay in a nutrient-limited soil: a case of exploitative competition? *Oecologia, 113*, 530–536. http://dx.doi.org/10.1007/s004420050406.
Moreira, H., Pereira, S. I. A., Marques, A. P. G. C., Rangel, A. O. S. S., & Castro, P. M. L. (2016). Mine land valorization through energy maize production enhanced by the application of plant growth promoting rhizobacteria and arbuscular mycorrhizal fungi. *Environment Science Pollution Research, 23*, 6940–6950. http://dx.doi.org/10.1007/s11356-015-5914-4.
Muehe, E. M., Weigold, P., Adaktylou, I. J., Planer-Friedrich, B., Kraemer, U., Kappler, A., & Behrens, S. (2015). Rhizosphere microbial community composition affects cadmium and zinc uptake of the metal-hyperaccumulating plant *Arabidopsis halleri*. *Applied and Environmental Microbiology*. http://dx.doi.org/10.1128/aem.03359–14.
Oline, D. K. (2006). Phylogenetic comparisons of bacterial communities from serpentine and nonserpentine soils. *Applied and Environmental Microbiology, 72*, 6965–6971.
Pal, A., Dutta, S., Mukherjee, P. K., & Paul, A. K. (2005). Occurrence of heavy metal-resistance in microflora from serpentine soil of Andaman. *Journal of Basic Microbiology, 45*, 207–218. http://dx.doi.org/10.1002/jobm.200410499.
Panagos, P., Van Liedekerke, M., Yigini, Y., & Montanarella, L. (2013). Contaminated sites in Europe: Review of the current situation based on data collected through a European network. *Journal of Environmental and Public Health*. http://dx.doi.org/10.1155/2013/158764.

Pereira, S. I. A., Barbosa, L., & Castro, P. M. L. (2015). Rhizobacteria isolated from a metal-polluted area enhance plant growth in zinc and cadmium-contaminated soil. *International Journal of Environmental Science and Technology, 12*, 2127–2142. http://dx.doi.org/10.1007/s13762-014-0614-z.

Pereira, S. I. A., & Castro, P. M. L. (2014). Diversity and characterization of culturable bacterial endophytes from *Zea mays* and their potential as plant growth-promoting agents in metal-degraded soils. *Environmental Science and Pollution Research, 21*, 14110–14123. http://dx.doi.org/10.1007/s11356-014-3309-6.

Pereira, S. I., Pires, C., Henriques, I., Correia, A., Magan, N., & Castro, P. M. (2015). Assessment of rhizospheric culturable bacteria of *Phragmites australis* and *Juncus effusus* from polluted sites. *Journal of Basic Microbiology*. http://dx.doi.org/10.1002/jobm.201500010.

Petriccione, M., Di Patre, D., Ferrante, P., Papa, S., Bartoli, G., Fioretto, A., & Scortichini, M. (2013). Effects of *Pseudomonas fluorescens* seed bioinoculation on heavy metal accumulation for *Mirabilis jalapa* phytoextraction in smelter-contaminated soil. *Water, Air, and Soil Pollution, 224*.

Petrisor, I. G., Dobrota, S., Komnitsas, K., Lazar, I., Kuperberg, J. M., & Serban, M. (2004). Artificial inoculation-perspectives in tailings phytostabilization. *International Journal of Phytoremediation, 6*, 1–15. http://dx.doi.org/10.1080/16226510490439918.

Phieler, R., Merten, D., Roth, M., Büchel, G., & Kothe, E. (2015). Phytoremediation using microbially mediated metal accumulation in *Sorghum bicolor*. *Environmental Science and Pollution Research, 22*, 19408–19416. http://dx.doi.org/10.1007/s11356-015-4471-1.

Podile, A. R., & Kishore, G. K. (2006). Plant growth-promoting rhizobacteria. In S. S. Gnanamanickam (Ed.), *Plant-associated bacteria* (pp. 195–230). Dordrecht: Springer Netherlands. http://dx.doi.org/10.1007/978-1-4020-4538-7_6.

Proctor, J. (1971). The plant ecology of serpentine: II. Plant response to serpentine soils. *The Journal of Ecology, 59*, 397–410.

Proctor, J. (1999). Toxins, nutrient shortages and droughts: the serpentine challenge. *Trends in Ecology & Evolution, 14*, 334–335. http://dx.doi.org/10.1016/S0169-5347(99)01698-5.

Raaijmakers, J. M., Vlami, M., & de Souza, J. T. (2002). Antibiotic production by bacterial biocontrol agents. *Antonie van Leeuwenhoek, 81*, 537–547.

Rajkumar, M., Ae, N., & Freitas, H. (2009). Endophytic bacteria and their potential to enhance heavy metal phytoextraction. *Chemosphere, 77*, 153–160. http://dx.doi.org/10.1016/j.chemosphere.2009.06.047.

Rani, A., Shouche, Y. S., & Goel, R. (2008). Declination of copper toxicity in pigeon pea and soil system by growth-promoting *Proteus vulgaris* KNP3 strain. *Current Microbiology, 57*, 78–82. http://dx.doi.org/10.1007/s00284-008-9156-2.

Reed, M. L. E., & Glick, B. R. (2005). Growth of canola (*Brassica napus*) in the presence of plant growth-promoting bacteria and either copper or polycyclic aromatic hydrocarbons. *Canadian Journal of Microbiology, 51*, 1061–1069. http://dx.doi.org/10.1139/w05-094.

Robinson, B. H., Bañuelos, G., Conesa, H. M., Evangelou, M. W. H., & Schulin, R. (2009). The phytomanagement of trace elements in soil. *Critical Reviews in Plant Sciences, 28*, 240–266.

Rodriguez, H., & Fraga, R. (1999). Phosphate solubilizing bacteria and their role in plant growth promotion. *Biotechnology Advances, 17*, 319–339. http://dx.doi.org/10.1016/S0734-9750(99)00014-2.

Rylott, E. L. (2014). Endophyte consortia for xenobiotic phytoremediation: The root to success? *Plant Soil, 385*, 389–394. http://dx.doi.org/10.1007/s11104-014-2296-2291.

Sangthong, C., Setkit, K., & Prapagdee, B. (2016). Improvement of cadmium phytoremediation after soil inoculation with a cadmium-resistant *Micrococcus* sp. *Environmental Science and Pollution Research, 23*, 756–764. http://dx.doi.org/10.1007/s11356-015-5318-5.

Schippers, A., Hallmann, R., Wentzien, S., & Sand, W. (1995). Microbial diversity in uranium mine waste heaps. *Applied and Environmental Microbiology, 61*, 2930–2935.

Schlegel, H. G., Cosson, J. P., & Baker, A. J. M. (1991). Nickel-hyperaccumulating plants provide a niche for nickel-resistant bacteria. *Botanica Acta, 104*, 18–25. http://dx.doi.org/10.1111/j.1438-8677.1991.tb00189.x.

Sessitsch, A., Kuffner, M., Kidd, P., Vangronsveld, J., Wenzel, W. W., Fallmann, K., & Puschenreiter, M. (2013). The role of plant-associated bacteria in the mobilization and phytoextraction of trace elements in contaminated soils. *Soil Biology and Biochemistry, 60*, 182–194. http://dx.doi.org/10.1016/j.soilbio.2013.01.012.

Sessitsch, A., & Puschenreiter, M. (2008). Endophytes and rhizosphere bacteria of plants growing in heavy metal-containing soils. In P. Dion, & C. Nautiyal (Eds.), *Microbiology of extreme soils* (pp. 317–332). Springer Berlin Heidelberg. http://dx.doi.org/10.1007/978-3-540-74231-9.

Sheng, X.-F., Xia, J.-J., Jiang, C.-Y., He, L.-Y., & Qian, M. (2008). Characterization of heavy metal-resistant endophytic bacteria from rape (*Brassica napus*) roots and their potential in promoting the growth and lead accumulation of rape. *Environmental Pollution, 156*, 1164–1170. http://dx.doi.org/10.1016/j.envpol.2008.04.007.

Shilev, S., Fernández, A., Benlloch, M., & Sancho, E. (2006). Sunflower growth and tolerance to arsenic is increased by the rhizospheric bacteria *Pseudomonas fluorescens*. In J. L. Morel, G. Echevarria, & N. Goncharova (Eds.), *Phytoremediation of metal-contaminated soils* (pp. 315–326). Netherlands: Springer.

Smolders, E., Buekers, J., Oliver, I., & McLaughlin, M. J. (2004). Soil properties affecting toxicity of zinc to soil microbial properties in laboratory-spiked and field-contaminated soils. *Environmental Toxicology and Chemistry/SETAC, 23*, 2633–2640.

Sprocati, A. R., Alisi, C., Segre, L., Tasso, F., Galletti, M., & Cremisini, C. (2006). Investigating heavy metal resistance, bioaccumulation and metabolic profile of a metallophile microbial consortium native to an abandoned mine. *Science of the Total Environment, 366*, 649–658. http://dx.doi.org/10.1016/j.scitotenv.2006.01.025.

Stoppel, R., & Schlegel, H. G. (1995). Nickel-resistant bacteria from anthropogenically nickel-polluted and naturally nickel-percolated ecosystems. *Applied and Environmental Microbiology, 61*, 2276–2285.

Sturz, A. V., Christie, B. R., & Nowak, J. (2000). Bacterial endophytes: Potential role in developing sustainable systems of crop production. *Critical Reviews in Plant Sciences, 19*, 1–30. http://dx.doi.org/10.1080/07352680091139169.

Taghavi, S., Garafola, C., Monchy, S., Newman, L., Hoffman, A., Nele Weyens, … van der Lelie, D. (2009). Genome survey and characterization of endophytic bacteria exhibiting a beneficial effect on growth and development of poplar trees. *Applied and Environmental Microbiology, 75*, 748–757. http://dx.doi.org/10.1128/aem.02239-08.

Tanimoto, E. (2005). Regulation of root growth by plant hormones—roles for auxin and gibberellin. *Critical Reviews in Plant Sciences, 24*, 249–265. http://dx.doi.org/10.1080/07352680500196108.

Thijs, S., Langill, T., & Vangronsveld, J. (2017). The bacterial and fungal microbiota of hyperaccumulator plants: Small organisms, large influence. In A. Cuypers, & J. Vangronsveld (Eds.), *Phytoremediation* (Vol. 83, pp. 43–86).

Thijs, S., Sillen, W., Rineau, F., Weyens, N., & Vangronsveld, J. (2016). Towards an enhanced understanding of plant-microbiome interactions to improve phytoremediation: Engineering the metaorganism. *Frontiers in Microbiology, 7*, 341. http://dx.doi.org/10.3389/fmicb.2016.00341.

Turgay, O. C., Gormez, A., & Bilen, S. (2012). Isolation and characterization of metal resistant-tolerant rhizosphere bacteria from the serpentine soils in Turkey. *Environmental Monitoring and Assessment, 184*, 515–526. http://dx.doi.org/10.1007/s10661-011-1984-z.

Unterbrunner, R., Wieshammer, G., Hollender, U., Felderer, B., Wieshammer-Zivkovic, M., Puschenreiter, M., & Wenzel, W. W. (2007). Plant and fertiliser effects on rhizodegradation of crude oil in two soils with different nutrient status. *Plant and Soil, 300*, 117–126. http://dx.doi.org/10.1007/s11104-007-9394-2.

Vangronsveld, J., Herzig, R., Weyens, N., Boulet, J., Adriaensen, K., Ruttens, A., ... Mench, M. (2009). Phytoremediation of contaminated soils and groundwater: lessons from the field. *Environmental Science and Pollution Research International, 16*, 765–794. http://dx.doi.org/10.1007/s11356-009-0213-6.

Vangronsveld, J., Sterckx, J., Van Assche, F., & Clijsters, H. (1995). Rehabilitation studies on an old non-ferrous waste dumping ground: effects of revegetation and metal immobilization by beringite. *Journal of Geochemical Exploration, 52*, 221–229. http://dx.doi.org/10.1016/0375-6742(94)00045-D.

Vessey, J. K. (2003). Plant growth promoting rhizobacteria as biofertilizers. *Plant and Soil, 255*, 571–586. http://dx.doi.org/10.1023/a:1026037216893.

Visioli, G., Vamerali, T., Mattarozzi, M., Dramis, L., & Sanangelantoni, A. M. (2015). Combined endophytic inoculants enhance nickel phytoextraction from serpentine soil in the hyperaccumulator *Noccaea caerulescens*. *Frontiers in Plant Science, 6*. http://dx.doi.org/10.3389/fpls.2015.00638.

Wehner, J., Antunes, P. M., Powell, J. R., Mazukatow, J., & Rillig, M. C. (2010). Plant pathogen protection by arbuscular mycorrhizas: A role for fungal diversity? *Pedobiologia, 53*, 197–201. http://dx.doi.org/10.1016/j.pedobi.2009.10.002.

Weissenhorn, I., Leyval, C., Belgy, G., & Berthelin, J. (1995). Arbuscular mycorrhizal contribution to heavy metal uptake by maize (*Zea mays* L.) in pot culture with contaminated soil. *Mycorrhiza, 5*, 245–251. http://dx.doi.org/10.1007/BF00204957.

Wenzel, W. W. (2009). Rhizosphere processes and management in plant-assisted bioremediation (phytoremediation) of soils. *Plant and Soil, 321*, 385–408.

Weyens, N., Truyens, S., Dupae, J., Newman, L., Taghavi, S., van der Lelie, D., ... Vangronsveld, J. (2010). Potential of the TCE-degrading endophyte *Pseudomonas putida* W619-TCE to improve plant growth and reduce TCE phytotoxicity and evapotranspiration in poplar cuttings. *Environmental Pollution, 158*, 2915–2919. http://dx.doi.org/10.1016/j.envpol.2010.06.004.

Weyens, N., van der Lelie, D., Taghavi, S., Newman, L., & Vangronsveld, J. (2009). Exploiting plant–microbe partnerships to improve biomass production and remediation. *Trends in Biotechnology, 27*, 591–598. http://dx.doi.org/10.1016/j.tibtech.2009.07.006.

Whiting, S. N., Leake, J. R., McGrath, S. P., & Baker, A. J. M. (2001a). Assessment of Zn mobilization in the rhizosphere of *Thlaspi caerulescens* by bioassay with non-accumulator plants and soil extraction. *Plant and Soil, 237*, 147–156.

Whiting, S. N., Leake, J. R., McGrath, S. P., & Baker, A. J. M. (2001b). Zinc accumulation by *Thlaspi caerulescens* from soils with different Zn availability: A pot study. *Plant and Soil, 236*, 11–18.

Wojcik, M., Gonnelli, C., Selvi, F., Dresler, S., Rostański, A., Vangronsveld, J., & Schat, H. (2017). Metallophytes of serpentine and calamine soils – Their unique ecophysiology and potential for phytoremediation. In A. Cuypers, & J. Vangronsveld (Eds.), *Phytoremediation* (Vol. 83, pp. 1–42).

Zaidi, S., Usmani, S., Singh, B. R., & Musarrat, J. (2006). Significance of *Bacillus subtilis* strain SJ-101 as a bioinoculant for concurrent plant growth promotion and nickel accumulation in *Brassica juncea*. *Chemosphere, 64*, 991–997. http://dx.doi.org/10.1016/j.chemosphere.2005.12.057.

Zhang, W. H., Huang, Z., He, L. Y., & Sheng, X. F. (2012). Assessment of bacterial communities and characterization of lead-resistant bacteria in the rhizosphere soils of metal-tolerant *Chenopodium ambrosioides* grown on lead-zinc mine tailings. *Chemosphere, 87*, 1171–1178. http://dx.doi.org/10.1016/j.chemosphere.2012.02.036.

Zheng, Y.-K., Qiao, X.-G., Miao, C.-P., Liu, K., Chen, Y.-W., Xu, L.-H., & Zhao, L.-X. (2016). Diversity, distribution and biotechnological potential of endophytic fungi. *Annals of Microbiology, 66*, 529–542. http://dx.doi.org/10.1007/s13213-015-1153-7.

CHAPTER FOUR

Mycorrhiza-Assisted Phytoremediation

Laura Coninx, Veronika Martinova and Francois Rineau[1]
Hasselt University, Diepenbeek, Belgium
[1]Corresponding author: E-mail: francois.rineau@uhasselt.be

Contents

Advances in Botanical Research, Volume 83
ISSN 0065-2296
http://dx.doi.org/10.1016/bs.abr.2016.12.005

Abstract

Soil contamination is a major environmental problem affecting human health worldwide. A sustainable solution for this contamination is phytoremediation, a technique that has gotten more and more attention over the past decade. In this chapter, we will discuss how mycorrhizal fungi (arbuscular mycorrhizal fungi, ericoid mycorrhiza and ectomycorrhiza) can assist in phytoremediation. Our aim is to evaluate the benefits and drawbacks of mycorrhiza-assisted phytoremediation, review examples from literature, and provide recommendations for future studies. Both inorganic and organic pollution and factors influencing the effect of mycorrhization will be addressed. Furthermore, molecular mechanisms associated with the beneficial effects of mycorrhizal inoculation will be discussed. Since there are a number of studies (34) reporting quantitative data on the effect of mycorrhizal fungi on organic soil pollution, a meta-analysis was performed to quantify the influence of mycorrhiza on the degradation of organics. Although the effects of mycorrhization are highly variable, it is clear that mycorrhizae have the potential to enhance the phytoremediation efficiency of organic as well as inorganic pollution. Mycorrhiza-assisted phytoremediation is thus a valuable strategy for the remediation of contaminated areas and should be further studied.

1. WHAT ARE MYCORRHIZA?

In late 19th-century Kingdom of Prussia, Alex Bernard Frank was appointed to investigate the occurrence of truffles to advance their cultivation. This endeavour gave rise to his classical paper, where he described the union of tree roots with fungal mycelium, which made up for a new organ (Frank, 2005). To describe this structure, he used the term 'mycorrhiza', literally translating from Greek as 'fungus root'. Nowadays, mycorrhizas can be defined as the association between a plant root and a fungus, which serves fundamentally as a water and nutrient transfer interface (Smith & Read, 2008). However, mycorrhizas are also known to be formed between hyphal fungi and the underground organs of many lower land plants. Mycorrhizal associations are generally described as mutualisms, i.e., symbioses where both partners benefit from the partnership. The term symbiosis has been recommended as more accurate, as mycorrhizal associations have been shown not to be exclusively mutualistic (Smith & Read, 2008). It has been suggested that mycorrhiza functions rather on a mutualistic—parasitic continuum (Johnson, Graham, & Smith, 1997; Neuhauser & Fargione, 2004). There can be a big variation in the development and function of mycorrhiza. These variations depend on the identity of both partners, time and environmental conditions, as well as on the measures of fitness and nutritional benefits taken into account in the experiments (Chagnon, Bradley, Maherali, & Klironomos, 2013). The outcome of the association can change over the life span of even a single plant–fungus relationship (Smith & Read, 2008).

Ever since mycorrhiza were described, they were gradually gaining attention in biology and ecology. Nowadays their importance at plant and ecosystem level is well recognized (van der Heijden, Martin, Selosse, & Sanders, 2015; Smith & Read, 2008). It is estimated that 86% of all angiosperms are mycorrhizal, with mycorrhizal associations being common in most terrestrial ecosystems. Exceptions include saline, extremely cold or disturbed habitats (Brundrett, 2009; van der Heijden et al., 2015).

It is now established that colonization of land by the first plants would not have been possible without mycorrhizas, as supported by both fossil and molecular evidence (Fortin, Plenchette, & Piché, 2009). Bryophyte fossils from Devonian were found to be colonized by arbuscular mycorrhizal-like fungi, which are thought to have a crucial role in acquiring nutrients from the early soils and aiding their development (Fortin et al., 2009; Johnson & Gehring, 2011; Smith & Read, 2008).

1.1 Types of Mycorrhizal Associations

Mycorrhizas can be nowadays broadly categorized into arbuscular mycorrhizas, ectomycorrhizas, orchid mycorrhizas and ericoid mycorrhizas. In the past, mycorrhizal fungi were classified morphologically according to the relative situation of the fungi on roots, namely into endomycorrhizal, ectomycorrhizal and ectendomycorrhizal fungi (Brundrett, 2004; Smith & Read, 2008). According to the old classification, arbuscular, ericoid, and orchid mycorrhizas fall into 'endomycorrhizal' associations, as the fungus colonizes the plant intracellularly (Brundrett, 2004). However, they have distinct anatomy and are present in different host and fungus taxonomical groups (Brundrett, 2004). Ectendomycorrhizas, arbutoid and monotropoid mycorrhizas are not considered in this chapter, as they are not known to significantly contribute to phytoremediation.

Arbuscular mycorrhizal fungi (AMF) (Fig. 1) are the most common mycorrhizal association and form mutualistic relationships with over 74% of flowering plants (Brundrett, 2009) and with over 80% of vascular plants (Peterson, Massicotte, & Melville, 2004). The AMF plant hosts range from nonvascular plants (e.g., some liverworths), some gymnosperm families (e.g., Cupressaceae, Araucariaceae, Podocarpaceae) (Brundrett, 2009) and include *Gingko biloba*, and the majority of angiosperm families

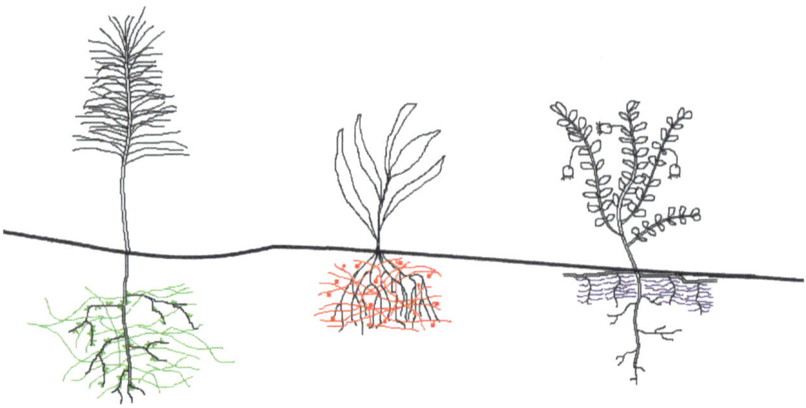

Figure 1 Schematic drawings of plants colonized by ECM (in green; with dichotomous root tips), AMF (in red; with vesicles) and ERM (blue; with root hairs). *AMF*, arbuscular mycorrhizal fungi; *ECM*, ectomycorrhizal fungi; *ERM*, ericoid mycorrhizae.

(Peterson et al., 2004). Most species of the economically important families Leguminosae, Rosaceae, Gramineae, Compositae and Labiaceae present AMF (Newman & Reddell, 1987). Genera that associate with AMF can present another type of mycorrhiza, either simultaneously or at different times (Smith & Read, 2008).

AMF are generally regarded as obligate biotrophs, i.e., they depend on finding a suitable host to complete their life cycle (Goltapeh, Danesh, Prasad, & Varma, 2008). Taxonomically, they belong to the phylum Glomeromycota, with approximately 230 described species and 29 genera (Johnson & Gehring, 2011; Oehl, Sieverding, Palenzuela, Ineichen, & Alves da Silva, 2011; Schüßler & Walker, 2010). These associations are most commonly characterized by the presence of arbuscules (modified hyphae that are the primary structures for resource exchange), intraradical hyphae, extraradical mycelium, and large spores (Peterson et al., 2004). AMF are thought to reproduce asexually, however advances in molecular methods have allowed the discovery of genes coding for meiotic tools in *Glomus* (Halary et al., 2011). Some fungal species form vesicles, intraradical structures with storage function (Johnson & Gehring, 2011; Peterson et al., 2004). AMF, in contrast with the other types of mycorrhizal fungi, are aseptate, which means that they lack regular cross walls and nuclei can flow along the length of hyphae relatively freely (Johnson & Gehring, 2011; Smith & Read, 2008). These nuclei are thought to be exchanged between different individuals by fusion of hyphae (anastomosis) and there is ongoing research investigating whether

AMF are heterokaryotic or homokaryotic (Pawlowska & Taylor, 2004; Smith & Read, 2008). These intriguing organisms are still a challenge for the scientific community as they present high genetic diversity, as shown with multiple ribosomal ribonucleic acid (rRNA) gene sequences from even a single spore, making it hard to develop genetic markers for AMF species identification (Lindahl et al., 2013; Rodriguez, Clapp, & Dodd, 2004; Smith & Read, 2008).

Ectomycorrhizal fungi (ECM) (Fig. 1) are also widespread in their distribution but associate with only 3% of vascular plant families. They are estimated to form the symbiosis with approximately 6000 plant species (Brundrett, 2009; van der Heijden et al., 2015) and they are the dominant mycorrhizal type in boreal and temperate forests. These ecosystems are often dominated by a few plant species, which in turn can harbour great ECM diversity. With the aid of next-generation sequencing, 600–1000 different ECM were discovered in beech-dominated temperate forest stands in France (Buée et al., 2009).

ECM species diversity is estimated to be much higher than the one of AMF, with at about 20,000 known ECM species (Tedersoo, May, & Smith, 2010), the majority belonging to Basidiomycota (252 genera) and the rest to Ascomycota (84 genera) (van der Heijden et al., 2015; Rinaldi, Comandini, & Kuyper, 2008). The ECM symbiosis is thought to have been derived multiple times independently from saprophytic lineages (Jayachandran & Fisher, 2008) through an extensive loss of lignocellulose-degrading genes, which is supported by molecular studies and recently published ECM genomes (Martin et al., 2011; Plett & Martin, 2011). Even though some ECM can be cultured without a host in laboratory conditions (unlike AMF), they depend on their host to acquire carbohydrates under field conditions (van der Heijden et al., 2015). Morphologically, ECM are characterized by the presence of a sheath or mantle of hyphae surrounding the root. Lateral projections of the mantle may penetrate between epidermal cells of the root in what is called the Hartig net (Peterson et al., 2004).

Orchid mycorrhizas associate with hosts from the family Orchidaceae, with approximately 20,000–35,000 plant species. These mycorrhizas are thought to be more host-specific, with up to 10 fungal taxa found per plant (Jacquemyn, Honnay, Cammue, Brys, & Lievens, 2010). Orchid seedlings are achlorophyllous and therefore completely dependent on the mycorrhiza to provide carbohydrates and nutrients. Fungi forming these mycorrhizal associations are typically Basidiomycetes. However, there is some recent evidence that several species of tropical achlorophyllous epiphytes form mycorrhizal associations with species in the Glomeromycota (Jayachandran & Fisher, 2008). Orchid mycorrhizas are characterized by hyphal coils in cortical cells of root-like tissue or the collar of the shoot. The fungus penetrates through passage cells of the exodermis and hyphae ramify through the cortical tissue by intracellular penetration (Peterson et al., 2004; Smith & Read, 2008).

As the name suggests, ericoid mycorrhizae (ERM) (Fig. 1) can be found in plants in the Ericaceae family, with genera such as *Erica, Calluna* and *Vaccinium.* These species are typical for nutrient-poor heathlands (Smith & Read, 2008). The fungal partners have been thought to be exclusively Ascomycetes, with Helotiales being a well-known order of ericoid fungi. However, it has been shown that Sebacinales, which are Basidiomycetes, are also forming these associations (Selosse et al., 2007). Some of these fungi have been found to form ECM associations with other host plants and fungi in the Ascomycota.

As for the morphology of ERM, the fungi colonize the epidermal fine hair root cells of their hosts, without mantle or Hartig net development (Jayachandran & Fisher, 2008). Coils of hyphae are formed in these cells and the fungi do not spread from cell to cell (Perotto, Girlanda, & Martino, 2002; Peterson et al., 2004).

1.2 Costs and Benefits of the Association

Improved plant mineral nutrition was the first benefit attributed to mycorrhizal fungi. The mycelium that extends from plant roots can bridge through nutrient depletion zones of poorly mobile elements, most importantly phosphorus (P). Absorptive surface area is greatly increased by the hyphal network in the rhizosphere, making nutrient and water acquisition more efficient. Mycorrhizal plants therefore invest less energy into extension of the root system than nonmycorrhizal plants (Smith & Read, 2008). While in the case of AMF, improved P acquisition is the physiological benefit that has been emphasized, ECM have been studied extensively for their nitrogen (N)-acquiring capacities, due to the generally N limited conditions in temperate forests (Courty et al., 2010). As ECM have at least partially conserved their capacity to degrade organic matter, they are able to access N and other minerals trapped within (Fortin et al., 2009). There can be various additional benefits to the plants derived from the association, as some authors report improved reproductive success of AMF plants due to an increase in shoot and seed P concentration (Nuortila, Kytöviita, &

Tuomi, 2004) and enhanced flower yield (Asrar, Abdel-Fattah, & Elhindi, 2012).

As mentioned earlier, the increased contact with the soil leads to improved water acquisition for mycorrhizal plants, which is also facilitated by the capacity of the fine hyphae to extract water from pores inaccessible to plant roots. Indeed, it has been shown that mycorrhizal plants are more resistant to drought stress than nonmycorrhizal plants. Moreover, the fungal partner is capable of inducing quicker closure of the leaf stomata, preventing plant wilting (Fortin et al., 2009) and increasing plant osmolyte levels such as proline (Garg & Chandel, 2012). In a pot experiment of Sarwat et al. (2016), AMF inoculation of *Brassica juncea* leads to a 63.47% increase in its proline content and overall alleviated symptoms of drought stress caused by sodium chloride (NaCl) treatment.

Studies with many economically important crops demonstrated that mycorrhizal plants were more resistant to pathogen attacks (Nagy & Fossdal, 2013; Pozo & Azcón-Aguilar, 2007; Sen, 2001). Mycorrhizal fungi present in the rhizosphere and on the root surfaces compete with pathogens and the presence of diverse microflora aids in maintaining the number of pathogens low (Fortin et al., 2009). The ECM mantle present on colonized forest trees also presents a physical barrier to the advance of pathogens and ECM were shown to be capable of producing antibiotics and/or increase the production of plant antibiotics (Schelkle & Peterson, 1997). However, it has been suggested that the in vitro protection of ECM plants is rather due to the acidification of the medium instead of production of antibiotics (Schelkle & Peterson, 1997). Furthermore, mycorrhizal fungi can also increase plant's resistance at physiological level. It has been shown that AMF can lead to greater resistance to herbivory through the action of jasmonic acid (Fortin et al., 2009). Moreover, the overall benefits in alleviation of environmental stress and improved mineral nutrition also contribute to reduced sensitivity of mycorrhizal plants to pathogen attacks.

Nevertheless, there are costs of colonization that can be high, as in AMF up to 20% of the host's fixed carbon (C) can be delivered to the microbial symbiont (Smith & Read, 2008). For ECM, it has been estimated that at about 20%–25% of photosynthetic products are delivered to the fungus (Nehls, 2008). Mycorrhizal fungi increase the drain of C to the rhizosphere as they speed up respiration, C retention in the mycelium and allocation to the roots (Olsson, Rahm, & Aliasgharzad, 2010). Nonetheless, under nutrient-limiting conditions, mycorrhizal crop plants typically exhibit a performance advantage over noncolonized individuals. However, under nutrient-saturated conditions occurring in high-input agricultural systems, the relative advantages are reduced while the C costs remain, and the performance of colonized plants can fall below that of noncolonized plants (Habte & Osorio, 2001; Jayachandran & Fisher, 2008). Stable 13C labelling can be used to determine the C flow and partitioning between the plant host and fungus. In a study with *Glomus intraradices* (AMF) (=*Rhizophagus irregularis*) and *Trifolium subterraneum* (subterranean clover), P fertilization and shading of the plant reduced the amount of C allocated to fungal mycelium as fungal storage lipids (Olsson et al., 2010).

As for environmental stresses, mycorrhizae have been shown to improve plant growth and survival when facing drought, salt stress and pollution due to heavy metals or organic pollutants (Asrar et al., 2012; Jayachandran & Fisher, 2008; Shaoxia, Tingting, & Runjin, 2011). Improved resistance to xenobiotics is an important benefit of the association, gaining more and more attention in the scientific community and in general society. The mechanisms involved in resistance to heavy metal and organic pollutants will be described in later sections.

1.3 Mycorrhiza in Ecosystems

Considering that substantial quantities of plant C are allocated to mycorrhizal fungi, and that in turn plants can largely depend on the fungal symbionts for their nutrition, mycorrhizal fungi play a significant role in mediating many ecosystem processes and have been increasingly considered a major actor in terrestrial C cycle, nutrient cycles and affecting plant communities (Dickie et al., 2013).

As proposed by Read (1993), it is generally considered that major mycorrhizal types change along plant primary succession in temperate and boreal forests, with early successional habitats with high nutrient substrates dominated by nonmycorrhizal plants, followed by an AMF-dominated vegetation, which is then changed into ECM-dominated forests with AMF understory, subsequently altered into ERM understory forests and finally end as ERM-dominated ecosystems. This model has been since challenged, as it has been reported that the main mycorrhizal types can be present in any of the stages of ecosystem development (Dickie et al., 2013).

There is increasing amount of evidence, which supports that mycorrhizal fungi are important players in the global terrestrial C cycle, especially when considering that mycorrhizal plants are dominant in most terrestrial ecosystems. Mycorrhizae constitute a considerable sink of C into soil ecosystems, as plants allocate significant amounts of their photosynthates to their fungal partners (Finlay, 2008; van der Heijden et al., 2015). Furthermore, mycorrhizal fungi can affect the rates of root decomposition (Fortin et al., 2009). They contain chitin, which makes them difficult to degrade and glomalin, a glycoprotein secreted by AMF hyphae, has been shown to be quite recalcitrant as well (van der Heijden et al., 2015; Rillig, Wright, & Eviner, 2002). The significance of glomalin as a C sink might be higher than previously thought, with estimates as high as one-third of soil-sequestered C to be held as this glycoprotein (Fortin et al., 2009; van der Heijden et al., 2015; Johnson & Gehring, 2011). The extensive hyphal networks in the soil also significantly contribute to stabilizing soil aggregates. One gram of grassland can contain as much as 100 m of AMF hyphae (Johnson & Gehring, 2011) and glomalin has a role in aiding to bind soil particles.

It is now well established that AMF are important for plant P nutrition, especially in nutrient-poor habitats. In some experiments, AMF have been shown to contribute up to 90% of plant P. AMF extraradical hyphae secrete extracellular phosphatases and their narrow diameter allows them to place these enzymes in pores that are too small for root hairs (Leake et al., 2004). AMF have been also shown to immobilize substantial amounts of N in their mycelia, however their importance for plant N uptake is less extensively studied (van der Heijden et al., 2015). The N and P acquired by ECM and ERM can account for up to 80% of plant N and P nutrition. In temperate and boreal forest ecosystems, ECM contribute to the N and P cycles through mobilization of these elements from soil organic matter (SOM) and minerals (van der Heijden et al., 2015). It has been reported that ECM are capable of producing a wide range of extracellular and cell wall-bound enzymes. ECM fungi have been demonstrated to mobilize P from SOM mainly in conditions of P deficiency and they were shown to contribute to soil phosphomonoesterase activity at the same rate as saprophytic fungi (Courty et al., 2010). However, there is a wide range of variability in the capacity to mobilize nutrients in different ECM species and even in different strains. So far, in vitro studies have been limited, as many species-rich ECM genera such as *Cortinarius* or *Tomentella* are challenging to grow in laboratory conditions. To overcome this problem, some authors developed enzyme assays to study catabolic activity of ECM morphotypes directly from ECM root tips (Courty et al., 2010).

As mentioned earlier, AMF association lies on a mutualism–parasitism continuum and the reported effects of AMF on plant biomass range from positive to negative (van der Heijden et al., 2015; Hoeksema et al., 2010; Nuortila et al., 2004). Plant species with more thick roots are usually considered more mycorrhiza dependent than species with fine roots, such as grasses (Dostálek, Pánková, Münzbergová, & Rydlová, 2013; Habte & Osorio, 2001; van der Heijden et al., 2015). It has also been observed that the origin of AMF influences the outcome of the association (Yao, Zhu, Hu, & Li, 2008), with the combination of both local plant host and fungus being the one yielding the highest plant biomass productivity (Klironomos, 2003).

In natural conditions, neighbouring mycorrhizal plants are frequently connected with the extraradical mycelium, creating a so-called mycorrhizal network (MN) (Johnson & Gehring, 2011; Leake et al., 2004). In this way, in temperate forests, for example, trees can be interconnected with ECM mycelium, while the understory plants may be connected with AMF, ERM, or even orchid MN (van der Heijden et al., 2015). Preexisting MNs in the soil were reported to aid seedling establishment (van der Heijden, 2004; van der Heijden et al., 2015) and the importance of MN for plant communities and geochemical processes is being intensively investigated.

At plant community level, AMF identity and diversity has been shown to influence plant species composition through mediating competitive interactions (Francis & Read, 1994; van der Heijden, Boller, Weimken, & Sanders, 1998) and affecting the dynamics of plant succession (Dostálek et al., 2013). AMF can alter plant community composition, as different plant species present different growth responses to different AMF species, even when the AMF are collected from the same habitat. In the study of van der Heijden et al. (1998), codominant grassland species *Bromus erectus* and *Festuca ovina* differed significantly in the shoot biomass when inoculated with different individual AMF species or their mixture. Furthermore, in nutrient-poor grasslands, it has been shown that perennial forbs, which were also the rarest species in the study of Dostálek et al. (2013), are as well the most mycorrhiza dependent. In the same study, suppression of AMF with a fungicide also resulted in a decrease of plant species richness.

2. PRACTICAL APPLICATIONS OF MYCORRHIZAL FUNGI

Soil contamination is a serious environmental issue threatening human health on a global scale. Over the last decades a vast number of remedial technologies have been developed. Conventional remediation techniques include confinement, excavation and physicochemical stabilization (Vangronsveld et al., 2009; Witters et al., 2009). However, these are not applicable to large areas and tend to destroy the biological activity in the soil by irreversibly changing the soil's physicochemical properties and disturbing the indigenous soil microflora. Cost, time and logistical concerns further strengthen the need for alternative, low-cost techniques that are environmentally sound (Mahar et al., 2016; McGrath, Zhao, & Lombi, 2001; Witters et al., 2009).

Phytoremediation is such an alternative remediation technique. As defined by Vangronsveld et al. (2009), phytoremediation is 'the use of plants and associated microorganisms to remove, contain, inactivate, or degrade harmful environmental contaminants'. An overview of the most common soil phytoremediation techniques is given in Table 1.

It is evident that mycorrhizal fungi, being the most prevalent beneficial organisms associated with plants (Smith & Read, 1997a), can be applied for phytoremediation purposes. We will further discuss this potential of mycorrhizal fungi for the remediation of metals and organic contaminants on the following pages.

Table 1 Overview of Phytoremediation Techniques (Ali et al., 2013; Vangronsveld et al., 2009)

Name	Description	Purpose
Rhizodegradation	The breakdown of organic contaminants in the rhizosphere by microorganisms	Site decontamination
Phytoextraction	The uptake of contaminants via plant roots and their translocation to accumulation in aboveground biomass	Site decontamination
Phytostabilization	The use of plants for stabilizing contaminants in contaminated soils	Site stabilization (risk reduction)
Phytodegradation	The degradation of organic contaminants by plants	Site decontamination
Phytovolatilization	The uptake of pollutants from soil by plants, their conversion to volatile form and subsequent release into the atmosphere	Contamination transfer (from soil to atmosphere)

3. PHYTOREMEDIATION OF METAL-CONTAMINATED SOILS USING MYCORRHIZAL PLANTS

Metals in the environment originate from both natural and anthropogenic sources. The most important natural sources of metal pollution are volcanic activity, erosion and the weathering of minerals (Ali, Khan, & Sajad, 2013; Fulekar, Singh, & Bhaduri, 2009; Wuana & Okieimen, 2011). Anthropogenic activities that lead to metal contamination include inappropriate waste disposal, mining, industrial activities, agricultural use of metals and metal-containing compounds, and so on (Ali et al., 2013; McGrath et al., 2001; Wuana & Okieimen, 2011). It is predominantly these anthropogenic activities that have affected large areas of land all over the globe. It has been estimated that in the European Union 3.5 million sites (Perez, 2012, pp. 2—22) and in America 600,000 sites (McKeehan, 2000) have potentially been contaminated with heavy metals. Furthermore one-sixth of the total agricultural land in China is polluted with heavy metals (Liu, Wu, & Banks, 2005). This widespread contamination of metals poses a serious risk for human health since some metals are mutagenic, teratogenic and carcinogenic (Ali et al., 2013; Mahar et al., 2016). Furthermore, metals can function as endocrine disruptors and some metals induce neurological and behavioural changes (Ali et al., 2013; Mahar et al., 2016). Given both the extent of the contamination and the risk to human health, there is an obvious need to remediate the contaminated sites. As described earlier a suitable remediation option for soil contamination is phytoremediation. However, metals are immutable and therefore several of the low-cost phytoremediation options that are available for the remediation of organic contamination, such as phytodegradation and rhizodegradation, are not applicable to metal-contaminated soils (Vangronsveld et al., 2009). Moreover, another phytoremediation technique, phytovolatilization can only be used for some metals like mercury (Hg) and selenium (Se) (Ali et al., 2013). Remediation options that remain are phytostabilization and phytoextraction (Vangronsveld et al., 2009). As stated previously, phytostabilization and phytoextraction require the use of plants. Consequently, the survival of plants on metalliferous soil is crucial for the success of a phytoremediation application. The selected plant species (or ecotypes) must thus be selected with care.

A group of plants considered for use in phytoremediation are the metallophytes (chapter: Metallophytes of Serpentine and Calamine Soils — Their Unique Ecophysiology and Potential for Phytoremediation by Wójcik et al., 2017). These plants have the ability to survive and reproduce on metal-polluted soils (Alford, Pilon-Smits, & Paschke, 2010;

Baker, Ernst, van der Ent, Malaisse, & Ginocchio, 2010). However, a lot of metallophytes are characterized by a slow growth and a small size (Ali et al., 2013; Cabral, Soares, Giachini, & Siqueira, 2015; Prasad, 2003), which is not desirable for a phytoremediation application. Some metallophytes can be further classified as metal hyperaccumulator plants (Alford et al., 2010). Hyperaccumulator plants can accumulate relative high amounts of metals, while their growth remains unaffected (Alford et al., 2010; Miransari, 2011), i.e., they can concentrate metals at levels that would be toxic to nonaccumulators. These plants can be used in phytoextraction applications (Alford et al., 2010). In most cases, hyperaccumulator plants are only able to accumulate one metal (Vangronsveld et al., 2009), while metal-polluted soils often contain a mixture of metals (Krznaric et al., 2010). This mixed pollution is not only challenging for phytoremediation with hyperaccumulator plants but also for metal-tolerant plants in general as plant metal tolerance mechanisms are usually metal specific. There is little evidence for cotolerance (tolerance to several different metals) (Hall, 2002; Schat, Llugany, & Bernhard, 2000). Another group of metallophytes, the metal excluders are also considered suitable for phytoremediation. These plants accumulate metals from the soil into their roots but restrict metal transport to other parts of the plant. Such plants cannot be applied for phytoextraction, instead they can be useful in phytostabilisation techniques (Sheoran, Sheoran, & Poonia, 2011). Other plants used for phytoextraction of metals are high biomass-producing nonhyperaccumulator plants (Vangronsveld et al., 2009). In comparison with hyperaccumulating plants, these plants accumulate a lower concentration of metals (Ali et al., 2013). Yet, they produce more aboveground biomass so that the absolute metal concentrations are similar to hyperaccumulators' concentrations (Ali et al., 2013). To summarize, there are a lot of different plant species and ecotypes mentioned in literature with the potential for use in phytoremediation (phytoextraction or phytostabilization). Nonetheless is it challenging to establish revegetation in contaminated areas and develop an efficient phytoremediation application (Cabral et al., 2015). This is extremely difficult for sites severely contaminated with heavy metals and sites contaminated with a mixed pollution.

Strategies to improve phytoremediation efficiency are the use of soil amendments (to increase or decrease metal availability) and the use of the plants' associated microorganisms (bacteria and mycorrhiza). Numerous studies report on the potential of plant-associated endophytic and rhizosphere bacteria for use in phytoremediation (chapters: Bio- and Phytoremediation of Pesticide Contaminated Environments: A Review by Eevers, White, Vangronsveld, & Weyens, 2017; The Bacterial and

Fungal Microbiota of Hyperaccumulator Plants: Small Organisms, Large Influence by Thijs, Langill, & Vangronsveld, 2017). It has been observed that plant-associated bacteria are capable to improve plant fitness on metalliferous soils (Ma, Rajkumar, Zhang, & Freitas, 2016; Santoyo, Moreno-Hagelsieb, Orozco-Mosqueda, & Glick, 2016; Vivas, Barea, Biro, & Azcon, 2006). Furthermore, there is also ample and clear evidence that mycorrhiza has the capacity to assist in phytoremediation techniques (Bhargava, Carmona, Bhargava, & Srivastava, 2012; Cabral et al., 2015; Colpaert, Wevers, Krznaric, & Adriaensen, 2011; Jentschke & Godbold, 2000; Meharg & Cairney, 2000). We will not discuss the use of amendments and plant-associated bacteria in detail, as it is beyond the scope of this chapter Instead, we will focus on the potential of mycorrhizal fungi, which are among soil microorganisms, the only ones functioning as a direct connection between soil and roots (Leyval, Turnau, & Haselwandter, 1997).

3.1 Mycorrhiza in Metalliferous Environments

It has been suggested that the diversity of mycorrhizal fungi found in metalliferous areas is in sharp contrast with the abundance of plant species present there (Macnair, 1993; Meharg, 2003). Data supporting this observation are usually case specific and vary depending on which metal(s) are present, the soil characteristics, the severity of the pollution, which type(s) of mycorrhizal fungi are investigated, which technique is used to analyze the fungal diversity in the soil and so on. This makes it challenging to estimate the ratio of plant: fungal species abundance in metal-polluted areas. Schneider, Bundschuh, and Araújo do Nascimento (2016) investigated a lead (Pb)-contaminated spodosol (a nutrient-poor, sandy soil) on AMF diversity and plant species richness. They reported that plant diversity was significantly correlated with fungal diversity in the Pb-contaminated area. In this area the fungal diversity was also higher than the plant diversity, there were at least two times more fungal species than plant species. A higher diversity of mycorrhizal fungi in metal-contaminated areas could imply that mycorrhizal fungi are inherently more capable in dealing with metals compared to vascular plants (Meharg, 2003). Khan, Kuek, Chaudhry, Khoo, and Hayes (2000) further suggested that plants colonizing metal-contaminated soils probably have a selective advantage if they are mycorrhized. This is supported by the fact that plants, traditionally viewed as nonmycorrhizal, become mycorrhized when colonizing soils subjected to major environmental perturbations (Colpaert, 2008; Wu, Isobe, & Ishii, 2004).

Mycorrhiza can assist in phytoremediation either by making metals more available for uptake by the plants or by reducing metal toxicity in their host plants (Bhargava et al., 2012; Cabral et al., 2015; Colpaert et al., 2011; Meharg & Cairney, 2000). However, the phytoremediation enhancement capacity is very variable depending on the fungal species and ecotype (Cabral et al., 2015; Colpaert et al., 2011; Krznaric, Verbruggen, Vangronsveld, & Colpaert, 2009; Orlowska, Przybylowicz, Orlowski, Turnau, & Mesjase-Przybylowicz, 2011; Toler, Morton, & Cumming, 2005). The fungal isolate selected for use in a phytostabilization project will therefore play a crucial role in the success of the application. Furthermore, selecting a fungal isolate for use in phytoremediation will also determine which plant species can be selected for the application. Evidently, to be able to exert a beneficial effect on phytoremediation efficiency, mycorrhizal fungi must first establish a mycorrhizal symbiosis. This in turn depends on the survival of the fungus itself (Colpaert, 2008). Therefore the ability of the fungus to survive in metal-contaminated areas is a prerequisite for its use in phytoremediation applications.

3.2 Mechanisms Involved in Metal Detoxification and Tolerance in Mycorrhizal Fungi

To guarantee their survival in metalliferous soils, mycorrhizal fungi can rely on various extracellular and intracellular defence mechanisms (Bellion, Courbot, Jacob, Blaudez, & Chalot, 2006; Cabral et al., 2015; Daghino, Martino, & Perotto, 2016). Extracellular mechanisms, such as chelation and cell wall binding, focus on the prevention of metal entry. While intracellular mechanisms including binding to nonprotein thiols and transport into intracellular compartments reduce the metal concentration in the cytosol (Bellion et al., 2006; Cabral et al., 2015; Daghino et al., 2016). Fig. 2 provides a schematic overview of these mechanisms. Additionally, there are also antioxidative detoxification systems that aim to counteract the accumulation of reactive oxygen species (ROS) and mechanisms that focus on the repair of metal-damaged biomolecules (Bellion et al., 2006; Cabral et al., 2015; Daghino et al., 2016; chapter: Molecular and Cellular Aspects of Contaminant Toxicity in Plants: The Importance of Sulphur and Associated Signalling Pathways by Hendrix, Schröder, Keunen, Huber, & Cuypers, 2017).

Chelation is a first defence mechanism of mycorrhizal fungi against toxic metal concentrations (Fig. 2). This mechanism is both metal and species dependent (Bellion et al., 2006; Meharg, 2003). A number of different organic molecules, excreted by fungi and able to chelate metals, are mentioned in literature (Table 2). In contrast, there are also a number of studies reporting the opposite metal solubilization from metal-containing compounds due to the exudation of organic molecules by mycorrhizal fungi

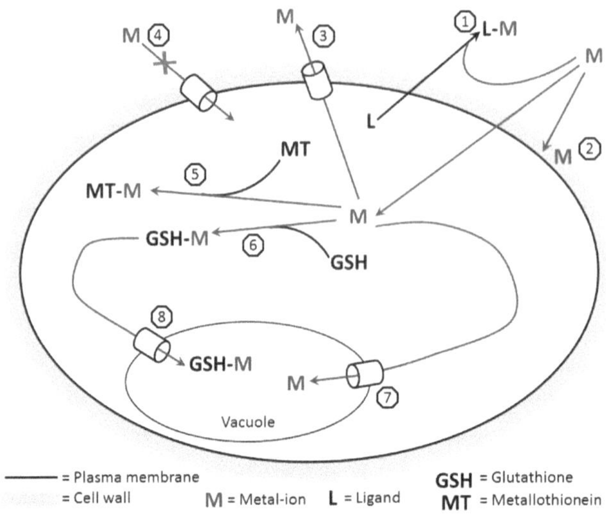

Figure 2 Schematic overview of the mechanisms involved in metal detoxification in mycorrhizal fungi. 1 = extracellular chelation by excreted ligands; 2 = biosorption of metals to the cell wall; 3 = enhanced efflux via transporters located at the plasma membrane; 4 = reduced uptake via transporters located at the plasma membrane; 5 and 6 = intracellular chelation (e.g., by metallothioneins, glutathione, organic acids, amino acids and compound-specific chaperones); 7 = subcellular compartmentation via transporters (into the vacuole or other internal cell compartments); 8 = vacuolar compartmentation of a complex (e.g., the GSH–M complex). *Adapted from Bellion, M., Courbot, M., Jacob, C., Blaudez, D., & Chalot, M. (2006). Extracellular and cellular mechanisms sustaining metal tolerance in ectomycorrhizal fungi.* FEMS Microbiology Letters, *254(2), 173–181.*

(Table 2). It can be concluded that although extracellular chelation of metals by excretion of ligands is an important defence strategy against metal toxicity, not all mycorrhizal fungi will apply this strategy (Table 2). Another extracellular mechanism preventing the entry of metals into the cell is cell wall binding or biosorption (Fig. 2) (Bellion et al., 2006; Gadd, 1993). In contrast with chelation of metals by excretion of organic molecules, this mechanism does not depend on the metabolic activity of the host (Bellion et al., 2006). It has been observed in a large number of mycorrhizal species (Table 2).

Even though the extracellular mechanisms described earlier can be effective in preventing metals from entering fungal cells, this is not always sufficient. Therefore, fungi require also intracellular metal detoxification mechanisms (Fig. 1). These intracellular mechanisms rely on transporter proteins and intracellular complexation by proteins (Bellion et al., 2006).

Table 2 Possible Mechanisms of Features Observed in Mycorrhizal Fungi Useful for the Phytoremediation of Metals

Feature Description	ECM (Ectomycorrhizal Fungi)	AMF (Arbuscular Mycorrhizal Fungi)	ERM (Ericoid Mycorrhizae)
Reduction of metal availability	Production of chelating agents (Bellion et al., 2006; Jentschke & Godbold, 2000)	Production of chelating agents (Cabral et al., 2015; Göhre & Paszkowski, 2006)	Production of chelating agents (Meharg, 2003)
Reduced metal transfer to plant host	Subcellular compartmentalization (Bellion et al., 2006; Blaudez, Botton, & Chalot, 2000; Colpaert et al., 2005; Jentschke & Godbold, 2000)	Subcellular compartmentalization (Cabral et al., 2015; Göhre & Paszkowski, 2006; Kaldorf, Kuhn, Schröder, Hildebrandt, & Bothe, 1999)	Subcellular compartmentalization (Khouja et al., 2013; Meharg, 2003)
	Binding, sorption, and incorporation to cell wall (Bellion et al., 2006; Brown & Wilkins, 1985; Colpaert et al., 2005; Galli, Schüepp, & Brunold, 1994; Wasserman, Mineo, Majumdar, & Vantyne, 1987)	Binding, sorption, and incorporation to cell wall (Cabral et al., 2015; Galli et al., 1994; Göhre & Paszkowski, 2006; Gonzalez-Chavez, D'Haen, Vangronsveld, & Dodd, 2004)	Binding, sorption, and incorporation to cell wall (Bradley et al., 1981)
	Active efflux and/or reduced influx (Bellion et al., 2006; Colpaert et al., 2005; Jentschke & Godbold, 2000)	Active efflux and/or reduced influx (Bruins, Kapil, & Oehme, 2000; Göhre & Paszkowski, 2006; Meier et al., 2012)	Active efflux and/or reduced influx (Khouja et al., 2013; Sharples et al., 2001)
	Intracellular chelation (e.g., metallothionein, glutathione) (Bellion et al., 2006; Colpaert et al., 2005)	Intracellular chelation (e.g., metallothionein, glutathione) (Cabral et al., 2015; Gonzalez-Guerrero, Cano, Azcon-Aguilar, & Ferrol, 2007; Lanfranco, Bolchi, Ross, Ottonello, & Bonfante, 2002; Meier et al., 2012)	Intracellular chelation (e.g., metallothionein, glutathione) (Vallino, Drogo, Abba, & Perotto, 2005)

(Continued)

Table 2 Possible Mechanisms of Features Observed in Mycorrhizal Fungi Useful for the Phytoremediation of Metals—cont'd

Feature Description	ECM (Ectomycorrhizal Fungi)	AMF (Arbuscular Mycorrhizal Fungi)	ERM (Ericoid Mycorrhizae)
	Metal transformation (e.g., oxidation, reduction, methylation) (Gadd, 1993)	Metal transformation (e.g., oxidation, reduction, methylation) (Gadd, 1993; Li et al., 2016; Meier et al., 2012)	Metal transformation (e.g., oxidation, reduction, methylation) (Gadd, 1993)
		Production and secretion of glomalin, which sequesters metals (Cabral 2015; Gonzalez-Chavez, Carrillo-Gonzalez, Wright, & Nichols, 2004; Göhre & Paszkowski, 2006; Khan, 2005)	
		Metal storage in arbuscules and vesicles (Cabral et al., 2015; Kaldorf et al., 1999; Whitfield, Richards, & Rimmer, 2004)	
		Metal retention in spores (Cabral et al., 2015; Gonzalez-Guerrero et al., 2001; Ferrol et al., 2009)	
Stimulation of plants' own metal defence mechanisms	Changing hormonal status of the plant (via production of phytohormones or by altering phytohormone levels in the plant) (Jentschke & Godbold, 2000; Vamerali, Bandiera, & Mosca, 2010)	Changing hormonal status of the plant (via production of phytohormones or by altering phytohormone levels in the plant) (Vamerali et al., 2010)	Changing hormonal status of the plant (via production of phytohormones or by altering phytohormone levels in the plant) (Vamerali et al., 2010)
		Increased production of glutathione and phytochelatins (Degola et al., 2015)	

Increased metal availability in the soil	Production of chelation agents (Fomina, Alexander, Colpaert, & Gadd, 2005; Sheoran et al., 2011; Vamerali et al., 2010)	Production of chelation agents (Vamerali et al., 2010; Sheoran et al., 2011)	Production of chelation agents (Fomina et al., 2005; Haselwandter et al., 1992; Sheoran et al., 2011; Vamerali et al., 2010)
	Metal transformations (Sheoran et al., 2011)	Metal transformations (Sheoran et al., 2011)	Metal transformations (Sheoran et al., 2011)
Increased metal concentration in plants		Passive or active transport from fungal cells to plant cells (Göhre & Paszkowski, 2006)	
Increased translocation from roots to shoots	Mediated by improved growth by improving the nutrient acquisition and water uptake (Sell et al., 2005; Vamerali et al., 2010)	Mediated by improved growth by improving the nutrient acquisition and water uptake (Kaldorf et al., 1999; Leyval et al., 1997; Usman & Mohamed, 2009; Vamerali et al., 2010; Wang et al., 2005)	Mediated by improved growth by improving the nutrient acquisition and water uptake (Vamerali et al., 2010)
Reduced metal toxicity in the plant			Extracellular enzyme release (superoxide dismutase) (Vallino et al., 2009)

Metal transporter proteins can alleviate metal stress by transporting metals out of the cell or into intracellular compartments (Fig. 1; Table 2). Furthermore, the uptake of metals via specific transporter systems located in the plasma membrane can be downregulated (Fig. 2; Table 2). Other proteins, such as metallothioneins or glutathione, are known to function as intracellular metal chelators (Table 2). Once chelated these metal complexes can be transported as well.

Besides these extracellular and intracellular metal-specific mechanisms that prevent metal toxicity, there are also 'secondary mechanisms' that do not deal with the metals directly but with their effects (Bellion et al., 2006). These mechanisms include various antioxidative and repair mechanisms (Bellion et al., 2006; Hall, 2002).

3.3 Beneficial Effects of Mycorrhization on Metal-Exposed Plants

There is a lot of evidence that suggests that mycorrhizal fungi are not only capable of surviving on metalliferous soils but also manage to improve the growth of their host plants in these harsh environments (Bhargava et al., 2012; Cabral et al., 2015; Colpaert et al., 2011; Jentschke & Godbold, 2000; Meharg & Cairney, 2000). However, the mechanisms responsible for these beneficial effects have proven difficult to resolve and are often correlating with the metal tolerance mechanism of the fungus itself. It is likely that the mechanisms involved are quite diverse given the observation that there is a large variation in response of mycorrhized plants to metals. This variation is observed between different fungal species, between different ecotypes of one fungal species and in response to different metals within a species or ecotype (Cabral et al., 2015; Colpaert et al., 2011; Krznaric et al., 2009; Orlowska et al., 2011).

In addition to plant growth promotion, it is observed that mycorrhizal fungi have other qualities beneficial for phytoremediation as well. These include the increase in metal availability in the soil, the increase in metal transfer from soil to roots and a higher translocation from roots to shoots. Although these effects are not beneficial for the plant, they may be quite useful for phytoextraction purposes (Göhre & Paszkowski, 2006; Jentschke & Godbold, 2000; Phieler, Merten, Roth, Büchel, & Kothe, 2015; Wang, Shi, Xu, Wang, & Li, 2013). We will further discuss these mycorrhizal properties and their possible mechanisms for ECM, AMF and ERM. We will focus on these three mycorrhizal types as they are known to play an important role in the survival of plants in harsh environments, such as metalliferous soils (Meharg, 2003). Therefore, they are the most promising to study for use in phytoremediation. General attributes of the mycorrhizal symbiosis to phytoremediation that are not specific for metals will not be discussed in detail. Examples of such general contributions to plant survival are the increased acquisition of water and nutrients, the increase in SOM, the prevention of predation by pathogenic fungi or nematodes and so on (Peterson et al., 2004). Although the useful effects of mycorrhiza on plants in metal-polluted environments have been described in several papers, the mechanisms behind these features are often poorly understood. An overview of these features is given in Table 2. Based on literature, possible mechanisms responsible for these effects are hypothesized.

3.4 Ectomycorrhiza and Their Role in the Phytoremediation of Heavy Metals

ECM are the dominant group of root symbionts of pioneer trees that attempt to colonize extremely contaminated sites (Colpaert, 2008). Even on severely metal-polluted sites the ECM symbiotic association persists, suggesting that the tree's adapted ECM symbiont helps the tree survive toxic effects of metal pollution (Colpaert, 2008; Wilkinson & Dickinson, 1995). Although woody plants are often not considered as primary colonizers of metalliferous soils (Schat et al., 2000), it has been observed that on a number of contaminated sites pioneer tree species are able to establish small mycorrhized pioneer populations. Pioneer species, for which this has been observed, include birches, willows, poplars and pines (Adriaensen, Vrålstad, Noben, Vangronsveld, & Colpaert, 2005; Colpaert, Muller, Lambaerts, Adriaensen, & Vangronsveld, 2004; Vrålstad, Myhre, & Schumacher, 2002). These observations illustrate the potential of ECM for primary colonization and subsequently for phytoremediation.

An example of such an ECM that can be useful for phytostabilization purposes is *Suillus bovinus*. Adriaensen, van der Lelie, Van Laere, Vangronsveld, and Colpaert (2003) studied a zinc (Zn)-tolerant *S. bovinus* ecotype and observed that is extremely effective in protecting its host *Pinus sylvestris* from Zn stress and improving its nutrient acquisition in a pot experiment. Fungal isolates of the Zn-tolerant ecotype were isolated from polluted soils and were compared with Zn-sensitive isolates originating from a nonpolluted control area. It was concluded that association with a Zn-adapted ecotype was probably a key strategy for the survival of pine trees colonizing Zn-contaminated soils. The mechanism responsible for this Zn tolerance trait was further investigated by Ruytinx et al. (2013), who established that this adaptive tolerance to Zn was the result of an increased export of Zn out of the fungal cells. The increased export was also responsible for a lower Zn influx in the host plant, hereby protecting the host plant from Zn stress. A study by Baum, Hrynkiewicz, Leinweber, and Meißner (2006) reports on the results from a pot experiment with *Salix x dasyclados* inoculated with *Paxillus involutus*. Two metal-contaminated soils and two *P. involutus* strains were tested. The two soils were contaminated with different metal concentrations and had a different metal extractability. One *P. involutus* strain was isolated from a metal-contaminated site, while the other experimental strain originated from a noncontaminated area. Inoculation with both fungal strains had a positive effect on plant growth and biomass production. Yet, differences in metal content in the plants were observed depending on both the experimental soil and the *P. involutus* strain that was tested. Only the *P. involutus* strain isolated from the metal-contaminated area was able to increase the extractability of cadmium (Cd). It was concluded that inoculation with *P. involutus* can be used as a strategy to increase the phytoextraction of metals. However, it should be kept in mind that the beneficial effect of inoculation will vary depending on the metal extractability of the soil, the fungal strain and probably a number of other factors as well. In a similar experiment, Sell, Kayser, Schulin, and Brunner (2005) inoculated *Salix viminalis* and *Populus canadensis* with three well-known ECM fungi, *Hebeloma crustuliniforme*, *P. involutus* and *Pisolithus tinctorius*. They found a highly significant increase in uptake and translocation of Cd from a sterilized Cd-contaminated soil for the association of *P. Canadensis* with *P. involutus*. Compared to the control, the Cd concentration in inoculated poplar increased nearly 100% and also the translocation from the roots to the leaves was significantly enhanced. Similar effects were not observed in association with *S. viminalis* or by inoculation with the other fungi. Therefore, it can be concluded that also the choice of the host plant is a determinant of the beneficial effect of ECM inoculation. Another pot experiment conducted by Dos Santos Utmazian et al. (2007) investigated Cd and Zn accumulation in the metal-accumulating willow species (*Salix smithiana* and *Salix caprea*). Plants were grown on metal-contaminated soil and inoculated with *Cadophora finlandica*. No beneficial effect of the mycorrhization was observed.

These examples of studies focussing on ECM and their potential to facilitate phytostabilization or phytoextraction clearly illustrate the importance of carefully selecting a specific fungal strain and plant clone. Testing another fungal strain or plant clone of the same species can have a very different result. Additionally, it is also important to use soil originating from the contaminated area one wants to remediate in experiments, because the specific soil characteristics (not only the metal concentration) will influence the effect of the inoculation. Furthermore, it will make data from pot experiments more translatable to field experiments.

3.5 Arbuscular Mycorrhizal Fungi and Their Role in Phytoremediation of Heavy Metals

AMF are ubiquitous and form symbiotic associations with the majority of terrestrial plants (Leung, Wu, Cheung, Ye, & Wong, 2010). Consequently, they represent an important part of the soil microbiome (Hildebrandt, Regvar, & Bothe, 2007). This omnipresence even persists in highly degraded sites (Gaur & Adholeya, 2004; Khan, 2005; Schneider, Stürmer, Guilherme, de Souza Moreira, & Soares, 2013; Turnau, Ryszka, Gianinazzi-Pearson, & van Tuinen, 2001) and numerous studies indicate that AMF are able to reduce metal toxicity in their host plants (Cabral et al., 2015; Meier, Borie, Bolan, & Cornejo, 2012). AMF are therefore considered to be the most important type of mycorrhiza for phytoremediation applications. Cabral et al. (2015, review article) already discussed the mechanisms and major benefits of the application of AMF for the phytoremediation of metals. Furthermore, they provided a summary of field studies investigating these benefits of AMF. Therefore, in the next paragraphs, we will only discuss examples of more recent studies investigating the potential of AMF inoculation for phytoremediation of metals.

Hristozkova, Geneva, Stancheva, Boychinova, and Djonovab (2016) recently published a paper describing their findings on the contributions of two *Claroideoglomus claroideum* isolates and one isolate for the attenuation of Cd and Pb exposure on *Calendula officinalis* development. All fungal isolates were originating from soils highly contaminated with metals. In the pot experiment, no Cd or Pb was detected in marigold flowers of the mycorrhized plants, while the nonmycorrhized plants had a much lower flower weight and higher Cd and Pb levels. The three tested mycorrhizal strains also stimulated the accumulation of important secondary metabolites in pot marigold flowers and, consequently, enhanced the plant's antioxidant capacity. Pot marigold is a valuable medicinal crop, due to its biologically active compounds and therefore this study not only demonstrated the potential of AMF for phytostabilization but also for contributing to pot marigold bioactive compounds synthesis. Another study focussing on the effects of AMF inoculation on Cd uptake was carried out by Li et al. (2016).

Oryza sativa plants were inoculated with *Rhizophagus intraradices* or *Glomus mosseae* (*Funneliformis mosseae*). A short-term hydroponic experiment was performed in which the plants were exposed to six Cd concentrations. Results indicate that AMF inoculation significantly decreased Cd concentrations in shoots and roots of Cd-exposed plants. This decrease in Cd was significantly higher in the roots of plants inoculated with *G. mosseae* compared to plants inoculated with *R. intraradices*. Furthermore, AMF colonization reduced the concentrations and relative abundance of Cd in inorganic and water-soluble form in shoots and roots of plants exposed to high Cd concentrations. This implies that Cd was converted into inactive, less toxic forms by the AMF. Moreover, the subcellular distribution of Cd was investigated. Results showed that in inoculated plants exposed to high levels of Cd, the concentrations of Cd in the cell wall fractions were significantly reduced. It was concluded that (1) AMF can enhance *O. sativa* resistance to Cd through reduced Cd uptake, transformation of Cd to inactive chemical forms and modification of subcellular Cd distribution and that (2) the choice of AMF species will determine the effect of the AMF inoculation. Conversely, Sheikh-Assadi, Khandam-Mirkohi, Alemardan, and Moreno-Jiménez (2015) reported a higher accumulation of Pb and Cd in roots of AMF inoculated *Limonium sinuatum* in a pot experiment. Plants were inoculated with a mixture of *G. mosseae* and *G. intraradices* and exposed to different Cd and Pb concentrations. *L. sinuatum* accumulated Cd and Pb in the roots and translocated very little to the shoots. Total Cd and total Pb accumulated in the roots was nearly two and three times higher in AMF-inoculated plants compared to nonmycorrhized plants. Furthermore, it was observed that mycorrhized plants had a higher metal tolerance. Therefore, it was concluded that mycorrhization alleviated metal toxicity in the plants and that *L. sinuatum* inoculated with AMF could be useful as a Cd or Pb controlling agent for phytoremediation.

These two examples clearly illustrate that inoculating different plant species with the same AMF species and exposing them to the same metal will not always have the same effect. In these studies, plants were exposed to Cd and inoculated with *G. mosseae*. Cd uptake was reduced in *O. sativa*, while Cd accumulation was promoted in roots of *L. sinuatum*. Another factor that can influence the effect of AMF inoculation on plants exposed to metals is the sex of the plant. This was described by Chen et al. (2015). In this study the effect of *G. mosseae* inoculation was investigated in *Populus cathayana* male and female plants exposed to Pb. Although symbiosis with *G. mosseae* did not significantly affect the growth and biomass allocation of both sexes, male and female plants did absorb more phosphorus (P) in the roots. It was observed that AMF inoculation promoted Pb uptake and accumulation in female plant roots but not in the roots of males. This increased Pb accumulation in females had no obvious negative effect on plant growth. Consequently, it was concluded that inoculated female *P. cathayana* plants are more suited for Pb phytoextraction than male plants.

Finally, it is also important to mention that not all studies, which investigate the effect of AMF inoculation on plants exposed to metals, report a significant beneficial effect of AMF symbiosis. Doubková and Sudová (2016) investigated the effect of *G. intraradices* inoculation on *Agrostis capillaris* exposed to a polymetallic contamination. Certain positive effects of AMF inoculation on their host plants were observed. These effects include an increase in root P concentrations and a more efficient photosynthetic apparatus in mycorrhized plants. However, these beneficial effects were not translated in an improved plant growth. Therefore, it was concluded that there are better alternatives for the revegetation of metal-contaminated sites than the use of *A. capillaris* inoculated with *G. intraradices*.

To conclude, there is a lot of literature available on the effects of AMF inoculation on plants exposed to metals. Although not all studies report beneficial effects of inoculation, there is a clear consensus that at least some AMF—plant symbioses are useful for the development of new phytoremediation strategies.

3.6 Ericoid Mycorrhiza and Their Role in Phytoremediation of Heavy Metals

ERM are especially suited for the remediation of metal-polluted, acidic, nutrient-poor soils, as members of the Ericales (plants associated with this type of mycorrhiza) are dominant in these unhospitable environments. Acidic soil conditions increase the availability of potentially toxic metals, which makes it even more challenging for organisms to survive and reproduce in these areas (Daghino et al., 2016). Although metal tolerance has been observed for plant species and ecotypes in the Ericales, their survival in these polluted soils has been mainly attributed to their association with ERM (Daghino et al., 2016). It was already reported in 1981 by Bradley et al. that ERM are able to confer metal resistance upon their host plants. It was observed that inoculation of *Calluna vulgaris* with copper (Cu)-tolerant *Hymenoscyphus ericae* (=*Pezizella ericae*) isolates had a positive effect on plant growth in the presence of Cu and Zn. It was concluded that *C. vulgaris* mycobionts confer Cu and Zn resistance in their host plants

(Bradley, Burt, & Read, 1981, 1982). Both Cu and Zn resistances are constitutive in ERM of *C. vulgaris* isolated from mining and uncontaminated sites. There is no evidence of evolution of Cu and Zn resistance in *H. ericae* (Bradley et al., 1981, 1982). Sharples, Meharg, Chambers, and Cairney (2000, 2001) further investigated this metal resistance in *H. ericae*. It was again confirmed that the resistance to Cu is constitutive. Additionally, also the resistance to arsenate was investigated and the results indicate adaptive rather than constitutive resistance to arsenate. The mechanism responsible for this arsenate resistance was investigated. Results suggest that intracellular arsenate ($H_2AsO_4^-$) is reduced to arsenous acid (H_3AsO_3) and actively transported out of the cells. This arsenate resistance mechanism is also beneficial for the plant, as the fungus acts as a filter, ensuring low-plant arsenic (As) levels, while maintaining a sufficient supply of P to the plant. Furthermore, efflux of H_3AsO_3 also limits the reabsorption of As from the soil. This research of Bradley et al. (1981, 1982) and Sharples et al. (2000, 2001) illustrates that ERM fungi can have constitutive resistance or develop adaptive tolerance to specific metals. However, whether the fungus has a constitutive resistance or an adaptive tolerance will not determine if the inoculation has a beneficial effect on the host plant. The major determinant of the effect of fungal inoculation on the host seems to be the mechanism responsible for the tolerance.

Gibson and Mitchell (2004) investigated two isolates of *H. ericae* and two dark ericoid mycobionts. It was demonstrated that these ERM isolates were capable to solubilize certain metal phosphates, including Zn phosphate and calcium (Ca) phosphate. This suggests that these fungi have the potential for use in the phytoremediation of sites contaminated with high levels of metal-bound phosphate. Not only by increasing the metal availability but also by improving the P nutritional status of host plants with metal phosphates as sole P sources. Another example of an ERM fungus capable to survive in metalliferous soil was described by Martino, Perotto, Parsons, and Gadd (2003). They examined the capacity of *Oidiodendron maius* to mobilize insoluble Zn oxide and Zn phosphate. This was analyzed for isolates originating from Zn-polluted and nonpolluted sites. It was observed that isolates from polluted sites mobilized less Zn in comparison with isolates from nonpolluted sites. The enhanced solubilization observed in the isolates from nonpolluted sites was achieved through increased excretion of citric and malic acid. The production of these Zn-chelating acids was also induced when Zn was present. An enhanced production of citric and malic acid is undesirable in polluted soils and mine spoils because most of the metals there are present in insoluble forms. Therefore, it can be concluded that the isolates from the polluted sites produce less chelating acids and that this is part of their survival strategy for growth in metal-contaminated areas. Obviously, such a lower metal bioavailability will also benefit plants growing on these metalliferous soils.

To conclude, there are a number of studies illustrating the potential of ERM fungi for phytoremediation. At present, *H. ericae* and *O. maius* remain the most studied ERM in respect to metal tolerance. More studies are needed to further explore the potential of ERM fungi and to expand the research to other ERM species as well.

3.7 Perspectives and Recommendations for Future Studies Investigating the Use of Mycorrhiza for the Phytoremediation of Heavy Metals

At present, the view that mycorrhizae have the capacity to increase the efficiency of phytoremediation is generally accepted. The last two decades great progress has been made in our understanding of mycorrhizal fungi and their potential role in the phytoremediation of metals. However, there is still a lot of work to do. First of all, the mechanisms involved in these beneficial effects remain largely unknown. Secondly, despite the fact that a considerable amount of literature is available on metal tolerance in plants, this is not yet the case for mycorrhizal fungi. Therefore not only the mechanisms in known tolerant fungi should be studied further but also the search for other tolerant mycorrhizae should continue. As indicated by Alford et al. (2010), in 2010 only 10% of the known hyperaccumulator plant species have had their rhizosphere examined and most of this research has focused on rhizospheric bacteria, overlooking potentially interesting mycorrhizal fungi. It is clear that by improving our understanding of the rhizosphere and its unique microorganisms in polluted areas, we will be able to develop more and better strategies for phytoremediation. Furthermore, this improved understanding of the rhizosphere will also help translating results from laboratory experiments to field applications (Wenzel, 2009).

For future studies focussing on these matters, it is advisable to include enough fungal isolates in their experiments to compensate for the large phenotypic diversity that can be present within one mycorrhizal species. Additionally, these studies of organisms adapted to contaminated sites will enable researchers to gain insights into microevolutionary processes.

Another concept that is also worthwhile to further explore is the application of mycorrhizal helper bacteria for mycorrhiza-assisted phytoremediation. It has been reported that there are a lot of bacterial strains capable to promote AMF and ECM symbioses (Frey-Klett, Garbay, & Tarkka, 2007; Garbaye, 1994; Khan, 2005). Logically, they have the potential to be applied for phytoremediation purposes. Finally, it should also be considered that tolerance mechanisms present in plants and mycorrhizal fungi might not always function independently. An idea that was already suggested by Meharg (2003).

While AMF are traditionally considered the most important fungi for phytoremediation, more and more research is illustrating the potential of ECM and ERM for phytoremediation as well. The type of mycorrhiza that can be selected for a phytoremediation application will therefore depend predominantly on the plant species that is used. Though it should be kept in mind that mycorrhizal fungi cannot be regarded as a sole solution for the detoxification of contaminated areas but as a phytoremediation enhancement strategy.

4. PHYTOREMEDIATION OF ORGANIC CONTAMINANTS USING MYCORRHIZAL PLANTS

4.1 Characteristics of Soil Organic Contamination

Soil organic contaminants were in 2006 the most frequent pollutant in Europe. In total 59% of the sites reported as contaminated by the European Environment Agency were so by an organic pollutant, even more than for heavy metals (37%) (Valentin, Nousiainen, & Mikkonen, 2013). Organic contamination covers a wide diversity of molecules that can be classified in different families based on their chemical composition: oil hydrocarbons (alkanes and alkenes), chlorinated compounds (polychlorinated biophenyls or PCBs), monomeric aromatic hydrocarbons (such as benzene, toluene, ethylbenzene, xylene referred to as BTEX), polycyclic aromatic hydrocarbons (PAH) and nitroaromatics [such as 2,4,6-trinitrotoluene (TNT), atrazine or some pesticides] (Valentin et al., 2013). Some sites can be contaminated with several organic pollutants at once: for example, both soil hydrocarbons and PAH come from fuel and often co-occur in polluted sites. Organic pollutants and heavy metals are as well very often found together in polluted soils (40% of hazardous waste sites on the

Environmental Protection Agency's National Priority are co-contaminations between heavy metals and organics) (Olaniran, Balgobind, & Pillay, 2013).

Contrary to heavy metals, however, organic contaminants can be chemically degraded by soil organisms. Therefore, phytodegradation has been extensively studied in the last 40 years, with lots of focus on soil microorganisms (Van Hamme, Singh, & Ward, 2003). Mycorrhizas are no exception. A synthesis of the literature on the subject shows that, among the 80 articles that included both mycorrhizas and the name of an organic contaminant, 35% focused on PAHs, 23% on petroleum hydrocarbons, 19% on oil and diesel, 10% on PCBs, 8% on nitroaromatics, 4% on monoaromatics and 6% on other types (lignite or creosote) (Table 3).

Mycorrhizas offer two theoretical advantages for phytoremediation over bacteria and saprophytic fungi. Contrary to bacteria, which are the most studied in phytoremediation, they use oxidative mechanisms that are induced by nutrient limitation, and which do not require a preexposition to the pollutant (Barr & Aust, 1994). Saprophytic fungi are limited by the C availability, and therefore often are limited to the topsoil, where litter still contains enough easy C sources. Mycorrhizal fungi, however, are provided with energy from the plant photosynthates and are therefore active at greater soil depths (Lindahl et al., 2007). Moreover, saprophytic fungi are usually difficult to cultivate and have a limited half-life in soil (White, Ross, Gent, Eitzer, & Mattina, 2006). Nevertheless, the results of mycorrhizal inoculation in phytodegradation trials in growth chambers and in the field had very variable outcomes from a slowdown to an acceleration of the process. Hence the aim of the following paragraphs is (1) to describe the organic contaminants and the reasons for their toxicity, (2) to summarize the effects of organic contaminants on mycorrhizal communities, (3) to provide a synthesis on the mycorrhizal mechanisms beneficial for phytoremediation, and (4) to quantify the mean and variation of the effect of mycorrhizas on phytoremediation.

4.2 The Organic Contaminants

As of 2014, 2.5 million sites were potentially polluted in Europe, of which 340.000 were highly likely contaminated (EEA, 2015). Among those, mineral oil was the most common organic contaminant (23%), then PAH (10%), monoaromatic hydrocarbons (10%), chlorinated hydrocarbons (8%) or other sources (cyanides, explosives… 16%) (EEA, 2015). All these

Table 3 Summary of the Results of a Meta-analysis of 34 Papers Dealing With Remediation of Organics by Mycorrhizal Fungi

Citation	Interactions With Bacteria, Yeast, Fungi or Earthworms	Effects of Pollution on Mycorrhizal Performance	Effects of Pollution on Mycorrhizal Functioning	Effects of Pollution on Mycorrhizal Communities	AMF	ECM	ERM	Petroleum Hydrocarbons	Oil	Diesel	PAHs	Nitroaromatics	Monoaromatics	Pesticides and PCBs	Other
Alarcón et al. (2008)	1	1			1			1							
Antibus and Linkins (1978)				1		1			1						
Aranda et al. (2013)			1		1						1				
Bell et al. (2014)	1			1		1		1							
Binet, Portal, and Leyval (2000)		1			1						1				
Cabello (1999)		1			1				1						
Cairney and Meharg (2003)															
Calonne et al. (2014)			1		1						1				
Chatarpaul, Chakravarty, and Subramaniam (1989)		1				1								1	
Chiapusio et al. (2007)	1	1		1	1						1				
Corgié et al. (2003)	1		1		1						1				
Corgié, Fons, Beguiristain, and Leyval (2006)	1		1		1						1				
Criquet et al. (2000)			1		1						1				
Danielson and Visser (1989)		1		1		1			1						
De Paula, De Siqueira, and Soares (2007)					1										1
Dong, Gu, Guo, Xun, and Liu (2014)	1	1			1			1							
Driai et al. (2015)			1		1					1					

Xun, Xie, Liu, and Guo (2015)	1	1		1		1					
Franco-Ramírez, Ferrera-Cerrato, Varela-Fregoso, Pérez-Moreno, and Alarcón (2007)			1	1			1				
Gao et al. (2010)			1	1					1		
Gao, Li, Ling, and Zhu (2011)				1					1		
Gao et al. (2014)		1		1		1					
Gaspar, Cabello, Cazau, and Pollero (2002)	1			1					1		
Genney et al. (2004)			1		1				1		
Gunderson, Knight, and Van Rees (2007)		1			1			1			
Heinonsalo, Jørgensen, Haahtela, and Sen (2000)	1				1	1					
Hernández-Ortega et al. (2012)		1		1				1			
Huang and Wang (2013)			1		1					1	
Huang, Zhao, and Luan (2007)			1		1					1	
Ibáñez et al. (2011)			1	1							1
Iffis et al. (2014)	1			1		1					
Joner and Leyval (2001)				1					1		
Joner and Leyval (2003)				1					1		
Joner et al. (2001)	1		1	1					1		
Joner et al. (2006)					1				1		

(*Continued*)

Table 3 Summary of the Results of a Meta-analysis of 34 Papers Dealing With Remediation of Organics by Mycorrhizal Fungi—cont'd

Citation	Interactions With Bacteria, Yeast, Fungi or Earthworms	Effects of Pollution on Mycorrhizal Performance	Effects of Pollution on Mycorrhizal Functioning	Effects of Pollution on Mycorrhizal Communities	AMF	ECM	ERM	Petroleum Hydrocarbons	Oil	Diesel	PAHs	Nitroaromatics	Monoaromatics	Pesticides and PCBs	Other
Khalvati, Bartha, Dupigny, and Schröder (2009)		1	1		1								1		
Kirk, Moutoglis, Klironomos, Lee, and Trevors (2005)			1		1			1							
Koivula, Salkinoja-Salonen, Peltola, and Romantschuk (2004)			1			1					1				
Kuo, Juang, Yang, Kuo, and Wu (2014)		1			1			1							
Laatikainen and Heinonen-Tanski (2002)						1								1	
Langer, Syafruddin, Steinkellner, Puschenreiter, and Wenzel, (2010)		1			1				1						
Linkins and Antibus (1978)			1			1			1						
Lu and Lu (2015)			1		1						1				
Lu, Lu, Peng, Wan, and Liao (2014)	1				1									1	
Lumini et al. (1994)	1	1			1										1

Małachowska-Jutsz and Kalka (2010)	1				1			1							
Małachowska-Jutsz, Rudek, and Janosz (2011)	1				1					1					
Meharg and Cairney (2000)						1		1	1	1	1	1	1		1
Meharg, Dennis, Cairney (1997)						1						1			
Nicolotti and Egli (1998)		1	1	1		1			1						
Nie et al. (2011)		1	1	1	1			1							
Noori, Maivan, and Alaie (2014)		1			1			1							
Oliveira, Castro, Dodd, and Vosatka (2005)	1	1	1		1										1
Powell et al. (2009)					1									1	
Rabie (2004)	1	1	1		1						1				
Redondo-Gómez, Petenello, and Feldman (2014)		1			1					1					
Richter, Warner, and Stephens (2003)						1									1
Robertson, Kennedy, Massicotte, and Rutherford (2010)	1			1		1	1				1				
Robertson, Rutherford, and Massicotte (2011)				1		1	1	1							
Robson, Diane Knight, Farrell, and Germida (2004)		1						1							

(Continued)

Table 3 Summary of the Results of a Meta-analysis of 34 Papers Dealing With Remediation of Organics by Mycorrhizal Fungi—cont'd

Citation	Interactions With Bacteria, Yeast, Fungi or Earthworms	Effects of Pollution on Mycorrhizal Performance	Effects of Pollution on Mycorrhizal Functioning	Effects of Pollution on Mycorrhizal Communities	AMF	ECM	ERM	Petroleum Hydrocarbons	Oil	Diesel	PAHs	Nitroaromatics	Monoaromatics	Pesticides and PCBs	Other
Saparrat, Balatti, Arambarri, and Martínez (2014)															
Sarand et al. (1998)	1					1		1							
Savin, Purcell, Daigh, and Manfredin (2006, pp. 49–51)					1									1	
Schnabel and White (2001)					1									1	
Tang, Chen, Huang, and Tian (2009)		1			1					1					
Teng et al. (2010)	1	1			1									1	
Thion et al. (2012)				1	1						1				
Trappe, Molina, and Castellano (1984)					1									1	
Verdin et al. (2006)		1	1		1						1				
White et al. (2006)					1							1			
Wu, Jacobson, and Hausner (2008)	1				1						1				
Wu, Zhang, Huang, and Christie (2008)	1		1		1							1			
Wu, Zhang, Huang, Shan, et al. (2008)					1						1				
Wu et al. (2011)		1			1						1				

Xiao, Chen, Si and Wu (2012)		1	1		1						1				
Xun et al. (2015)	1	1	1		1			1							
Yergeau et al. (2015)	1	1		1		1		1							
Yu et al. (2011)	1	1	1		1										
Zhou, Cébron, Béguiristain, and Leyval (2009)			1		1						1				
Zhou, Zhou, and Xiang (2013)		1			1						1				
Total	25	31	28	10	56	21	2	18	8	7	28	6	3	8	5
Percentage of number of studies	31%	39%	35%	13%	70%	26%	3%	23%	10%	9%	35%	8%	4%	10%	6%

Effect of AMF (arbuscular mycorrhizal fungi) on plant biomass was calculated as (mycorrhizal plant weight−nonmycorrhizal plant weight)/nonmycorrhizal plant weight. Effect of AMF on pollutant removal was calculated as (pollutant concentration in the mycorrhized treatment−pollutant concentration in the nonmycorrhized treatment)/pollutant concentration in the mycorrhized treatment. Plant weight was either total dry weight (DW), shoot DW or total FW. In case of multiple inoculation (e.g., bacteria + AMF treatments), we compared plant + other inoculant with plant + other inoculant + AMF.

ECM, ectomycorrhizal fungi; *ERM*, ericoid mycorrhizae; *PAHs*, polycyclic aromatic hydrocarbons.

pollutants can be sorted in four families based on their chemical structure. Aliphatics are linear hydrocarbons found in oils and diesel. They all have a linear, aliphatic structure but present different degrees of unsaturation. They have many similar properties with lipids. Their rate of degradation is negatively correlated with the length of the aliphatic chain, as well as the number of double bonds (Robertson, McGill, Massicotte, & Rutherford, 2007). They represent up to 25% of crude oil composition in mass (Robertson et al., 2007). Because they involve relatively low energy binding, aliphatic compounds are usually degraded first by microorganisms. PAHs are neutral, nonpolar molecules coming from the combustion of organic polymers (fuel, wood...) and are therefore ubiquitous. They are composed of three to ten aromatic rings, with most of the PAHs having three to six of them arranged in different patterns. Because of the high stability of the aromatic rings and their insolubility, they are very recalcitrant to degradation. Hence, as a general rule, larger PAHs are harder to degrade and more toxic than small ones. On top of that, PAH degradation is influenced by other factors, such as production of toxic or dead-end metabolites, metabolite repression and the lack of cometabolic or inducer substrates (Van Hamme et al., 2003). The most widespread PAHs are naphthalene, benzo[a]pyrene or phenanthrene. Monoaromatic hydrocarbons are, as their name suggests, composed of one aromatic ring, with or without substitutions. They are naturally found in the vicinity of natural gas and petroleum deposits and are detected as well in the fumes of forest fires (Valentin et al., 2013). Well-known examples of this class are BTEX, phenol or aniline. Nitroaromatics are aromatic groups with one or more nitro substitutions. They are found in explosives such as TNT or 2-nitrophenol but also in many pesticides (parathion) and herbicides (fluorodiphen) (Ju & Parales, 2010). They are often present as spills in explosive manufacturing, testing and decommissioning sites (Thijs et al., 2014). Their catabolism is analogous to the general oxidative pathways of aromatic hydrocarbons but with extra steps to accommodate the nitro group (Ju & Parales, 2010). Organochlorides are organic molecules with at least one covalent binding with a chlorine atom. They are mostly found in pesticides (e.g., dichlorodiphenyltrichloroethane, DDT), insulators (PCBs) or also dioxins. Though they have been a major environmental issue in the 1990s, they appear to become less of a concern for soil pollution recently (EEA, 2015).

All of these organic contaminants are toxic to living organisms through an unspecific mode of action, which involves partitioning into the cell membrane, thereby disrupting its integrity and increasing its permeability

(Robertson et al., 2007). The cell wall therefore plays an important role in the protection of membranes from the contaminant, and fungi could be more tolerant than bacteria to higher organic pollutant concentrations thanks to the specific structure of their cell wall (Blakely, Neher, & Spongberg, 2002). On top of this direct toxic effect, organic contaminants also have indirect effects through the disturbance of soil properties and plant–fungus interactions. Indeed, because of their high hydrophobicity, these molecules will alter the availability of water, and consequently soluble nutrients and oxygen. It has been show that hydrocarbon pollution reduces water porosity (filling fine water retention pores), which causes faster water-logging, followed by faster oxygen depletion and retarded decomposition by oxidative enzymes (Trofimov & Rozanova, 2003). The organic pollutant may as well sequester to SOM, which decreases its bioavailability (Robertson et al., 2007). Finally, as the pollutant ages in the soil, it gets harder and harder to degrade, either through sorption to organic matter or by incorporation to the organic matter by the unwanted action of soil enzymes (Robertson et al., 2007).

4.3 Effects of Organic Contaminants on Mycorrhizal Communities

Organic contamination can be considered as a disturbance, because of its direct toxic effects and of its indirect lowering effect on water and nutrient availability for the soil organisms. Hence, it should theoretically select tolerant species in the community and eliminate the other ones. Empirical data show that most of the time organic contamination indeed leads to a decrease in the diversity of the mycorrhizal community. However, even though we know the consequences of organic pollution on soil communities relatively well, surprisingly few studies addressed their influence on mycorrhizal ones. We can however distinguish some general trends out of this limited amount of literature. As a general rule, in vitro abilities of the mycorrhizal fungus to tolerate the contaminant are correlated with field abundance in contaminated sites. It was, for example, found that pioneer species (such as the ones belonging to the genus *Laccaria*) were resistant to hydrocarbon contamination, while more late-stage species (*Amanita, Tricholoma*) were sensitive (Nicolotti & Egli, 1998). In another study, another ECM pioneer species *Thelephora terrestris* proved to be the best inoculant to improve pine growth on petroleum-contaminated tailings (Danielson & Visser, 1989). However, the inoculant was quickly replaced by local mycorrhizal communities, which were dominated by ascomycetes (Danielson & Visser, 1989).

More precise trends were observed by Bell et al. (2014), who found that both Agaricomycetes and Pezizomycetes (the latter being mostly represented by *Sphaerosporella brunnea*, a pioneer ECM species) had higher relative abundance in highly hydrocarbon-contaminated treatments, while Dothideomycetes were very negatively affected. In the longer term, however, the pollutant does not appear to be a factor of structuration of the fungal community anymore (Thion, Cébron, Beguiristain, & Leyval, 2012).

Organic contamination does not affect only mycorrhizal community structure but also mycorrhization levels, with higher pollution reducing colonization. This has been observed in both AMF (Chiapusio, Pujol, Toussaint, Badot, & Binet, 2007) and ECM (Antibus & Linkins, 1978). Usually the immediate effect of the organic pollutant on colonization decreases with time (Antibus & Linkins, 1978).

4.4 Mechanisms Contributing to the Phytoremediation of Organic Pollutants

Mycorrhizal fungi contribute to the phytoremediation of organics thanks to different mechanisms. The first one is a direct degradation of the xenobiotic. This is done in several steps, most of them catalyzed by enzymes. As observed with bacteria, the first one consists of increasing contaminant solubility, which can be done through the secretion of surfactants (Van Hamme et al., 2003). Production of surfactant in mycorrhizal fungi has not been investigated in relation with pollution, but we know that AMF produce large quantities of amphiphilic peptides, called glomalins, which chemically function as surfactants (Rillig & Mummey, 2006). Among basidiomycete ECMs, hydrophobins and other proteins called repellents have also the potential to behave as surfactants (Kershaw & Talbot, 1998). Influence of both glomalins and hydrophobins on organic pollutant degradation has however never been estimated. The second step depends on the chemical family of the pollutant. Aliphatics are structurally related to lipids and are therefore suspected to be degraded using lipases (Margesin, Zimmerbauer, & Schinner, 1999, pp. 859–863). Indeed Li, Huang, and Wei (2008) report that the mycorrhizal species *Laccaria bicolor* could degrade up to 40% of diesel in vitro, and that the degraded compounds were mostly 16–22C alkenes. PAHs, which are structurally similar to lignin and soil polyphenols, are degraded with the pathways inherited from white- and brown-rot fungi. Hence the contaminants are degraded through a nonspecific free radical-based mechanism, usually by class II peroxidases (manganese (Mn), lignin and versatile peroxidases) or laccases (Barr & Aust, 1994). Class II peroxidases are not common in the most investigated mycorrhizal species (Kohler et al., 2015) but can be found within the genera *Cortinarius*, *Lactarius* and *Russula* (Bödeker, Nygren, Taylor, Olson, & Lindahl, 2009). In opposite, members of the laccase gene family appear to be most of the time numerous and over-expressed in ECM fungi degrading organic matter (Shah et al., 2015). They can lead to significant degradation of polyphenols especially in the presence of redox mediators (Burke & Cairney, 2002). Finally, recent results showed that several ECM species relied on unspecific nonenzymatic reactions based on the Fenton reaction to degrade organic matter, including polyphenols (Rineau et al., 2012; Shah et al., 2015). It is therefore possible that this mechanism, already mentioned 20 years ago (Burke & Cairney, 1998) but until now often overlooked, could significantly contribute to degradation of P by mycorrhizal fungi. Much less literature is available on this subject for AMF, probably because they are believed to have much lower litter degradation abilities than ECMs (Smith & Read, 1997a, 1997b). Finally, even though degradation of soil pollutants by ERM has almost not been investigated, screening of their enzymatic activities suggests that they have the potential to contribute significantly, having higher activity and diversity of ligninolytic enzymes than ECMs (Read, Leake, & Perez-Moreno, 2004).

Degradation of nitroaromatics is proceeded in two steps: first, a reduction of the nitro- or chlorine substitution with a nitrate reductase, followed by a further oxidative attack by ligninolytic enzymes (Genney, Alexander, Killham, & Meharg, 2004). Several ECM have been found to degrade nitroaromatics effectively, though the mechanisms involved were not investigated in-depth (Meharg et al., 1997). Finally, organochlorines are degraded in a similar way as nitroaromatics, except that the initial phase consists in the removal of the chlorine substitutions by dehalogenases (Lal et al., 2010). Eight out of thirteen ECM fungi have been found to degrade lower chlorinated PCBs in vitro, while degradation of higher PCBs was achieved by only one strain (Meharg & Cairney, 2000). The occurrence frequency of dehalogenase genes (or enzymatic activities) in mycorrhizas has, to our knowledge, not been investigated yet.

The second mechanism by which mycorrhizal fungi can contribute to phytoremediation is a passive one and comes from the surface and adsorption ability of its mycelium. A large proportion of the plant roots are mycorrhizal. Poplar and willow, the most common used trees in phytoremediation, have between 60% and 80% of their root system colonized by ECMs and up to 10% by AMFs (Bucking, 2011). The fungal mycelium extends by several orders of magnitude the surface area of contact between the soil and the roots, thereby maximizing the potential for remediation and the range of soil volume remediated. A significant amount of pollutants can therefore be adsorbed to the mycelial surface, with up to three times higher partition coefficient in mycelium than in root (Gao, Cheng, Ling, & Huang, 2010). The ECMs in particular, by making a sheath around the absorbing tip of the root, represent a barrier for the pollutant (Bucking, 2011). Logically, the adsorption potential is directly correlated with the amount of extraradical hyphae. Hence, one would expect high adsorption capacities of long distance ECMs (the ones making cords) (Agerer, 2001) and in proportion lower amounts for certain AMFs.

Mycorrhizal fungi are interacting with many other soil organisms in the rhizosphere. The metabolic activity of symbiotic fungi alters the rhizosphere properties (e.g., decreasing the pH by the secretion of organic acids or secretion of ammonium, NH_4^+), which selects for a different bacterial community than bulk soil. The functional profile of bacteria in the mycorrhizosphere had, for example, a different functional profile than the one from the bulk soil (Uroz et al., 2007). In particular, mycorrhizal fungi are reported to stimulate the growth of bacterial species with higher pollutant degradation abilities than in the bulk soil, often regardless of the contaminant. For example, the soil in the vicinity of the ECM tips was enriched in Pseudomonads, which were able to degrade xylene, toluene and harboured catalytic marker genes (Sarand et al., 1998). In the same way, PAH degradation was higher at close distance from the AMF root, where a higher proportion of PAH-degrading bacteria was found (Corgié, Joner, & Leyval, 2003). Similar results were found around AMF propagules: in a heavily petroleum-contaminated soil, there was a specific and diverse community of bacteria shaping up, consisting of Pseudomonads, Sphingomonas, Massilia and Methylobacterium (Iffis, St-Arnaud, & Hijri, 2014). It is therefore not surprising that consortia made of a mycorrhizal fungus plus a rhizosphere bacterial strain performed better in the field than a mycorrhizal fungus alone, whether it was an AMF—bacteria (Małachowska-Jutsz & Kalka, 2010) or an AMF—bacteria—saprophytic fungus association (Alarcón, Davies, Autenrieth, & Zuberer, 2008). Moreover, mycorrhizal fungi (just as saprophytic fungi) can improve bacterial motility in soils, thanks to the long, linear structure of the mycelium — sometimes called bacterial highways (Simon et al., 2015) — increasing significantly the accessibility of pollutants for the prokaryotes.

Mycorrhizae do not only influence the response to the pollutant by the surrounding microbial community but also affect the host one. Indeed, the mycorrhizal symbiosis is known to stimulate host-oxidative stress response, which alleviates in part the toxic effect of the pollutant on the plant (Finlay, 2008). The AMF fungi, for example, increased peroxidase (PX), superoxide dismutase (SOD) and ascorbate peroxidase (APX) in the root, which was associated with a protection (no lipid peroxidation) from phenol exposure (Ibáñez, Medina, & Agostini, 2011). This pattern of increasing root antioxidant response associated with decrease in lipid peroxidation seems a common feature of mycorrhizas whether it is for PX (Criquet, Joner, Leglize, & Leyval, 2000; Driai, Verdin, Laruelle, Beddiar, & Sahraoui, 2015; Ibáñez et al., 2011; Xun et al., 2015; Yu, Wu, Wu, & Wong, 2011), catalase (CAT) (Xun et al., 2015) or APX (Ibáñez et al., 2011). SOD, however, was triggered as well in the study of Ibáñez et al. (2011) and Xun et al. (2015) but not in the study of Driai et al. (2015).

4.5 Consequences for Remediation Organic Contaminants

There is a lot of literature on the effect of mycorrhizal fungi on remediation of organics from soil with a wide variation between the outcomes of the experiment from positive to detrimental. We were therefore interested in quantifying the influence of mycorrhiza on pollutant degradation as well as on plant growth in polluted soils and in addition to estimate the range of variation of these parameters. For that, we performed a meta-analysis. We could find 34 articles with quantitative data on the effect of mycorrhizal fungi on soil organic pollutant concentration or increased plant growth by mycorrhizal fungi, which provided a table of 93 observations. Most of the studies focused on AMF (78% of the observations), the remaining ones being on ECM (there was no exploitable data on ERMs). Observations were obtained on alfalfa, ryegrass and corn (18%, 13% and 8% of the total, respectively) for AMF and pine (19%) for ECMs, the remaining observations being shared with 18 other plant species. *G. mosseae*, *G. intraradices* and *Glomus etunicatum* were the most studied AMF species (27%, 12% and 9%, respectively), while the data covered a total of 24 mycorrhizal species. A majority of the observations were obtained in greenhouse or growth chambers (43% and 35%, respectively), with some field experiments as well (20%).

Since most of the data were obtained on AMF, we could analyze more in detail the distribution of data around the mean effect on pollutant concentration. In average, the presence of AMF reduced pollutant concentration by 6% only, with values normally distributed and centred at a reduced pollutant concentration of 11% (Fig. 3A). Effects of AMF on plant biomass in contaminated soils were also positive, with an increase of 29% of the shoot biomass in average, with values normally distributed as well around a median of 5% (Fig. 3B). There were less studies involving ECM, but results showed that their inoculation improved in average plant biomass in polluted soils and decreased soil contaminant concentration in case of PAH, but not in case of hydrocarbons, though the latter is represented by only one observation (Fig. 4A and B). It has been indeed previously noted that ECM inoculation sometimes slows down remediation, probably because of increased nutrient scavenging in the rhizosphere by the symbiotic fungi (Joner, Leyval, & Colpaert, 2006). The reasons why this phenomenon does not happen everytime are yet to be clarified but may depend on nutrient and water

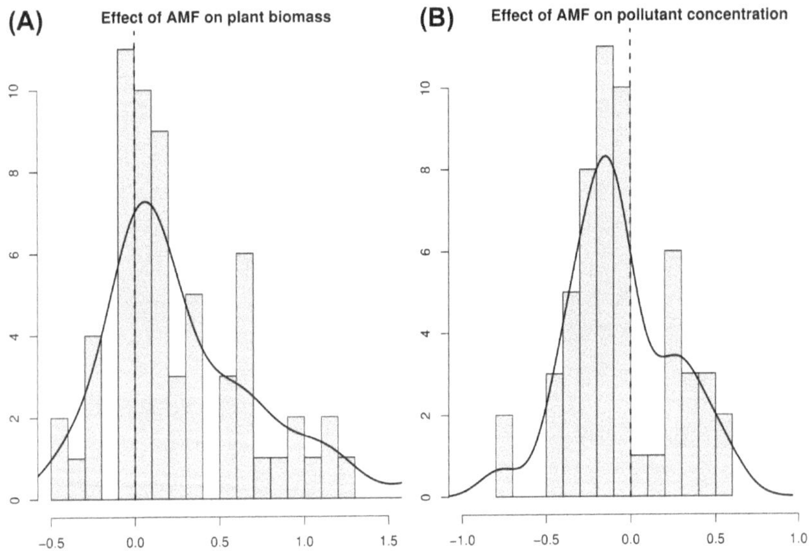

Figure 3 Density histogram of the effects of AMF (arbuscular mycorrhizal fungi) inoculation on plant biomass (A) and soil pollutant concentration (B). In y-axis is the number of observations. *Data come from Table 1.*

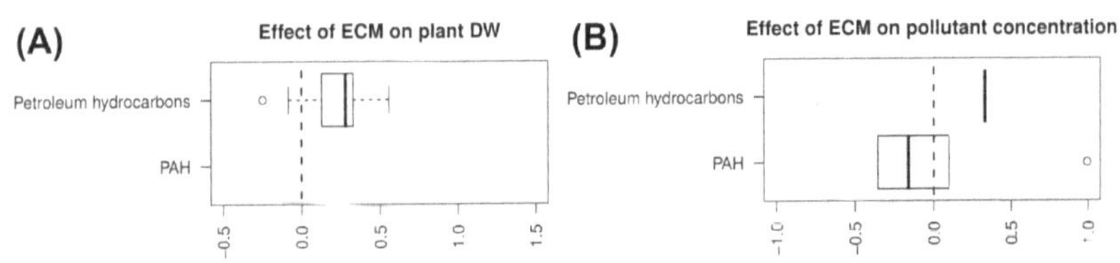

Figure 4 Box-and-whiskers plot of the effects of ECM (ectomycorrhizal fungi) inoculation on plant biomass (A) and soil pollutant concentration (B). *Data come from Table 1.*

availability. When detailing the results of the mycorrhizal effect per pollutant, inoculation always moderately improved the plant biomass (by 5%–50% except for paracetamol, 95% but represented by one value only, Fig. 5A) and the degradation of the contaminant (−18% for PCBs, −14% for PAHs, −10% for petroleum hydrocarbons, Fig. 5B) in average. The exception was for pesticides, where mycorrhizal treatments shoed in average lower pollutant remediation (Fig. 5B). This was true for bulk soil but not in rhizosphere, where about 80% of the initial DDT was left in the soil as dichlorodiphenyldichloroethylene (DDE) (Wu, Zhang, Huang, & Christie, 2008). In the study of White et al. (2006), five of the zucchini cultivars/AMF inocula combinations strongly reduced the DDE dissipation from soil. The DDE is much more recalcitrant than DDT and therefore the rates of degradation and number of species able to degrade it could be expected to be lower, which could explain these low figures. We also compared the differences between phytoremediation by single AMF species and a mix of several ones. AMF species mixtures had in general on average no effect on plant growth, while single species improved it. On the other hand, AMF consortia had a stronger effect on pollutant removal than single species (−21% and −10% in median values, respectively) (Fig. 6A and B). Consortia offer indeed higher phytoremediation potential because it increases both the probability of competition and synergistic interactions between symbionts for the degradation of the pollutant, while the effects

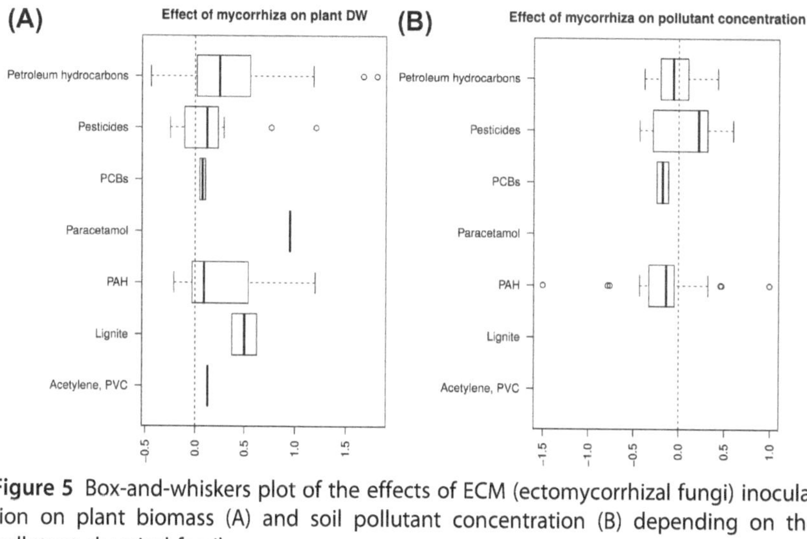

Figure 5 Box-and-whiskers plot of the effects of ECM (ectomycorrhizal fungi) inoculation on plant biomass (A) and soil pollutant concentration (B) depending on the pollutant chemical family.

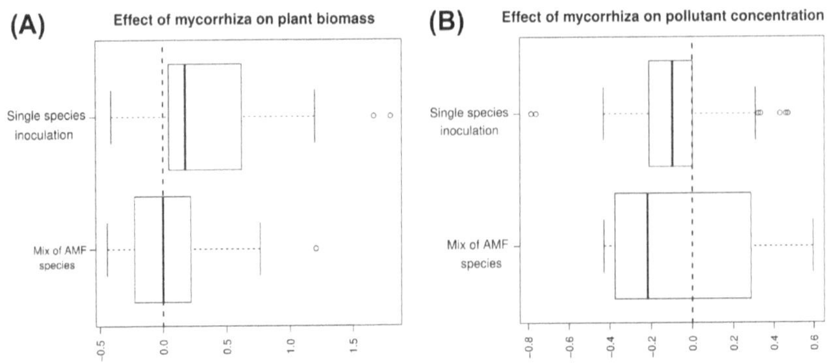

Figure 6 Box-and-whiskers plot of the effects of single mycorrhiza inoculation versus consortia with other soil biota on plant biomass (A) and soil pollutant concentration (B).

of mixtures on plant growth could be explained by a higher belowground allocation to sustain a higher proportion of AMF.

Finally, comparison between several AMF species shows that *G. intraradices* had a larger effect on pollutant removal than *G. mosseae* for petroleum hydrocarbons, while the latter had a wider range of effects on plant biomass (Fig. 7). This can be explained by the larger amount of extraradical hyphae that *G. intraradices* produces as compared with *G. mosseae* (=*F. mosseae*)

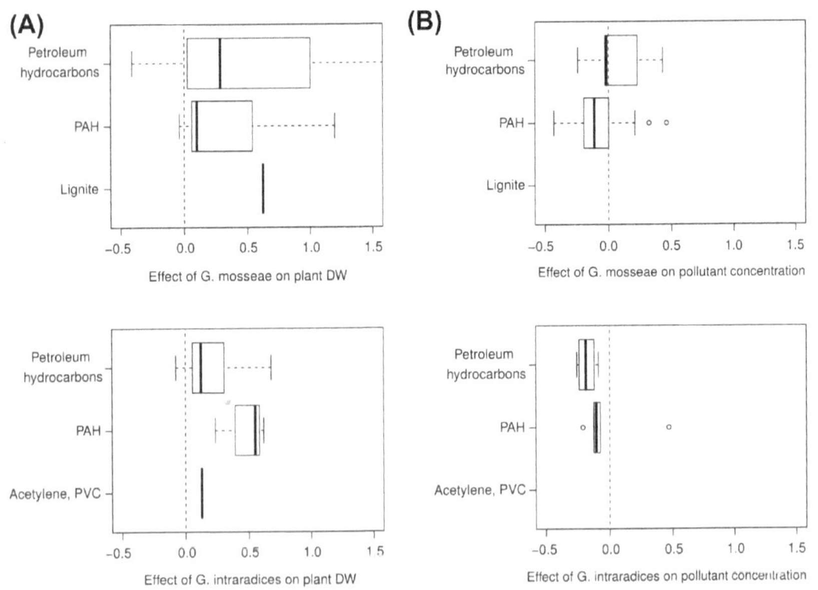

Figure 7 Box-and-whiskers plot of the effects of *Glomus mosseae* and *Glomus intraradices* on plant biomass (A) and soil pollutant concentration (B).

(Avio, Pellegrino, Bonari, & Giovannetti, 2006), increasing its remediation potential (as it is often dependent on the amount of extraradical mycelium, see above).

Important is that, despite the important soil diversity, the inoculum does have a significant effect on plant survival (Yergeau et al., 2015).

4.6 Conclusions on the Benefit of Mycorrhization on Soil Organic Pollutant Remediation

In average, mycorrhizal inoculation moderately improves both plant growth in polluted soils and degradation of the organic pollutants. This effect is however highly variable, depending on the contaminant, the mycorrhizal species, the plant type, the interaction with the surrounding soil organisms, and the soil conditions. It is likely that nutrient scavenging by mycorrhizal mycelium is responsible for the lower degradation seen in some studies.

5. GENERAL CONCLUSION

Soil contamination is a major issue in Europe, especially when soil use is under growing pressure for urbanization food, feed and fibre production. Phytoremediation is a potentially valuable solution to clean up those soils using plants. Their roots are colonized by rhizosphere microorganisms, which contribute significantly to the decontamination of the soil. Mycorrhizal fungi are symbionts that are at the interface between plant roots and soil. They are a potent partner for the plant to mitigate the effects of soil pollution, thanks to the variety of their mechanisms to tolerate heavy metals, as well as their enzymatic potential to degrade complex organic molecules, notwithstanding their ability to uptake nutrients and reduce pathogen infections in stressed environments. The three major types of mycorrhizas (arbuscular, ecto- and ericoid) have been investigated for their potential to enhance phytoremediation outcomes, though the use of ericoid mycorrhizas remains anecdotal. An overview of the work carried out so far on this subject shows that the effect of mycorrhization on phytoremediation has in average a positive effect, though with tremendous variability. This could be explained by different factors, from the complexity of microbial interactions in the rhizosphere (mycorrhizal fungi may compete for some essential nutrients with saprophytic fungi, bacteria or may be affected by pathogens) to the functional diversity of these fungi. The field of mycorrhizal ecology certainly requires substantial experimental research to improve our engineering of phytoremediation solutions.

APPENDIX

List of abbreviations

AMF	arbuscular mycorrhizal fungi
APX	Ascorbate peroxidase
As	Arsenic
BTEX	Benzene, toluene, ethylbenzene, xylene
C	Carbon
Ca	Calcium
CAT	catalase
Cd	Cadmium
DDE	Dichlorodiphenyldichloroethylene
DDT	Dichlorodiphenyltrichloroethane
DW	Dry weight
ECM	Ectomycorrhiza
ERM	Ericoid mycorrhiza
$H_2AsO_4^-$	Arsenate
H_3AsO_3	Arsenous acid
Hg	Mercury
MN	Mycorrhizal network
Mn	Manganese
N	Nitrogen
NaCl	Sodium chloride
NH_4^+	Ammonium
PAH	Polycyclic aromatic hydrocarbons
Pb	Lead
PCBs	Polychlorinated biophenyls
PX	Peroxidase
ROS	Reactive oxygen species
rRNA	Ribosomal ribonucleic acid
Se	Selenium
SOD	Superoxide dismutase
SOM	Soil organic matter
TNT	2,4,6-Trinitrotoluene
Zn	Zinc

List of species and genera mentioned throughout the chapter

Agrostis capillaris
Amanita
Brassica juncea
Bromus erectus
Cadophora finlandica
Calendula officinalis
Calluna vulgaris
Claroideoglomus claroideum
Cortinarius
Festuca ovina
Gingko biloba
Glomus etunicatum
Glomus intraradices = *Rhizophagus irregularis*
Glomus mosseae = *Funneliformis mosseae*
Hebeloma crustuliniforme
Hymenoscyphus ericae = *Pezizella ericae*
Laccaria
Laccaria bicolor
Lactarius
Limonium sinuatum
Oidiodendron maius
Oryza sativa
Paxillus involutus
Pinus sylvestris
Pisolithus tinctorius
Populus canadensis
Populus cathayana
Rhizophagus intraradices
Russula
Salix caprea
Salix smithiana
Salix viminalis
Salix x dasyclados
Sphaerosporella brunnea
Suillus bovinus
Thelephora terrestris
Tomentella
Tricholoma
Trifolium subterraneum

REFERENCES

Adriaensen, K., van der Lelie, D., Van Laere, A., Vangronsveld, J., & Colpaert, J. V. (2003). A zinc-adapted fungus protects pines from zinc stress. *New Phytologist, 161*, 549–555.

Adriaensen, K., Vrålstad, T., Noben, J. P., Vangronsveld, J., & Colpaert, J. V. (2005). Copper-adapted *Suillus luteus*, a symbiotic solution for pines colonizing Cu mine spoils. *Applied and Environmental Microbiology, 71*, 7279–7284.

Agerer, R. (2001). Exploration types of ectomycorrhizae: A proposal to classify ectomycorrhizal mycelial systems according to their patterns of differentiation and putative ecological importance. *Mycorrhiza, 11*(2), 107–114.

Alarcón, A., Davies, F. T., Jr., Autenrieth, R. L., & Zuberer, D. (2008). Arbuscular mycorrhiza and petroleum-degrading microorganisms enhance phytoremediation of petroleum-contaminated soil. *International Journal of Phytoremediation, 10*(November 2014), 251–263. Retrieved from http://www.ncbi.nlm.nih.gov/pubmed/19260211.

Alford, E. R., Pilon-Smits, E. A. H., & Paschke, M. W. (2010). Metallophytes – A view from the rhizosphere. *Plant and Soil, 337*(1), 33–50.

Ali, H., Khan, E., & Sajad, M. A. (2013). Phytoremediation of heavy metals — concepts and applications. *Chemosphere, 91*, 869–881.

Antibus, R., & Linkins, A. (1978). Ectomycorrhizal fungi of *Salix rotundifolia* Trautv. I. Impact of surface applied Prudhoe Bay crude oil on mycorrhizal structure and composition. *Arctic, 31*(3), 366–380. Retrieved from http://www.jstor.org/stable/10.2307/40508911.

Aranda, E., Scervino, J. M., Godoy, P., Reina, R., Ocampo, J. A., Wittich, R. M., & García-Romera, I. (2013). Role of arbuscular mycorrhizal fungus *Rhizophagus custos* in the dissipation of PAHs under root-organ culture conditions. *Environmental Pollution, 181*, 182–189. http://dx.doi.org/10.1016/j.envpol.2013.06.034.

Asrar, A., Abdel-Fattah, G., & Elhindi, K. (2012). Improving growth, flower yield, and water relations of snapdragon (*Antirhinum majus* L.) plants grown under well-watered and water-stress conditions using arbuscular mycorrhizal fungi. *Photosynthetica, 50*(2), 305–316.

Avio, L., Pellegrino, E., Bonari, E., & Giovannetti, M. (2006). Functional diversity of arbuscular mycorrhizal fungal isolates in relation to extraradical mycelial networks. *New Phytologist, 172*(2), 347–357.

Baker, A. J. M., Ernst, W. H. O., van der Ent, A., Malaisse, F., & Ginocchio, R. (2010). Metallophytes: The unique biological resource, its ecology and conservational status in Europe, central Africa and Latin America. In L. C. Batty, & K. B. Hallberg (Eds.), *Ecology of industrial pollution* (pp. 7–40). Cambridge: Cambridge University Press.

Barr, D., & Aust, S. (1994). Pollutant degradation by white rot fungi. *Reviews of Environmental Contamination and Toxicology, 138*, 49–72.

Baum, C., Hrynkiewicz, K., Leinweber, P., & Meißner, R. (2006). Heavy-metal mobilization and uptake by mycorrhizal and nonmycorrhizal willows (*Salix x dasyclados*). *Journal of Plant Nutrition and Soil Science, 169*, 516–522.

Bell, T. H., El-Din Hassan, S., Lauron-Moreau, A., Al-Otaibi, F., Hijri, M., Yergeau, E., & St-Arnaud, M. (2014). Linkage between bacterial and fungal rhizosphere communities in hydrocarbon-contaminated soils is related to plant phylogeny. *The ISME Journal, 8*(2), 331–343. http://dx.doi.org/10.1038/ismej.2013.149.

Bellion, M., Courbot, M., Jacob, C., Blaudez, D., & Chalot, M. (2006). Extracellular and cellular mechanisms sustaining metal tolerance in ectomycorrhizal fungi. *FEMS Microbiology Letters, 254*(2), 173–181.

Bhargava, A., Carmona, F. F., Bhargava, M., & Srivastava, S. (2012). Approaches for enhanced phytoextraction of heavy metals. *Journal of Environmental Management, 105*, 103–120.

Binet, P., Portal, J. M., & Leyval, C. (2000). Fate of polycyclic aromatic hydrocarbons (PAH) in the rhizosphere and mycorrhizosphere of ryegrass. *Plant and Soil, 227*(1–2), 207–213.

Blakely, J. K., Neher, D. A., & Spongberg, A. L. (2002). Soil invertebrate and microbial communities, and decomposition as indicators of polycyclic aromatic hydrocarbon contamination. *Applied Soil Ecology, 21*(1), 71–88.

Blaudez, D., Botton, B., & Chalot, M. (2000). Cadmium uptake and subcellular compartmentation in the ectomycorrhizal fungus *Paxillus involutus*. *Microbiology, 146*, 1109–1117.

Bödeker, I. T. M., Nygren, C. M. R., Taylor, A. F. S., Olson, Å., & Lindahl, B. D. (2009). ClassII peroxidase-encoding genes are present in a phylogenetically wide range of ectomycorrhizal fungi. *The ISME Journal, 3*(12), 1387–1395. http://dx.doi.org/10.1038/ismej.2009.77.

Bradley, R., Burt, A. J., & Read, D. J. (1981). Mycorrhizal infection and resistance to heavy metal toxicity in *Calluna vulgaris*. *Nature, 292*, 335–337.

Bradley, R., Burt, A. J., & Read, D. J. (1982). The biology of mycorrhiza in the Ericaceae. VIII. The role of mycorrhizal infection in heavy metal resistance. *New Phytologist, 91*, 669–682.

Brown, M. T., & Wilkins, D. A. (1985). Zinc tolerance of mycorrhizal Betula. *New Phytologist, 99*, 101–106.

Bruins, M., Kapil, S., & Oehme, F. (2000). Microbial resistance to metals in the environment. *Ecotoxicology and Environmental Safety, 45*, 198–207.

Brundrett, M. (2004). Diversity and classification of mycorrhizal associations. *Biological Reviews of the Cambridge Philosophical Society, 79*, 473–495.

Brundrett, M. C. (2009). Mycorrhizal associations and other means of nutrition of vascular plants: Understanding the global diversity of host plants by resolving conflicting information and developing reliable means of diagnosis. *Plant and Soil, 320*(1), 37–77.

Bucking, H. (2011). Ectomycoremediation: An eco-friendly technique for the remediation of polluted sites. In M. Rai, & A. Varma (Eds.), *Diversity and biotechnology of ectomycorrhizae*. Berlin: Springer-Verlag.

Buée, M., Reich, M., Murat, C., Morin, E., Nilsson, R. H., Uroz, S., & Martin, F. (2009). 454 pyrosequencing analyses of forest soils reveal an unexpectedly high fungal diversity. *New Phytologist, 184*, 449–456.

Burke, R. M., & Cairney, J. W. (2002). Laccases and other polyphenol oxidases in ecto- and ericoid mycorrhizal fungi. *Mycorrhiza*, 105–116.

Burke, R. M., & Cairney, J. W. G. (1998). Carbohydrate oxidases in ericoid and ectomycorrhizal fungi : A possible source of Fenton radicals during the degradation of lignocellulose. *New Phytologist, 139*, 637–645.

Cabello, M. N. (1999). Effectiveness of indigenous arbuscular mycorrhizal fungi (AMF) isolated from hydrocarbon polluted soils. *Journal of Basic Microbiology, 39*(2), 89–95.

Cabral, L., Soares, C. F. R. S., Giachini, A. J., & Siqueira, J. O. (2015). Arbuscular mycorrhizal fungi in phytoremediation of contaminated areas by trace elements: Mechanisms and major benefits of their applications. *World Journal of Microbiology and Biotechnology, 31*, 1655–1664.

Cairney, J. W. G., & Meharg, A. A. (2003). Ericoid mycorrhiza: A partnership that exploits harsh edaphic conditions. *European Journal of Soil Science, 54*(4), 735–740.

Calonne, M., Fontaine, J., Debiane, D., Laruelle, F., Grandmougin-Ferjani, A., & Lounès-Hadj Sahraoui, A. (2014). Impact of anthracene on the arbuscular mycorrhizal fungus lipid metabolism1. *Botany, 92*(2), 173–178. Retrieved from http://www.scopus.com/inward/record.url?eid=2-s2.0-84893837630&partnerID=40&md5=96d627e7277168557d1aa1834c08c187.

Chagnon, P.-L., Bradley, R. L., Maherali, H., & Klironomos, J. N. (2013). A trait-based framework to understand life history of mycorrhizal fungi. *Trends in Plant Science, 18*(9), 484–491.

Chatarpaul, L., Chakravarty, P., & Subramaniam, P. (1989). Studies in tetrapartite symbioses. I. Role of ecto- and endomycorrhizal fungi and Frankia on growth performance of Alnus incana. *Plant Soil, 118*, 145–150.

Chen, L., Hu, X., Yang, W., Xu, Z., Zhang, D., & Gao, S. (2015). The effects of arbuscular mycorrhizal fungi on sex-specific responses to Pb pollution in *Populus cathayana*. *Ecotoxicology and Environmental Safety, 113*, 460–468.

Chiapusio, G., Pujol, S., Toussaint, M. L., Badot, P. M., & Binet, P. (2007). Phenanthrene toxicity and dissipation in rhizosphere of grassland plants (*Lolium perenne* L. and *Trifolium pratense* L.) in three spiked soils. *Plant and Soil, 294*(1–2), 103–112.

Colpaert, J. V. (2008). Chapter 11 Heavy metal pollution and genetic adapdations in ectomycorrhizal fungi. *British Mycological Society Symposia Series, 27*, 157–173.

Colpaert, J. V., Adriaensen, K., Muller, L. A., Lambaerts, M., Faes, C., Carleer, R., & Vangronsveld, J. (2005). Element profiles and growth in Zn-sensitive and Zn-resistant Suilloid fungi. *Mycorrhiza, 15*, 628–634.

Colpaert, J. V., Muller, L. A. H., Lambaerts, M., Adriaensen, K., & Vangronsveld, J. (2004). Evolutionary adaptation to zinc toxicity in populations of Suilloid fungi. *New Phytologist, 162*, 549–559.

Colpaert, J. V., Wevers, J. H. L., Krznaric, E., & Adriaensen, K. (2011). How metal-tolerant ecotypes of ectomycorrhizal fungi protect plants from heavy metal pollution. *Annals of Forest Science, 68*, 17–24.

Corgié, S. C., Fons, F., Beguiristain, T., & Leyval, C. (2006). Biodegradation of phenanthrene, spatial distribution of bacterial populations and dioxygenase expression in the mycorrhizosphere of *Lolium perenne* inoculated with *Glomus mosseae*. *Mycorrhiza, 16*, 207–212.

Corgié, S. C., Joner, E. J., & Leyval, C. (2003). Rhizospheric degradation of phenanthrene is a function of proximity to roots. *Plant and Soil, 257*(1), 143–150.

Courty, P.-E., Buée, M., Diedhiou, A. G., Frey-Klett, P., Tacon, F. L., Rineau, F., … Garbaye, J. (2010). The role of ectomycorrhizal communities in forest ecosystem processes: New perspectives and emerging concepts. *Soil Biology and Biochemistry, 42*(5), 679–698.

Criquet, S., Joner, E., Leglize, P., & Leyval, C. (2000). Anthracene and mycorrhiza affect the activity of oxidoreductases in the roots and the rhizosphere of lucerne (*Medicago sativa* L.). *Biotechnology Letters, 22*(21), 1733–1737.

Daghino, S., Martino, E., & Perotto, S. (2016). Model systems to unravel the molecular mechanisms of heavy metal tolerance in the ericoid mycorrhizal symbiosis. *Mycorrhiza, 26*, 263–274.

Danielson, R. M., & Visser, S. (May 1989). Host response to inoculation and behaviour of introduced and indigenous ectomycorrhizal fungi of jack pine grown on oil-sands tailings. *Canadian Journal of Forest Research, 19*, 1412–1421.

De Paula, A. M., De Siqueira, J. O., & Soares, C. R. F. S. (2007). Contaminacao do solo com antraceno e creosoto e o crescimento vegetal e a colonizacao micorrizica pelo *Glomus etunicatum*. *Revista Brasileira de Ciencia do Solo, 31*(4), 805–811.

Degola, A., Fattorini, L., Bona, E., Sprimuto, C. T., Argese, E., Bertac, G., & di Toppi, L. S. (2015). The symbiosis between *Nicotiana tabacum* and the endomycorrhizal fungus *Funneliformis mosseae* increases the plant glutathione level and decreases leaf cadmium and root arsenic contents. *Plant Physiology and Biochemistry, 92*, 11–18.

Dickie, I. A., Martínez-García, L. B., Koele, N., Grelet, G.-A., Tylianakis, J. M., Peltzer, D. A., & Richardson, S. J. (2013). Mycorrhizas and mycorrhizal fungal communities throughout ecosystem development. *Plant and Soil, 367*, 11–39.

Dong, R., Gu, L., Guo, C., Xun, F., & Liu, J. (2014). Effect of PGPR *Serratia marcescens* BC-3 and AMF *Glomus intraradices* on phytoremediation of petroleum contaminated soil. *Ecotoxicology, 23*(4), 674–680.

Dos Santos Utmazian, M. N., Schweiger, P., Sommer, P., Gorfer, M., Strauss, J., & Wenzel, W. W. (2007). Influence of *Cadophora finlandica* and other microbial treatments on cadmium and zinc uptake in willows grown on polluted soil. *Plant, Soil and Environment, 53*, 158–166.

Dostálek, T., Pánková, H., Münzbergová, Z., & Rydlová, J. (2013). The effect of AMF suppression on plant species composition in a nutrient-poor dry grassland. *PLoS One, 8*, e80535.

Doubková, P., & Sudová, R. (2016). Limited impact of arbuscular mycorrhizal fungi on clones of *Agrostis capillaris* with different heavy metal tolerance. *Applied Soil Ecology, 99*, 78–88.

Driai, S., Verdin, A., Laruelle, F., Beddiar, A., & Sahraoui, A. L.-H. (May 2015). Is the arbuscular mycorrhizal fungus *Rhizophagus irregularis* able to fulfil its life cycle in the presence of diesel pollution? *International Biodeterioration and Biodegradation, 105*, 58–65.

EEA. (2015). *Progress in management of contaminated sites — European environment agency*. Retrieved from http://www.eea.europa.eu/data-and-maps/indicators/progress-in-management-of-contaminated-sites/progress-in-management-of-contaminated-1.

Eevers, N., White, J. C., Vangronsveld, J., & Weyens, N. (2017). Bio- and phytoremediation of pesticide contaminated environments: A review. In A. Cuypers, & J. Vangronsveld (Eds.), *Phytoremediation* (Vol. 83, pp. 277–318).

Ferrol, N., Gonzales-Guerrero, M., Valderaz, A., Benabdellah, K., & Azcon-Aguillar, C. (2009). Survival strategies of arbuscular mycorrhizal fungi in Cu-polluted environments. *Phytochemistry Reviews, 8*, 551–559.

Finlay, R. D. (2008). Ecological aspects of mycorrhizal symbiosis: With special emphasis on the functional diversity of interactions involving the extraradical mycelium. *Journal of Experimental Botany, 59*(5), 1115–1126.

Fomina, M. A., Alexander, I. J., Colpaert, J. V., & Gadd, G. M. (2005). Solubilization of toxic minerals and metal tolerance of mycorrhizal fungi. *Soil Biology and Biochemistry, 37*(5), 851–866.

Fortin, J. A., Plenchette, C., & Piché, Y. (2009). *Mycorrhizas: The new green revolution*. Québec City, CA: Editions MultiMondes.

Francis, R., & Read, D. J. (1994). The contributions of mycorrhizal fungi to the determination of plant community structure. *Plant and Soil, 159*, 11–25.

Franco-Ramírez, A., Ferrera-Cerrato, R., Varela-Fregoso, L., Pérez-Moreno, J., & Alarcón, A. (2007). Arbuscular mycorrhizal fungi in chronically petroleum-contaminated soils in Mexico and the effects of petroleum hydrocarbons on spore germination. *Journal of Basic Microbiology, 47*(5), 378–383.

Frank, B. (2005). On the nutritional dependence of certain trees on root symbiosis with belowground fungi (an english translation of a.b. frank's classic paper of 1885). *Mycorrhiza, 15*(4), 267–275.

Frey-Klett, P., Garbay, J., & Tarkka, M. (2007). The mycorrhiza helper bacteria revisited. *New Phytologist, 176*, 22–36.

Fulekar, M., Singh, A., & Bhaduri, A. M. (2009). Genetic engineering strategies for enhancing phytoremediation of heavy metals. *African Journal of Biotechnology, 8*, 529–535.

Gadd, G. (1993). Interactions of fungi with toxic metals. *New Phytologist, 124*, 25–60.

Galli, U., Schüepp, H., & Brunold, C. (1994). Heavy metal binding by mycorrhizal fungi. *Physiologia Plantarum, 92*, 364–368.

Gao, Y., Cheng, Z., Ling, W., & Huang, J. (2010). Arbuscular mycorrhizal fungal hyphae contribute to the uptake of polycyclic aromatic hydrocarbons by plant roots. *Bioresource Technology, 101*(18), 6895–6901. http://dx.doi.org/10.1016/j.biortech.2010.03.122.

Gao, Y., Li, Q., Ling, W., & Zhu, X. (2011). Arbuscular mycorrhizal phytoremediation of soils contaminated with phenanthrene and pyrene. *Journal of Hazardous Materials, 185*(2–3), 703–709. http://dx.doi.org/10.1016/j.jhazmat.2010.09.076.

Gao, Y. C., Guo, S. H., Wang, J. N., Li, D., Wang, H., & Zeng, D. H. (2014). Effects of different remediation treatments on crude oil contaminated saline soil. *Chemosphere, 117*(1), 486–493. http://dx.doi.org/10.1016/j.chemosphere.2014.08.070.

Garbaye, J. (1994). Helper Bacteria: A new dimension to mycorrhizal symbiosis. *New Phytologist, 128*(2), 197–210.

Garg, N., & Chandel, S. (2012). Role of arbuscular mycorrhizal (AM) fungi on growth, cadmium uptake, osmolyte, and phytochelatin synthesis in *Cajanus cajan* (L.) millsp. under NaCl and Cd stresses. *Journal of Plant Growth Regulation, 31*(3), 292–308.

Gaspar, M. L., Cabello, M. N., Cazau, M. C., & Pollero, R. J. (2002). Effect of phenanthrene and *Rhodotorula glutinis* on arbuscular mycorrhizal fungus colonization of maize roots. *Mycorrhiza, 12*(2), 55–59.

Gaur, A., & Adholeya, A. (2004). Prospects of arbuscular mycorrhizal fungi in phytoremediation of heavy metal contaminated soils. *Current Science, 86*, 528–534.

Genney, D. R., Alexander, I. J., Killham, K., & Meharg, A. A. (2004). Degradation of the polycyclic aromatic hydrocarbon (PAH) fluorene is retarded in a Scots pine ectomycorrhizosphere. *New Phytologist, 163*(3), 641–649.

Gibson, B. R., & Mitchell, D. T. (2004). Nutritional influences on the solibilization of metal phosphate by ericoid mycorrhizal fungi. *Mycological Research, 108*(8), 947–954.

Göhre, V., & Paszkowski, U. (2006). Contribution of the arbuscular mycorrhizal symbiosis to heavy metal phytoremediation. *Planta, 223*, 1115–1122.

Goltapeh, E. M., Danesh, Y. R., Prasad, R., & Varma, A. (2008). Mycorrhizal fungi: What we know and what should we know?. In *Mycorrhiza. State of the Art, genetics and molecular biology, eco-function, biotechnology, eco-physiology, structure and systematics*. Springer Berlin Heidelberg.

Gonzalez-Chavez, M. C., Carrillo-Gonzalez, R., Wright, S. F., & Nichols, K. A. (2004). The role of glomalin, a protein produced by mycorrhizal fungi, in sequestering potentially toxic elements. *Environmental Pollution, 130*, 317–323.

Gonzalez-Chavez, M. C., D'Haen, J., Vangronsveld, J., & Dodd, J. C. (2004). Copper sorption and accumulation by the extraradical mycelium of different *Glomus* spp. (arbuscular mycorrhizal fungi) isolated from the same polluted soil. *Plant and Soil, 240*, 287–297.

Gonzalez-Guerrero, M., Cano, C., Azcon-Aguilar, C., & Ferrol, N. (2007). GinMT1 encodes a functional metallothionein in *Glomus intraradices* that responds to oxidative stress. *Mycorrhiza, 17*, 327–335.

Gonzalez-Guerrero, M., Melville, L. H., Ferrol, N., Lott, J. N., Azcon-Aguilar, C., & Peterson, R. L. (2001). Ultrastructural localization of heavy metals in the extraradical mycelium and spores of the arbuscular mycorrhizal fungus *Glomus intraradices*. *Canadian Journal of Microbiology, 54*, 103–110.

Gunderson, J. J., Knight, J. D., & Van Rees, K. C. J. (2007). Impact of ectomycorrhizal colonization of hybrid poplar on the remediation of diesel-contaminated soil. *Journal of Environmental Quality, 36*(4), 927–934. Retrieved from http://www.ncbi.nlm.nih.gov/pubmed/17526871.

Habte, M., & Osorio, N. W. (2001). *Arbuscular Mycorrhizas: Producing and applying arbuscular mycorrhizal inoculum*. College of Tropical Agriculture and Human Resources, University of Hawaii at Manoa.

Halary, S., Malik, S.-B., Lildhar, L., Slamovits, C. H., Hijri, M., & Corradi, N. (2011). Conserved meiotic machinery in *Glomus* spp., a putatively ancient asexual fungal lineage. *Genome Biology and Evolution, 3*, 950–958.

Hall, J. (2002). Cellular mechanisms for heavy metal detoxification and tolerance. *Journal of Experimental Botany, 53*(366), 1–11.

Haselwandter, K., Dobernigg, B., Beck, W., Jung, G., Cansier, A., & Winkelmann, G. (1992). Isolation and identification of hydroxamate siferophores of ericoid mycorrhizal fungi. *Biometals, 5*(1), 51–56.

van der Heijden, M. G. A. (2004). Arbuscular mycorrhizal fungi as support systems for seedling establishment in grassland. *Ecology Letters, 7*, 293–303.
van der Heijden, M. G. A., Boller, T., Weimken, A., & Sanders, A. R. (1998). Different arbuscular mycorrhizal fungal species are potential determinants of plant community structure. *Ecology, 79*, 2082–2091.
van der Heijden, M. G. A., Martin, F. M., Selosse, M. A., & Sanders, I. R. (2015). Mycorrhizal ecology and evolution: The past, the present, and the future. *New Phytologist, 205*(4), 1406–1423.
Heinonsalo, J., Jørgensen, K. S., Haahtela, K., & Sen, R. (May 2000). Effects of *Pinus sylvestris* root growth and mycorrhizosphere development on bacterial carbon source utilization and hydrocarbon oxidation in forest and petroleum-contaminated soils. *46*, 451–464.
Hendrix, S., Schröder, P., Keunen, E., Huber, C., & Cuypers, A. (2017). Molecular and cellular aspects of contaminant toxicity in plants: The importance of sulphur and associated signalling pathways. In A. Cuypers, & J. Vangronsveld (Eds.), *Phytoremediation* (Vol. 83, pp. 223–276).
Hernández-Ortega, H. A., Alarcón, A., Ferrera-Cerrato, R., Zavaleta-Mancera, H. A., López-Delgado, H. A., & Mendoza-López, M. R. (2012). Arbuscular mycorrhizal fungi on growth, nutrient status, and total antioxidant activity of *Melilotus albus* during phytoremediation of a diesel-contaminated substrate. *Journal of Environmental Management, 95*(Suppl.), S319–S324. http://dx.doi.org/10.1016/j.jenvman.2011.02.015.
Hildebrandt, U., Regvar, M., & Bothe, H. (2007). Arbuscular mycorrhiza and heavy metal tolerance. *Phytochemistry, 68*, 139–146.
Hoeksema, J. D., Chaudhary, V. B., Gehring, C. A., Johnson, N. C., Karst, J., Koide, R. T., … Umbanhowar, J. (2010). A meta-analysis of context-dependency in plant response to inoculation with mycorrhizal fungi. *Ecology Letters, 13*(3), 394–407.
Hristozkova, M., Geneva, M., Stancheva, I., Boychinova, M., & Djonovab, E. (2016). Contribution of arbuscular mycorrhizal fungi in attenuation of heavy metal impact on *Calendula officinalis* development. *Applied Soil Ecology, 101*, 57–63.
Huang, Y., & Wang, J. (2013). Degradation and mineralization of DDT by the ectomycorrhizal fungi, *Xerocomus chrysenteron*. *Chemosphere, 92*(7), 760–764. http://dx.doi.org/10.1016/j.chemosphere.2013.04.002.
Huang, Y., Zhao, X., & Luan, S. (2007). Uptake and biodegradation of DDT by 4 ectomycorrhizal fungi. *Science of the Total Environment, 385*(1–3), 235–241.
Ibáñez, S. G., Medina, M. I., & Agostini, E. (2011). Phenol tolerance, changes of antioxidative enzymes and cellular damage in transgenic tobacco hairy roots colonized by arbuscular mycorrhizal fungi. *Chemosphere, 83*(5), 700–705.
Iffis, B., St-Arnaud, M., & Hijri, M. (2014). Bacteria associated with arbuscular mycorrhizal fungi within roots of plants growing in a soil highly contaminated with aliphatic and aromatic petroleum hydrocarbons. *FEMS Microbiology Letters, 358*(1), 44–54.
Jacquemyn, H., Honnay, O., Cammue, B. P. A., Brys, R., & Lievens, B. (2010). Low specificity and nested subset structure characterize mycorrhizal associations in five closely related species of the genus *Orchis*. *Molecular Ecology, 19*(18), 4086–4095.
Jayachandran, K., & Fisher, J. (2008). *Mycorrhizae: Sustainable agriculture and forestry*. Springer Science + Business Media B. V.
Jentschke, G., & Godbold, D. L. (2000). Metal toxicity and ectomycorrhizas. *Physiologia Plantarum, 109*, 107–116.
Johnson, & Gehring. (2011). *The rhizosphere: An ecological perspective*. Burlington, US: Academic Press, 2011. ProQuest ebrary. Web: 26 July 2016.
Johnson, N. C., Graham, J.-H., & Smith, F. A. (1997). Functioning of mycorrhizal associations along the mutualism-parasitism continuum. *New Phytologist, 135*(4), 575–585.
Joner, E. J., Johansen, A., Loibner, A. P., de la Cruz, M. A., Szolar, O. H., Portal, J. M., & Leyval, C. (2001). Rhizosphere effects on microbial community structure and dissipation

and toxicity of polycyclic aromatic hydrocarbons (PAHs) in spiked soil. *Environmental Science and Technology, 35*(13), 2773–2777.

Joner, E. J., & Leyval, C. (2001). Influence of arbuscular mycorrhiza on clover and ryegrass grown together in a soil spiked with polycyclic aromatic hydrocarbons. *Mycorrhiza, 10*(4), 155–159.

Joner, E. J., & Leyval, C. (2003). Rhizosphere gradients of polycyclic aromatic hydrocarbon (PAH) dissipation in two industrial soils and the impact of arbuscular mycorrhiza. *Environmental Science and Technology, 37*(11), 2371–2375.

Joner, E. J., Leyval, C., & Colpaert, J. V. (2006). Ectomycorrhizas impede phytoremediation of polycyclic aromatic hydrocarbons (PAHs) both within and beyond the rhizosphere. *Environmental Pollution, 142*(1), 34–38.

Ju, K.-S., & Parales, R. E. (2010). Nitroaromatic compounds, from synthesis to biodegradation. *Microbiology and Molecular Biology Reviews, 74*(2), 250–272. Retrieved from http://mmbr.asm.org/content/74/2/250.long.

Kaldorf, M., Kuhn, A. J., Schröder, W. H., Hildebrandt, U., & Bothe, H. (1999). Selective element deposits in maize colonized by a heavy metal tolerance conferring arbuscular mycorrhizal fungus. *Journal of Plant Physiology, 154*, 718–728.

Kershaw, M. J., & Talbot, N. J. (1998). Hydrophobins and repellents: Proteins with fundamental roles in fungal morphogenesis. *Fungal Genetics and Biology, 23*(1), 18–33. Retrieved from http://www.sciencedirect.com/science/article/pii/S1087184597910227.

Khalvati, M., Bartha, B., Dupigny, A., & Schröder, P. (2009). Arbuscular mycorrhizal association is beneficial for growth and detoxification of xenobiotics of barley under drought stress. *Journal of Soils and Sediments, 10*(1), 54–64.

Khan, A. G. (2005). Role of soil microbes in the rhizospheres of plants growing on trace metal contaminated soils in phytoremediation. *Journal of Trace Elements in Medicine and Biology: Organ of the Society for Minerals and Trace Elements (GMS), 18*, 355–364.

Khan, A. G., Kuek, C., Chaudhry, T. M., Khoo, C. S., & Hayes, W. J. (2000). Role of plants, mycorrhizae and phytochelators in heavy metal contaminated land remediation. *Chemosphere, 41*, 197–207.

Khouja, H. R., Abba, S., Lacercat-Didier, L., Daghino, S., Doillon, D., Richaud, P., … Blaudez, D. (2013). OmZnT1 and OmFET, two metal transporters from the metal-tolerant strain Zn of the ericoid mycorrhizal fungus *Oidiodendron maius*, confer zinc tolerance in yeast. *Fungal Genetics and Biology, 52*, 53–64.

Kirk, J. L., Moutoglis, P., Klironomos, J., Lee, H., & Trevors, J. T. (2005). Toxicity of diesel fuel to germination, growth and colonization of *Glomus intraradices* in soil and in vitro transformed carrot root cultures. *Plant and Soil, 270*(1), 23–30.

Klironomos, J. N. (2003). Variation in plant response to native and exotic arbuscular mycorrhizal fungi. *Ecology, 84*, 2292–2301.

Kohler, A., Kuo, A., Nagy, L. G., Morin, E., Barry, K. W., Buscot, F., … Martin, F. (2015). Convergent losses of decay mechanisms and rapid turnover of symbiosis genes in mycorrhizal mutualists. *Nature Genetics, 47*(4), 410–415. Retrieved from http://www.ncbi.nlm.nih.gov/pubmed/25706625.

Koivula, T. T., Salkinoja-Salonen, M., Peltola, R., & Romantschuk, M. (2004). Pyrene degradation in forest humus microcosms with or without pine and its mycorrhizal fungus. *Journal of Environmental Quality, 33*(1), 45–53. Retrieved from http://www.ncbi.nlm.nih.gov/entrez/query.fcgi?cmd=Retrieve&db=PubMed&dopt=Citation&list_uids=14964357.

Krznaric, E., Verbruggen, N., Vangronsveld, J., & Colpaert, J. V. (2009). Cd-tolerant *Suillus luteus*: A fungal insurance for pines exposed to Cd. *Environmental Pollution, 157*, 1581–1585.

Krznaric, E., Wevers, J. H., Cloquet, C., Vangronsveld, J., Vanhaecke, F., & Colpaert, J. V. (2010). Zn pollution counteracts Cd toxicity in metal-tolerant ectomycorrhizal fungi and their host plant, *Pinus sylvestris*. *Environmental Microbiology, 12*(8), 2133–2141.

Kuo, H. C., Juang, D.-F., Yang, L., Kuo, W.-C., & Wu, Y.-M. (2014). Phytoremediation of soil contaminated by heavy oil with plants colonized by mycorrhizal fungi. *International Journal of Environmental Science and Technology, 11*(6), 1661–1668.

Laatikainen, T., & Heinonen-Tanski, H. (2002). Mycorrhizal growth in pure cultures in the presence of pesticides. *Microbiological Research, 157*(2), 127–137. Retrieved from http://www.ncbi.nlm.nih.gov/pubmed/12002401.

Lal, R., Pandey, G., Sharma, P., Kumari, K., Malhotra, S., Pandey, R., … Oakeshott, J. G. (2010). Biochemistry of microbial degradation of hexachlorocyclohexane and prospects for bioremediation. *Microbiology and Molecular Biology Reviews, 74*(1), 58–80.

Lanfranco, L., Bolchi, A., Ross, E. C., Ottonello, S., & Bonfante, P. (2002). Differential expression of a metallothionein gene during the presymbiotic versus the symbiotic phase of an arbuscular mycorrhizal fungus. *Plant Physiology, 130*, 58–67.

Langer, I., Syafruddin, S., Steinkellner, S., Puschenreiter, M., & Wenzel, W. W. (2010). Plant growth and root morphology of *Phaseolus vulgaris* L. grown in a split-root system is affected by heterogeneity of crude oil pollution and mycorrhizal colonization. *Plant and Soil, 332*(1), 339–355.

Leake, J., Johnson, D., Donnelly, D., Muckle, G., Boddy, L., & Read, D. (2004). Networks of power and influence: The role of mycorrhizal mycelium in controlling plant communities and agroecosystem functioning. *Canadian Journal of Botany, 82*(8), 1016–1045.

Leung, H. M., Wu, F. Y., Cheung, K. C., Ye, Z. H., & Wong, M. H. (2010). Synergistic effects of arbuscular mycorrhizal fungi and phosphate rock on heavy metal uptake and accumulation by an arsenic hyperaccumulator. *Journal of Hazardous Materials, 181*, 497–507.

Leyval, C., Turnau, K., & Haselwandter, K. (1997). Effect of heavy metal pollution on mycorrhizal colonization and function: Physiological, ecological and applied aspects. *Mycorrhiza*, 7(3), 139–153.

Li, H., Luo, N., Zhang, L. J., Zhao, H. M., Li, Y. W., Cai, Q. Y., … Mo, C. H. (2016). Do arbuscular mycorrhizal fungi affect cadmium uptake kinetics, subcellular distribution and chemical forms in rice? *Science of the Total Environment, 571*, 1183–1190.

Li, X., Huang, Y., & Wei, Z. (2008). Degradation effect of two pure cultured ectomycorrhizal fungi on diesel. *Ying Yong Sheng Tai Xue Bao, 19*(7), 1579–1584.

Lindahl, B. D., Ihrmark, K., Boberg, J., Trumbore, S. E., Högberg, P., Stenlid, J., & Finlay, R. D. (2007). Spatial separation of litter decomposition and mycorrhizal nitrogen uptake in a boreal forest. *New Phytologist*, 611–620.

Lindahl, B. D., Nilsson, R. H., Tedersoo, L., Abarenkov, K., Carlsen, T., Kjoller, R., … Kauserud, H. (2013). Fungal community analysis by high-throughput sequencing of amplified markers – A user's guide. *New Phytologist, 199*(1), 288–299.

Linkins, A. E., & Antibus, R. K. (1978). Ectomycorrhizal fungi of *Salix rotundifolia* Trautv. II. Impact of surface applied Prudhoe Bay crude oil on mycorrhizal root respiration and cold acclimation. *31*(3), 381–391.

Liu, X. M., Wu, Q. T., & Banks, M. K. (2005). Effect of simultaneous establishment of *Sedum alfridii* and Zea mays on heavy metal accumulation in plants. *International Journal of Phytoremediation,* 7(1), 43–53.

Lu, Y. F., & Lu, M. (2015). Remediation of PAH-contaminated soil by the combination of tall fescue, arbuscular mycorrhizal fungus and epigeic earthworms. *Journal of Hazardous Materials, 285*, 535–541. http://dx.doi.org/10.1016/j.jhazmat.2014.07.021.

Lu, Y. F., Lu, M., Peng, F., Wan, Y., & Liao, M. H. (2014). Remediation of polychlorinated biphenyl-contaminated soil by using a combination of ryegrass, arbuscular mycorrhizal fungi and earthworms. *Chemosphere, 106*, 44–50. http://dx.doi.org/10.1016/j.chemosphere.2013.12.089.

Lumini, E., Bosco, M., Puppi, G., Isopi, R., Frattegiani, M., Buresti, E., & Favilli, F. (1994). Field performance of *Alnus cordata* loisel (Italian alder) inoculated with Frankia and VA-mycorrhizal strains in mine-spoil afforestation plots. *Soil Biology and Biochemistry, 26*(5), 659–661.

Ma, Y., Rajkumar, M., Zhang, C., & Freitas, H. (2016). Beneficial role of bacterial endophytes in heavy metal phytoremediation. *Journal of Environmental Management, 174*, 14–25.

Macnair, M. R. (1993). Tansley review no.49. The genetics of metal tolerance in vascular plants. *New Phytologist, 124*, 541–559.

Mahar, A., Wang, P., Ali, A., Awasthi, M. K., Lahori, A. H., Wang, Q., … Zhang, Z. (2016). Challenges and opportunities in the phytoremediation of heavy metals contaminated soils: A review. *Ecotoxicology and Environmental Safety, 126*, 111–121.

Małachowska-Jutsz, A., & Kalka, J. (2010). Influence of mycorrhizal fungi on remediation of soil contaminated by petroleum hydrocarbons. *Fresenius Environmental Bulletin, 19*(12 B), 3217–3223.

Małachowska-Jutsz, A., Rudek, J., & Janosz, W. (2011). The effect of Ribwort (*Plantago lanceolata*) and its myrorrhizas on the growth of microflora in soil contaminated with used engine oil. *Archives of Environmental Protection, 37*(1), 99–113.

Margesin, R., Zimmerbauer, A., & Schinner, F. (1999). *Soil lipase activity-Biodegradation of PAHs*.

Martin, F., Cullen, D., Hibbett, D., Pisabarro, A., Spatafora, J. W., Baker, S. E., & Grigoriev, I. V. (2011). Sequencing the fungal tree of life. *New Phytologist, 190*(4), 818–821.

Martino, E., Perotto, S., Parsons, R., & Gadd, G. M. (2003). Solubilization of insoluble inorganic zinc compounds by ericoid mycorrhizal fungi derived from heavy metal polluted sites. *Soil Biology and Biochemistry, 35*, 133–141.

McGrath, S. P., Zhao, F. J., & Lombi, E. (2001). Plant and rhizosphere processes involved in phytoremediation of metal-contaminated soils. *Plant and Soil, 232*, 207–214.

McKeehan, P. (2000). *Brownfields: The financial, legislative and social aspects of the redevelopment of contaminated commercial and industrial properties*. http://md3.csa.com/discoveryguide/brown/overview.php?SID=05c43ivvp4r0detrha3d9r5g.

Meharg, A. A. (2003). The mechanistic basis of interactions between mycorrhizal associations and toxic metal cations. *Mycological Research, 107*(11), 1253–1265.

Meharg, A. A., & Cairney, J. W. G. (2000). Ectomycorrhizas – Extending the capabilities of rhizosphere remediation? *Soil Biology and Biochemistry, 32*(11–12), 1475–1484.

Meharg, A. A., Dennis, G., & Cairney, J. W. G. (1997). Biotransformation of 2,4,6-trinitrotoluene (TNT) by ectomycorrhizal basidiomycetes. *Journal of Chemical Information and Modeling, 35*, 513–521.

Meier, S., Borie, F., Bolan, N., & Cornejo, P. (2012). Phytoremediation of metal-polluted soils by arbuscular mycorrhizal fungi. *Critical Reviews in Environmental Science and Technology, 42*, 741–775.

Miransari, M. (2011). Hyperaccumulatrs, arbuscular mycorrhizal fungi and stress of heavy metals. *Biotechnology Advances, 29*, 645–653.

Nagy, N. E., & Fossdal, C. G. (2013). Host responses in Norway spruce roots induced to the pathogen *Ceratocystis polonica* are evaded or suppressed by the ectomycorrhizal fungus *Laccaria bicolor*. *Plant Biology, 15*(1), 99–110.

Nehls, U. (2008). Mastering ectomycorrhizal symbiosis: The impact of carbohydrates. *Journal of Experimental Botany, 59*(5), 1097–1108.

Neuhauser, C., & Fargione, J. E. (2004). A mutualism-parasitism continuum model and its application to plant-mycorrhizae interactions. *Ecological Modelling, 177*, 337–352.

Newman, E. I., & Reddell, P. (1987). The distribution of mycorrhizas among families of vascular plants. *New Phytologist, 106*, 747.

Nicolotti, G., & Egli, S. (1998). Soil contamination by crude oil: Impact on the mycorrhizosphere and on the revegetation potential of forest trees. *Environmental Pollution, 99*(1), 37–43.

Nie, M., Wang, Y., Yu, J., Xiao, M., Jiang, L., Yang, J., … Li, B. (2011). Understanding plant-microbe interactions for phytoremediation of petroleum-polluted soil. *PLoS One, 6*(3).

Noori, A. S., Maivan, H. Z., & Alaie, E. (2014). *Leucanthemum vulgare* Lam. Germination, growth and mycorrhizal symbiosis under crude oil contamination. *International Journal of Phytoremediation, 16*(9), 962–970. http://dx.doi.org/10.1080/15226514.2013.810577. Retrieved from http://www.tandfonline.com/loi/bijp20.

Nuortila, C., Kytöviita, M. M., & Tuomi, J. (2004). Mycorrhizal symbiosis has contrasting effects on fitness components in *Campanula rotundifolia*. *New Phytologist, 164*, 543–553.

Oehl, F., Sieverding, E., Palenzuela, J., Ineichen, K., & Alves da Silva, G. (2011). Advances in glomeromycota taxonomy and classification. *IMA Fungus, 2*(2), 191–199.

Olaniran, A. O., Balgobind, A., & Pillay, B. (2013). Bioavailability of heavy metals in soil: Impact on microbial biodegradation of organic compounds and possible improvement strategies. *International Journal of Molecular Sciences, 14*(5), 10197–10228.

Oliveira, R. S., Castro, P. M. L., Dodd, J. C., & Vosatka, M. (2005). Synergistic effect of *Glomus intraradices* and *Frankia* spp. on the growth and stress recovery of *Alnus glutinosa* in an alkaline anthropogenic sediment. *Chemosphere, 60*, 1462–1470.

Olsson, P. A., Rahm, J., & Aliasgharzad, N. (2010). Carbon dynamics in mycorrhizal symbioses is linked to carbon costs and phosphorus benefits. *FEMS Microbiology Ecology, 72*(1), 125–131.

Orlowska, E., Przybylowicz, W., Orlowski, D., Turnau, K., & Mesjase-Przybylowicz, J. (2011). The effect of mycorrhiza on the growth and elemental composition of Ni-hyperaccumulating plant *Berkheya coddii* Roessler. *Environmental Pollution, 159*, 3730–3738.

Pawlowska, T. E., & Taylor, J. W. (2004). Organization of genetic variation in individuals of arbuscular mycorrhizal fungi. *Nature, 427*(6976), 733–737.

Perez, J. (2012). *The soil remediation industry in Europe: The recent past and future perspectives.*

Perotto, S., Girlanda, M., & Martino, E. (2002). Ericoid mycorrhizal fungi: Some new perspectives on old acquaintances. *Plant and Soil, 244*(1–2), 41–53.

Peterson, R. L., Massicotte, H. B., & Melville, L. H. (2004). *Mycorrhizas: Anatomy and cell biology*. NRC Research Press, ISBN 0-660-19087-7.

Phieler, R., Merten, D., Roth, M., Büchel, G., & Kothe, E. (2015). Phytoremediation using microbially mediated metal accumulation in *Sorghum bicolor*. *Environmental Science and Pollution Research International, 22*, 19408–19416.

Plett, J. M., & Martin, F. (2011). Blurred boundaries: Lifestyle lessons from ectomycorrhizal fungal genomes. *Trends in Genetics, 27*(1), 14–22.

Powell, J. R., Campbell, R. G., Dunfield, K. E., Gulden, R. H., Hart, M. M., Levy-Booth, D. J., … Antunes, P. M. (2009). Effect of glyphosate on the tripartite symbiosis formed by *Glomus intraradices*, *Bradyrhizobium japonicum*, and genetically modified soybean. *Applied Soil Ecology, 41*(1), 128–136.

Pozo, M. J., & Azcón-Aguilar, C. (2007). Unraveling mycorrhiza-induced resistance. *Current Opinion in Plant Biology, 10*(4), 393–398.

Prasad, M. N. V. (2003). Metal hyperaccumulation in plants—biodiversity prospecting for phytoremediation technology. *Electronic Journal of Biotechnology, 6*, 285–321.

Rabie, G. H. (2004). Using wheat-mungbean plant system and arbuscular mycorrhiza to enhance in-situ bioremediation. *International Journal of Food, Agriculture and Environment, 2*(2), 381–390.

Read, D. J. (1993). Mycorrhiza in plant communities. *Advances in Plant Pathology, 9*, 1–31.

Read, D. J., Leake, J. R., & Perez-Moreno, J. (2004). Mycorrhizal fungi as drivers of ecosystem processes in heathland and boreal forest biomes. *Canadian Journal of Botany, 82*, 1243–1263.
Redondo-Gómez, S., Petenello, M. C., & Feldman, S. R. (2014). Growth, nutrient status, and photosynthetic response to diesel-contaminated soil of a cordgrass, *Spartina argentinensis*. *Marine Pollution Bulletin, 79*(1–2), 34–38.
Richter, D. L., Warner, J. I., & Stephens, A. L. (2003). A comparison of mycorrhizal and saprotrophic fungus tolerance to creosote in vitro. *International Biodeterioration and Biodegradation, 51*(3), 195–202.
Rillig, M. C., & Mummey, D. L. (2006). Mycorrhizas and soil structure. *New Phytologist*, 41–53.
Rillig, M. C., Wright, S. F., & Eviner, V. T. (2002). The role of arbuscular mycorrhizal fungi and glomalin in soil aggregation: Comparing effects of five plant species. *Plant and Soil, 238*, 325–333.
Rinaldi, A., Comandini, O., & Kuyper, T. (2008). Ectomycorrhizal fungal diversity: Separating the wheat from the chaff. *Fungal Diversity, 33*, 1–45.
Rineau, F., Roth, D., Shah, F., Smits, M., Johansson, T., Canbäck, B., … Tunlid, A. (2012). The ectomycorrhizal fungus *Paxillus involutus* converts organic matter in plant litter using a trimmed brown-rot mechanism involving Fenton chemistry. *Environmental Microbiology, 14*(6), 1477–1487.
Robertson, S. J., Kennedy, N. M., Massicotte, H. B., & Rutherford, P. M. (2010). Enhanced biodegradation of petroleum hydrocarbons in the mycorrhizosphere of sub-boreal forest soils. *Environmental Microbiology Reports, 2*, 587–593.
Robertson, S. J., McGill, W. B., Massicotte, H. B., & Rutherford, P. M. (2007). Petroleum hydrocarbon contamination in boreal forest soils: A mycorrhizal ecosystems perspective. *Biological Reviews of the Cambridge Philosophical Society, 82*(2), 213–240.
Robertson, S. J., Rutherford, P. M., & Massicotte, H. B. (2011). Plant and soil properties determine microbial community structure of shared Pinus-Vaccinium rhizospheres in petroleum hydrocarbon contaminated forest soils. *Plant and Soil, 346*(1), 121–132.
Robson, D. B., Diane Knight, J., Farrell, R. E., & Germida, J. J. (2004). Natural revegetation of hydrocarbon-contaminated soil in semi-arid grasslands. *Canadian Journal of Botany, 82*(1), 22–30. Retrieved from http://www.scopus.com/inward/record.url?eid=2-s2.0-1842735085&partnerID=40&md5=763c2303c8876ef5bc0392046bc3945c.
Rodriguez, A., Clapp, J. P., & Dodd, J. C. (2004). Ribosomal RNA gene sequence diversity in arbuscular mycorrhizal fungi (glomeromycota). *Journal of Ecology, 92*(6), 986–989.
Ruytinx, J., Nguyen, H., Van Hees, M., Op De Beeck, M., Vangronsveld, J., Carleer, R., … Adriaensen, K. (2013). Zinc export results in adaptive zinc tolerance in the ectomycorrhizal basidiomycete *Suillus bovinus*. *Metallomics, 5*(9), 1225–1233.
Santoyo, G., Moreno-Hagelsieb, G., Orozco-Mosqueda, M. D., & Glick, B. R. (2016). Plant growth-promoting bacterial endophytes. *Microbiological Research, 183*, 92–99.
Saparrat, M. C., Balatti, P. A., Arambarri, A. M., & Martínez, M. J. (2014). *Coriolopsis rigida*, a potential model of white-rot fungi that produce extracellular laccases. *Journal of Industrial Microbiology and Biotechnology, 41*(4), 607–617.
Sarand, I., Timonen, S., Nurmiaho-Lassila, E.-L., Koivula, T., Haahtela, K., Romantschuk, M., & Sen, R. (1998). Microbial biofilms and catabolic plasmid harbouring degradative fluorescent pseudomonads in Scots pine mycorrhizospheres developed on petroleum contaminated soil. *FEMS Microbiology Ecology, 27*(2), 115–126.
Sarwat, M., Hashem, A., Ahanger, M. A., Abd_Allah, E. F., Alqarawi, A. A., Alyemeni, M. N., … Gucel, S. (2016). Mitigation of NaCl stress by arbuscular mycorrhizal fungi through the modulation of osmolytes, antioxidants and secondary metabolites in mustard (*Brassica juncea* L.) plants. *Frontiers in Plant Science, 7*, 869.

Savin, M. C., Purcell, L. C., Daigh, A., & Manfredin, A. (2006). *Mycorrhizal response in Soybean, corn, and cotton to glyphosate applications and P fertilization.*

Schat, H., Llugany, M., & Bernhard, R. (2000). Metal-specific patterns of tolerance, uptake and transport of heavy metals in hyperaccumulating and nonhyperaccumulating metallophytes. In N. Terry, & G. Banuelos (Eds.), *Phytoremediation of contaminated soil and water* (pp. 171–188). Boca Raton: Lewis.

Schelkle, M., & Peterson, R. (1997). Suppression of common root pathogens by helper bacteria and ectomycorrhizal fungi in vitro. *Mycorrhiza, 6*(6), 481–485.

Schnabel, W. E., & White, D. M. (2001). The effect of mycorrhizal fungi on the fate of Aldrin: Phytoremediation potential. *International Journal of Phytoremediation, 3*(April 2012), 221–241.

Schneider, J., Bundschuh, J., & Araújo do Nascimento, C. W. (2016). Arbuscular mycorrhizal fungi-assisted phytoremediation of a lead-contaminated site. *Science of the Total Environment, 572*, 86–97.

Schneider, J., Stürmer, S. L., Guilherme, L. R., de Souza Moreira, F. M., & Soares, C. R. (2013). Arbuscular mycorrhizal fungi in arsenic contaminated areas in Brazil. *Journal of Hazardous Materials, 262*, 1105–1115.

Schüßler, A., & Walker, C. (2010). *The glomeromycota. A species list with new families and new genera.* Gloucester, UK: A. Schüßler & C. Walker.

Sell, J., Kayser, A., Schulin, R., & Brunner, I. (2005). Contribution of ectomycorrhizal fungi to cadmium uptake of poplars and willows from a heavily polluted soil. *Plant and Soil, 277*(1), 245–253.

Selosse, M.-A., Setaro, S., Glatard, F., Richard, F., Urcelay, C., & Weiß, M. (2007). Sebacinales are common mycorrhizal associates of Ericaceae. *New Phytologist, 174*(4), 864–878.

Sen, R. (2001). Multitrophic interactions between a *Rhizoctonia* sp. and mycorrhizal fungi affect Scots pine seedling performance in nursery soil. *New Phytologist, 152*(3), 543–553.

Shah, F., Nicolás, C., Bentzer, J., Ellström, M., Smits, M., Rineau, F., ... Tunlid, A. (2015). Ectomycorrhizal fungi decompose soil organic matter using oxidative mechanisms adapted from saprotrophic ancestors. *New Phytologist, 209*(4).

Shaoxia, G., Tingting, H., & Runjin, L. (2011). Effects of arbuscular mycorrhizal (AM) fungi *Glomus mosseae* on characteristics of leaf development of *Paeonia suffruticosa* under salt stress. *African Journal of Microbiology Research, 5*, 714–719.

Sharples, J. M., Meharg, A. A., Chambers, S. M., & Cairney, J. W. G. (2000). Mechanism of arsenate resistance in the ericoid mycorrhizal fungus *Hymenoscyphus ericae. Plant Physiology, 124*, 1327–1334.

Sharples, J. M., Meharg, A. A., Chambers, S. M., & Cairney, J. W. G. (2001). Arsenate resistance in the ericoid mycorrhizal fungus *Hymenoscypus ericae. New Phytologist, 151*, 265–270.

Sheikh-Assadi, M., Khandam-Mirkohi, A., Alemardan, A., & Moreno-Jiménez, E. (2015). Mycorrhizal *Limonium sinuatum* (L.) mill. Enhances accumulation of lead and cadmium. *International Journal of Phytoremediation, 17*, 556–562.

Sheoran, V., Sheoran, A., & Poonia, P. (2011). Role of hyperaccumulators in phytoextraction of metals from contaminated mining sites: A review. *Critical Reviews in Environmental Science and Technology, 41*, 168–214.

Simon, A., Bindschedler, S., Job, D., Wick, L. Y., Filippidou, S., Kooli, W. M., ... Junier, P. (2015). Exploiting the fungal highway: Development of a novel tool for the in situ isolation of bacteria migrating along fungal mycelium. *FEMS Microbiology Ecology, 91*(11), 1–13.

Smith, S. E., & Read, D. J. (1997a). *Mycorrhizal symbiosis* (2nd ed.). Academic Press, ISBN 978-0-12-652840-4.

Smith, S. E., & Read, D. J. (1997b). 15-The roles of mycorrhizas in ecosystems. In *Mycorrhizal symbiosis* (2nd ed., pp. 409–452). Retrieved from http://www.sciencedirect.com/science/article/pii/B9780126528404500164.

Smith, S., & Read, D. (2008). *Mycorrhizal symbiosis*. Academic press Inc.

Tang, M., Chen, H., Huang, J. C., & Tian, Z. Q. (2009). AM fungi effects on the growth and physiology of Zea mays seedlings under diesel stress. *Soil Biology and Biochemistry, 41*(5), 936–940. http://dx.doi.org/10.1016/j.soilbio.2008.11.007.

Tedersoo, L., May, T., & Smith, M. (2010). Ectomycorrhizal lifestyle in fungi: Global diversity, distribution, and evolution of phylogenetic lineages. *Mycorrhiza, 20*(4), 217–263.

Teng, Y., Luo, Y., Sun, X., Tu, C., Xu, L., Liu, W., … Christie, P. (2010). Influence of arbuscular mycorrhiza and *Rhizobium* on phytoremediation by alfalfa of an agricultural soil contaminated with weathered PCBs: A field study. *International Journal of Phytoremediation, 12*(October 2014), 516–533.

Thijs, S., Langill, T., & Vangronsveld, J. (2017). The bacterial and fungal microbiota of hyperaccumulator plants: Small organisms, large influence. In A. Cuypers, & J. Vangronsveld (Eds.), *Phytoremediation* (Vol. 83, pp. 43–86).

Thijs, S., Van Dillewijn, P., Sillen, W., Truyens, S., Holtappels, M., D'Haen, J., … Vangronsveld, J. (2014). Exploring the rhizospheric and endophytic bacterial communities of *Acer pseudoplatanus* growing on a TNT-contaminated soil: Towards the development of a rhizocompetent TNT-detoxifying plant growth promoting consortium. *Plant and Soil, 385*(1–2), 15–36.

Thion, C., Cébron, A., Beguiristain, T., & Leyval, C. (2012). Long-term in situ dynamics of the fungal communities in a multi-contaminated soil are mainly driven by plants. *FEMS Microbiology Ecology, 82*(1), 169–181.

Toler, H. D., Morton, J. B., & Cumming, J. R. (2005). Growth and metal accumulation of mycorrhizal sorghum exposed to elevated copper and zinc. *Water, Air, and Soil Pollution, 164*, 155–172.

Trappe, J. M., Molina, R., & Castellano, M. (1984). Reactions of mycorrizal fungi and mycorrizal formation to pesticides. *Annual Review of Phytopathology, 22*(22), 331–359. Retrieved from http://www.annualreviews.org/doi/pdf/10.1146/annurev.py.22.090184.001555.

Trofimov, S. Y., & Rozanova, M. S. (2003). Transformation of soil properties under the impact of oil pollution. *Eurasian Soil Science, 36*, 82–87.

Turnau, K., Ryszka, P., Gianinazzi-Pearson, V., & van Tuinen, D. (2001). Identification of arbuscular mycorrhizal fungi in soils and roots of plants colonizing zinc wastes in southern Poland. *Mycorrhiza, 10*, 169–174.

Uroz, S., Calvaruso, C., Turpault, M. P., Pierrat, J. C., Mustin, C., & Frey-Klett, P. (2007). Effect of the mycorrhizosphere on the genotypic and metabolic diversity of the bacterial communities involved in mineral weathering in a forest soil. *Applied and Environmental Microbiology, 73*(9), 3019–3027.

Usman, A. R. A., & Mohamed, H. M. (2009). Effect of microbial inoculation and EDTA on the uptake and translocation of heavy metal by corn and sunflower. *Chemosphere, 76*, 893–899.

Valentin, L., Nousiainen, A., & Mikkonen, A. (2013). Introduction to organic contaminants in soil: Concepts and risks. In T. Vicent, et al. (Eds.), *The Handbook of Environmental Chemistry*. Springer-Verlag.

Vallino, M., Drogo, V., Abba, S., & Perotto, S. (2005). Gene expression of the ericoid mycorrhizal fungus *Oidiodendron maius* in the presence of high zinc concentrations. *Mycorrhiza, 15*(5), 333–344.

Vallino, M., Martino, E., Boella, F., Murat, C., Chiapello, M., & Perotto, S. (2009). Cu, Zn superoxide dismutase and zinc stress in the metal-tolerant ericoid mycorrhizal fungus *Oidiodendron maius* Zn. *FEMS Microbiology Letters, 293*, 48–57.

Vamerali, T., Bandiera, M., & Mosca, G. (2010). Field crops for phytoremediation of metal-contaminated land. A review. *Environmental Chemistry Letters, 8*, 1–17.

Van Hamme, J. D., Singh, A., & Ward, O. P. (2003). Recent advances in petroleum microbiology recent advances in petroleum microbiology. *Microbiology and Molecular Biology Reviews, 67*(4), 503–549.

Vangronsveld, J., Herzig, R., Weyens, N., Boulet, J., Adriaensen, K., Ruttens, A., … Mench, M. (2009). Phytoremediation of contaminated soils and groundwater: Lessons from the field. *Environmental Science and Pollution Research International, 16*, 765–794.

Verdin, A., Lounès-Hadj Sahraoui, A., Fontaine, J., Grandmougin-Ferjani, A., & Durand, R. (2006). Effects of anthracene on development of an arbuscular mycorrhizal fungus and contribution of the symbiotic association to pollutant dissipation. *Mycorrhiza, 16*(6), 397–405.

Vivas, A., Barea, J. M., Biro, B., & Azcon, R. (2006). Effectiveness of autochthonous bacterium and mycorrhizal fungus on *Trifolium* growth, symbiotic development and soil enzymatic activities in Zn contaminated soil. *Journal of Applied Microbiology, 100*, 587–598.

Vrålstad, T., Myhre, E., & Schumacher, T. (2002). Molecular diversity and phylogenetic affinities of symbiotic root-associated ascomycetes of the Helotiales in burnt and metal polluted habitats. *New Phytologist, 155*, 131–148.

Wang, F., Lin, X., & Yin, R. (2005). Heavy metal uptake by arbuscular mycorrhizas of *Elsholtzia splendens* and the potential for phytoremediation of contaminated soil. *Plant and Soil, 269*, 225–232.

Wang, F. Y., Shi, Z. Y., Xu, X. F., Wang, X. G., & Li, Y. J. (2013). Contribution of AM inoculationand cattle manure to lead and cadmium phytoremediation by tobacco plants. *Environmental Science. Processes and Impacts, 15*, 794–801.

Wasserman, J. L., Mineo, L., Majumdar, S. K., & Vantyne, C. (1987). Detection of heavy metals in oak mycorrhizae of northeastern Pennsylvania forests, using X-ray microanalysis. *Canadian Journal of Botany, 65*, 2622–2627.

Wenzel, W. W. (2009). Rhizosphere processes and management in plant-assisted bioremediation (phytoremediation) of soils. *Plant and Soil, 321*, 385–408.

White, J. C., Ross, D. W., Gent, M. P., Eitzer, B. D., & Mattina, M. I. (2006). Effect of mycorrhizal fungi on the phytoextraction of weathered p,p-DDE by *Cucurbita pepo*. *Journal of Hazardous Materials, 137*(3), 1750–1757.

Whitfield, L., Richards, A. J., & Rimmer, D. L. (2004). Relation-ships between soil heavy metal concentration and mycorrhizal colonization in *Thymus polytrichus* in northern England. *Mycorrhiza, 14*, 55–62.

Wilkinson, D. M., & Dickinson, N. M. (1995). Metal resistance in trees: The role of mycorrhizae. *Oikos, 72*(2), 298–300.

Witters, N., Van Slycken, S., Ruttens, A., Adriaensen, K., Meers, E., Meiresonne, L., … Vangronsveld, J. (2009). Short-rotation coppice of willow for phytoremediation of a metal-contaminated agricultural area: A sustainability assessment. *Bioenergy Research, 2*(144), 152.

Wójcik, M., Gonnelli, C., Selvi, F., Dresler, S., Rostański, A., Vangronsveld, J., et al. (2017). Metallophytes of serpentine and calamine soils — Their unique ecophysiology and potential for phytoremediation. In A. Cuypers, & J. Vangronsveld (Eds.), *Phytoremediation* (Vol. 83, pp. 1–42).

Wu, B. Y., Isobe, K., & Ishii, R. (2004). Arbuscular mycorrhizal colonization of the dominant plant species in primary successional volcanic deserts on the southeast slope of Mount Fuji. *Mycorrhiza, 14*, 391–395.

Wu, F. Y., Yu, X. Z., Wu, S. C., Lin, X. G., & Wong, M. H. (2011). Phenanthrene and pyrene uptake by arbuscular mycorrhizal maize and their dissipation in soil. *Journal of Hazardous Materials, 187*(1–3), 341–347.

Wu, L., Jacobson, A. D., & Hausner, M. (2008). Characterization of elemental release during microbe-granite interactions at T = 28°C. *Geochimica et Cosmochimica Acta, 72*(4), 1076–1095.

Wu, N., Zhang, S., Huang, H., & Christie, P. (2008). Enhanced dissipation of phenanthrene in spiked soil by arbuscular mycorrhizal alfalfa combined with a non-ionic surfactant amendment. *Science of the Total Environment, 394*(2–3), 230–236.

Wu, N., Zhang, S., Huang, H., Shan, X., Christie, P., & Wang, Y. (2008). DDT uptake by arbuscular mycorrhizal alfalfa and depletion in soil as influenced by soil application of a non-ionic surfactant. *Environmental Pollution, 151*(3), 569–575.

Wuana, R. A., & Okieimen, F. E. (2011). Heavy metals in contaminated soils: A review of sources, chemistry, risks and best available strategies for remediation. *ISRN Ecology*, 1–20.

Xiao, X., Chen, H., Si, C., & Wu, L. (2012). Influence of biosurfactant-producing strain *Bacillus subtilis* BS1 on the mycoremediation of soils contaminated with phenanthrene. *International Biodeterioration and Biodegradation, 75*, 36–42. http://dx.doi.org/10.1016/j.ibiod.2012.09.002.

Xun, F., Xie, B., Liu, S., & Guo, C. (2015). Effect of plant growth-promoting bacteria (PGPR) and arbuscular mycorrhizal fungi (AMF) inoculation on oats in saline-alkali soil contaminated by petroleum to enhance phytoremediation. *Environmental Science and Pollution Research International, 22*(1), 598–608.

Yao, Q., Zhu, H.-H., Hu, Y.-L., & Li, L.-Q. (2008). Differential influence of native and introduced arbuscular mycorrhizal fungi on growth of dominant and subordinate plants. *Plant Ecology, 196*, 261–268.

Yergeau, E., Bell, T. H., Champagne, J., Maynard, C., Tardif, S., Tremblay, J., & Greer, C. W. (December 2015). Transplanting soil microbiomes leads to lasting effects on willow growth, but not on the rhizosphere microbiome. *Frontiers in Microbiology, 6*, 1–14.

Yu, X. Z., Wu, S. C., Wu, F. Y., & Wong, M. H. (2011). Enhanced dissipation of PAHs from soil using mycorrhizal ryegrass and PAH-degrading bacteria. *Journal of Hazardous Materials, 186*(2–3), 1206–1217.

Zhou, X., Zhou, J., & Xiang, X. (2013). Impact of four plant species and arbuscular mycorrhizal (AM) fungi on polycyclic aromatic hydrocarbon (PAH) dissipation in spiked soil. *Polish Journal of Environmental Studies, 22*(4), 1239–1245. Retrieved from http://www.pjoes.com/pdf/22.4/Pol.J.Environ.Stud.Vol.22.No.4.1239-1245.pdf.

Zhou, X. B., Cébron, A., Béguiristain, T., & Leyval, C. (2009). Water and phosphorus content affect PAH dissipation in spiked soil planted with mycorrhizal alfalfa and tall fescue. *Chemosphere, 77*(6), 709–713. http://dx.doi.org/10.1016/j.chemosphere.2009.08.050.

CHAPTER FIVE

Phytoremediation and Phytomining: Status and Promise

Rufus L. Chaney*,[1] and Ilya A. Baklanov[§]
*USDA-Agricultural Research Service, Crop Systems and Global Change Laboratory, Beltsville, MD, United States
[§]University of Maryland, College Park, MD, United States
[1]Corresponding author: E-mail: rufuschaney@verizon.net

Contents

Abstract

Phytoremediation of inorganics is comprised of technologies to protect the environment from contaminated soils. This broad group of technologies includes phytoextraction (removal from soil), phytomining (accumulating economic metal value in plant biomass) and phytostabilization (limiting plant metals and soil metal uptake and/or bioavailability using plants and soil amendments). As agronomic technologies, selection of plant genotypes, management of soil properties, appropriate use of fertilizers and soil pH adjustment amendments are required for optimum effectiveness. Phytoextraction depends on the availability of plants which can accumulate and tolerate ~100-fold higher metal concentrations than tolerated by common crop species. Unless a plant can accumulate over 1% (10 g/kg) of a metal in dry shoot biomass, it is unlikely to remove metals rapidly enough to support economic remediation. For element cases where contamination is low (Cd), plants which accumulate over 1000 μg/g may be able to achieve adequate phytoextraction. In the case of Ni, the combination of extensive ultramafic soils rich in Ni, and natural hypernickelophores (accumulate over 10 g Ni/kg) with high harvestable biomass yield allows a phytomining technology to be profitable. Increasing knowledge of the biochemistry and genetic mechanisms used by hyperaccumulator plants portend the day that high biomass element hyperaccumulator plant genotypes can be constructed for soil remediation in all climatic zones.

Advances in Botanical Research, Volume 83
ISSN 0065-2296
http://dx.doi.org/10.1016/bs.abr.2016.12.006

1. INTRODUCTION

1.1 What Is Phytoremediation?

Phytotechnology is a subset of Agronomy in which plants and soils are managed to solve environmental problems. Using cover crops to reduce soil erosion has been practiced for over a century. Even using plants plus soil amendments to revegetate metal-contaminated landscapes is relatively old with reports in the 1960s and earlier. As researchers began to use plants and soil microbes to biodegrade soil xenobiotic contaminants (complex organic compounds including polycyclic aromatic hydrocarbons), and the concept of phytoextraction was reported, the larger concept of 'phytotechnologies' was adopted by many researchers and environmental managers. Phytotechnologies or Phytoremediation include a number of separate technologies, each of which is being studied to develop or improve potential value of the technology. Research is being conducted on both metals and xenobiotics. Other reviews on phytoremediation and hyperaccumulator plants have summarized available knowledge from varied points of view (Assunção, Schat, & Aarts, 2003; Baker & Brooks, 1989; Baker, McGrath, Reeves, & Smith, 2000; Brooks, 1998; Cappa & Pilon-Smits, 2013; Chaney et al., 2007; Chaney, Broadhurst, & Centofanti, 2010, chap. 17; Chaney et al., 2000; Dickinson, Baker, Doronila, Laidlaw, & Reeves, 2009; Ernst, 2005; Hanikenne & Nouet, 2011; Jaffré, Pillon, Thomine, & Merlot, 2013; Koopmans et al., 2008; Koopmans, Römkens, Song, Temminghoff, & Japenga, 2007; Lombi, Zhao, Dunham, & McGrath, 2001; McGrath, Zhao, & Lombi, 2001; Pilon-Smits, 2005; Pollard, Powell, Harper, & Smith, 2002; Rascio & Navari-Izzo, 2011; Robinson, Anderson, & Dickinson, 2015; Robinson, Bañuelos, Conesa, Evangelou, & Schulin, 2009; Robinson et al., 2003; Tang et al., 2012; Van der Ent, Baker, van Balgooy, & Tjoa, 2013; Van der Ent et al., 2015; Van der Ent, Baker, Reeves, Pollard, & Schat, 2012; Van Nevel, Mertens, Oorts, & Verheyen, 2007; Vangronsveld et al., 2009).

For over 60 years, research has been developing to show use of metal-tolerant plants and soil amendments to restore vegetation on metal toxic soils where mine wastes were dispersed or surrounding metal smelters. It started with using Al-tolerant and Mn-tolerant plants plus soil amendments to raise soil pH and add nutrients to establish vegetation on strongly acidic barren coal mine wastes (Foy, Chaney, & White, 1978; Sopper, 1993). Essentially all soils may become Al or Mn phytotoxic if pH drops low enough; these are 'natural' soil metal phytotoxicities. When soil pH drops below 5.4, soil microbes which normally oxidize Mn^{2+} to Mn^{4+} (which precipitates as MnO_2) are mostly inactive so that Mn^{2+} can accumulate in the exchangeable ion pool to levels which are phytotoxic. When soil pH drops below 5.2,

Al^{3+} is increasingly dissolved in soil solution and replaces Ca and Mg on the cation exchange complex and can be phytotoxic to most plant species. There are long-known naturally Mn-resistant and Al-resistant plants, and obvious soil amendments to raise soil pH to reverse Mn and Al phytotoxicity. Much research has been conducted in the last few decades to understand the fundamental processes in Al phytotoxicity and to develop natural or genetically modified crop plants resistant to the Al^{3+} in soil solution of acidic soils (Kochian, Pineros, & Hoekenga, 2005). Al phytotoxic soils are extensive; enough limestone to prevent these toxicities can be expensive and limit choices in developing countries. When cultivars of food crops are bred to tolerate strongly acidic soils, the potential for higher crop Cd requires separate attention (Chaney, 2015).

Other researchers studied plants which grew on highly metal-contaminated (mine and smelter contaminated) or metal-mineralized (natural mine deposits not dispersed by humans) soils. Bradshaw et al. characterized metal-tolerant ecotypes of some grasses which were remarkably more tolerant of soil metals than normal genotypes, and even developed metal-tolerant *Merlin* red fescue (*Festuca rubra* L.) for use in revegetation of Zn–Pb-contaminated soils (Bradshaw, Humphries, & Johnson, 1978; Smith & Bradshaw, 1979). The difficulty in using metal-tolerant grasses is the need to supply nitrogen needed for growth. Although a Zn-tolerant legume, *Anthyllis vulneraria*, has been discovered more recently (Mahieu et al., 2013), the effectiveness of in situ remediation using metal-tolerant grasses without raising soil pH makes this a less persistent solution for such soils.

The early research by Ernst et al. (Ernst, 1974, 1996) on metal-tolerant and metal-accumulating plants has been a guide to the older knowledge. As noted in review papers, high Zn levels in and tolerance by *Noccaea caerulescens* (J. Presl & C. Presl) F.K. Mey. was reported by Baumann (1885). Minguzzi and Vergnano (1948) reported the remarkable Ni accumulation and tolerance by *Alyssum bertolonii* Desv. Because methods of Ni analysis were less reliable at that time, few believed this plant could be alive with this much foliar Ni. Vergnano subsequently studied Ni phytotoxicity in crop plants while working on her Ph.D. in Scotland and reported that essentially all plants suffered visible Ni phytotoxicity symptoms and substantially reduced shoot yields at 50–100 mg Ni/kg dry leaves (Vergnano & Hunter, 1952; see also Kukier & Chaney, 2004). So the 10% NiO in *A. bertolonii* ash in the 1948 report seemed unlikely to other scientists who understood that crop plants suffered phytotoxicity with <100 mg/kg.

Others worked to develop methods to use plants to achieve biodegradation of organic compounds in contaminated soils in which it is generally accepted that plants support microbial communities as well as express biodegradative or biotransformative capabilities which can achieve remediation of xenobiotic contaminated soils. In the case of xenobiotics, only the compounds which persist long enough to comprise a persistent hazard have gotten significant attention, but much power has been demonstrated in the literature (e.g., Schnoor, Licht, McCutcheon, Wolfe, & Carreira, 1995).

Unusual plants associated with ore deposits was the early focus of work on 'botanical prospecting' (e.g., Cannon (1960)). The concept that rare metal-tolerant and metal-accumulating plants had evolved to survive on soils so rich in particular metals that normal plants could not grow there is the predecessor of bioindicator, botanical prospecting, geobotany, biogeochemistry, metal accumulation phylogeny, phytoextraction, phytomining, phytoremediation, agromining and related ideas. Cannon (1960), discussing botanical prospecting, included a list of species which occurred in locations where soils contained unusual concentrations of elements, including the reported hyperaccumulators of Zn, Ni and Se. Botanical prospecting was developing for decades and based on a report by Beath (1943), Cannon had used characteristic flowering Se accumulator plants (e.g., *Astragalus pattersoni* Gray ex Brand) to locate soils overlying uranium deposits during World War 2. Mineralized soils or ore deposits contained U, V and Se so that Se-tolerant *Astragalus* species were a bioindicator plant for U-mineralized deposits in western United States (Cannon, 1962). Other early important research on botanical prospecting and metal accumulation in unusual plants was summarized by Malyuga (1964). Wider recognition and appreciation of how unusual hyperaccumulator plants were followed in the report by Jaffré and Schmid (1974) about the 'hypernickelophore' *Psychotria douarrei* (Beauvis.) Däniker. Subsequently Brooks working with Jaffré, then additionally with Reeves and Baker and their students (Brooks, Lee, Reeves, & Jaffré, 1977) coined the term 'hyperaccumulator' to describe extreme accumulation of several elements by many plant species based on analysis of small pieces of leaves held by herbaria. Chaney relied on the papers of Brooks et al. and others to develop the concepts of phytoextraction and phytomining which he presented at a US-EPA conference in 1980 (Chaney, Hornick, & Sikora, 1981), and then in a book chapter (Chaney, 1983). Baker (1981) summarized the literature about accumulator and excluder plant species which introduced the hyperaccumulator idea more widely. The definition of Ni hyperaccumulation became that of Reeves (1992): "... it is suggested that a hyperaccumulator of nickel be defined as 'a plant in which a nickel concentration of at least 1000 μg/g has been recorded in the dry matter of any above-ground tissue in at least one specimen growing in its natural habitat'." Over time the definitions for hyperaccumulation of several elements have shifted as knowledge grew and the definitions of Van der Ent et al. (2012) are generally accepted today (see also chapter: Metallophytes of Serpentine and Calamine Soils – Their Unique Ecophysiology and Potential for Phytoremediation by Wójcik et al. (2017)). Although Van der Ent et al. (2012) suggested that only the term 'hyperaccumulator' (accumulate >1.0 g Ni/kg dry shoots) should be used to discuss Ni-accumulating plant species, the importance of 'hypernickelophores' (accumulate >10 g Ni/kg dry shoots) to phytomining technology caused Van der Ent et al. (2015) to recommend use of this more exclusive label for these remarkable species that are so important in the potential to develop commercial phytomining of soil Ni.

The concept of phytoextraction (Chaney, 1983) is simple, but often misunderstood. His original focus was on decontamination of hazardous contaminated soils at lower cost than the 'dig and haul' approach favored by the engineering community. For plants to absorb metals from soils and transport them to their shoots in high enough amounts to achieve eventual decontamination of the soil, ordinary crop plants simply do not work. But natural element hyperaccumulating species have been identified for a few elements (Table 1). The clear effect of soil pH on Zn and Cd accumulation by *N. caerulescens* was already reported by Rascio (1977). Phytoextraction plants must both absorb and translocate high concentrations of soil metals to harvestable plant shoots, and the shoots must tolerate high metal concentrations. He was able to propose this concept because hyperaccumulator plants had been reported in the literature for Zn, Ni and Se before 1980.

Table 1 Example Plant Species Which Hyperaccumulate Elements to Levels Usually Greater Than 100-Fold Levels Tolerated by Crop Species

Element	Plant Species	Max. Conc. (mg/kg dry wt)	Location Collected	References
As	*Pteris vittata*	22,300	Florida	Ma et al. (2001)
Cd	*Noccaea caerulescens*	2,910	France	Reeves et al. (2001)
Co	*Phyllantus favieri*	8,050	United States	van der Ent (unpubl.)
Mn	*Alyxia rubricaulis*	11,500	New Caledonia	Brooks, Trow, Veillon, and Jaffré (1981)
Ni	*Phyllanthus serpentinus*	38,100	New Caledonia	Kersten, Brooks, Reeves, and Jaffré (1979)
Se	*Astragalus racemosus*	14,900	Wyoming	Beath, Eppsom, and Gilbert (1937)
Tl	*Biscutella laevigata*	15,200	France	Anderson et al. (1999)
Zn	*Noccaea caerulescens*	39,600	Germany	Reeves and Brooks (1983)

As many others have since noted, it will not be useful to phytoextract Zn because it would take too long to deplete Zn in contaminated soils to reach regulatory levels (see below).

Unfortunately study of phytoextraction took a drastic wrong turn when Kumar, Dushenkov, Motto, and Raskin (1995) reported nutrient solution tests of Pb accumulation by *Brassica juncea* (L.) Czern. genotypes. The nutrient solutions they used during Pb uptake assays kept the Pb highly soluble by limiting phosphate and sulfate, and achieved over 1% of Pb in shoot dry matter in some *B. juncea* genotypes. This is now recognized as an extreme artifact. It was well known that if plants were P-deficient, Pb^{2+} could be absorbed and translocated much like Ca^{2+}, reaching high shoot Pb concentration and injuring plants (Miller & Koeppe, 1971). But Phytotech obtained patents for what they reported was all possible phytoextraction technologies based on these early defective studies. Pb uptake by plants is normally very limited because Pb is either insoluble in soil due to phosphate (promotes formation of very low solubility Pb pyromorphite) or pH (promotes adsorption) so that Pb phytotoxicity is never observed in nature (Koeppe, 1981).

After the patent was granted to Phytotech and they obtained investments, they had to show that they could attain high shoot Pb from real world contaminated soils. Because having enough phosphate to promote plant growth naturally limits Pb solubility, uptake and translocation, high yields are contrary to high shoot Pb levels. Subsequently Blaylock et al. (1997) reported that application of chelating agents (NTA, EDTA, DTPA, etc.) could cause dissolution of soil metals and induce accumulation in shoots of *B. juncea*. Many others have studied this process in the lab, greenhouse and field, and it is now recognized that addition of chelating agents to metal-contaminated soils to promote phytoextraction is both environmentally unacceptable and cost prohibitive. Chaney et al. (2014, chap. 15) used the EDTA concentration reported by Blaylock to be most effective to induce phytoextraction in their test system (10 mmol/kg soil), and the commercial bulk price of EDTA to derive an estimated cost of chemicals alone of $23,500/ha year. Others showed that only a small fraction of the metals dissolved by chelating agents was accumulated by the plants, and under leaching conditions, more metal chelates leached to groundwater than were accumulated by plants. In addition, if other metals were present, the combination of metal chelates caused phytotoxicity and that phytotoxicity persisted until the chelators were biodegraded so multiple cropping was not practical. The leaching of metal chelates is environmentally unacceptable and permits for chelator-induced phytoextraction without a contained system are not issued in the United States or Europe (see review by Nowack, Schulin, & Robinson, 2006).

Fundamental research to better understand chelator-induced Pb phytoextraction was reported by several research groups. Huang, Chen, Berti, and Cunningham (1997) and Wu, Hsu, and Cunningham (1999) showed that significant amounts of the chelators did reach the plant shoots, which was confirmed in detail by Schaider, Parker, and Sedlak (2006). Vassil, Kapulnik, Raskin, and Salt (1998) determined that enough EDTA had to be added that 'free' EDTA (not strongly chelated with metals, likely chelating Ca in soil solution) remained in the soil solution to harm the rhizodermal cell membranes and cause roots to leak the PbEDTA into cells where it flowed to the xylem with the transpiration stream. Others considered other chelating compounds which were more rapidly biodegradable than EDTA, and studied EDDS (ethylenediamine-disuccinic acid). But this was no more effective and is considerably more expensive (Meers et al., 2004; Meers, Ruttens, Hopgood, Samson, & Tack, 2005). Actually many of the tests of the effectiveness of chelators in inducing phytoextraction ignored the information reported by the Phytotech team. Blaylock et al. (1997) showed that it was best to grow *B. juncea* to early flowering stage (long days promote flowering to produce oilseeds) with high potential transpiration, and then add the chelator to the soil surface with irrigation water so that a large biomass and transpiration demand could absorb more of the Pb-EDTA from the soil solution. Otherwise the plants suffered adverse effects of the chelator addition immediately upon addition which limited shoot biomass in other research. The fundamental unacceptability of chelator-induced phytoextraction has been summarized by many review papers including Evangelou, Ebel, and Schaeffer (2007), Nowack et al. (2006) and Chaney et al. (2014, chap. 15).

There were attempts to find actual Pb-hyperaccumulating plant species because it is clear that there is a market for Pb removal from contaminated soils. Reeves and Brooks (1983) reported very high levels of Pb in shoots of a very small plant growing on Pb mine wastes in Italy, *Noccaea rotundifolium* (L.) Gaudin var. *cepaeifolium* (Wulf.) Rouy et Fouc. As discussed below regarding soil contamination of foliage, when plants are short and soil highly contaminated, leaf samples might appear to be evidence of hyperaccumulation but it is just soil particles adhering to leaf surfaces. One way to prevent that error in metal uptake studies is to cover the soil surface with plastic beads or washed sand. Huang and Cunningham (1996) used this approach to test phytoextraction of Pb from a soil with 2500 mg/kg; they found *N. rotundifolium* had little Pb uptake to shoots (79 mg/kg), similar to that of ragweed (*Ambrosia artemisiifolia* L.), the Pb-accumulating species their team reported to accumulate 75 mg/kg. Adding chelators greatly increased shoots Pb, but as noted above, that method is unacceptable.

For some elements the important risk from element-contaminated soils is the potential for children and wildlife to ingest the soil and absorb bioavailable elements from the ingested soil (see Ryan et al., 2004). In particular, the greatest risk from soil Pb and As is through soil ingestion, not plant uptake (with risk to food chains) or even toxicity to soil organisms. Thus any reduction in soil element concentration or bioavailability could reduce risk; dilution can significantly reduce risk for most soils. Several Pb and As phytoremediation demonstrations achieved clean-up goals of reduction in surface soil Pb concentration largely by mixing 'hotspot' high Pb soils with surrounding less contaminated, or deeper uncontaminated soil. For Pb and As, field projects of phytoremediation achieved reduction in surface soil Pb concentration (and hence estimated risk) more by tillage than by plant uptake, but was useful soil remediation and risk reduction. Phytostabilization is discussed below in the context of changing the chemical speciation of elements in soils to limit risk in addition to possible dilution by tillage to achieve significant risk reduction (Scheckel, Chaney, Basta, & Ryan, 2009).

1.2 True Hyperaccumulators Versus Imposters Due To Soil Contamination of Short Plants

True hyperaccumulators absorb elements from soils and translocate high amounts to shoots where the elements could be harvested to remove them from the field (phytoextraction), or to recover the element as a 'bio-ore' (phytomining). One key principle of hyperaccumulators is that the shoots have higher element concentration than do the roots; efficient translocation is required to achieve high shoot concentrations. Very few species are true hyperaccumulators in real soils, the basis for the fundamental definition of hyperaccumulators (e.g., Van der Ent et al., 2012). Hundreds of papers have reported high Cd or Ni accumulation in plant shoots when they were grown on soils spiked with soluble Cd or Ni salts. Others failed to recognize that their favorite plant did not actually reach hyperaccumulator status in leaves before phytotoxicity occurred. Every report of Pb phytoextraction is invalid due to the use of chelators, or soluble Pb in the absence of phosphate or sulfate which would precipitate Pb in nutrient solutions or soils. Every paper. Even the early papers which appeared to suggest Pb hyperaccumulation by *N. rotundifolium* was a result of soil contamination of plant foliage.

Soil contaminates all field-grown plants to a small or large extent. Soil contamination is readily demonstrated by analysis of elements which have very limited uptake and translocation due to their very low solubility in natural soils: Cr^{3+}, Fe^{3+}, Ti, Zr, Y, etc. This principle had been used to estimate soil on plants in the diet of grazing livestock (Cary, Grunes, Bohman, & Sanchirico, 1986). Soil contamination of plant shoots was elegantly illustrated by Cary and Kubota (1990) who sampled vegetation on a serpentine soil with over 1% Cr. By analysis of Ti in the soil and plant shoots, they could estimate the fraction of soil Cr in the plant shoot total Cr. Soil and shoot Ti analysis revealed that more than the total Cr in the plant shoot sample could be explained by the soil present based on Ti analysis. This can result if the particle size of the soil analyzed does not match the particle size of materials adhering to plant shoots. This was further demonstrated with vegetable crops where Fe in the shoots was predominantly soil particles (which could be nutritionally significant because soil Fe is less bioavailable to animals than is plant Fe) (Cary et al., 1994). Soil contamination is especially important in short leafy vegetables growing close to soils. We calculated the fraction of Fe in commercial spinach (*Spinacia oleracea* L.) which was actually soil. Based on the analysis of US commercial spinach samples by the US Food and Drug Administration Total Diet Study, mean spinach contains 24.6 mg Fe/kg fresh weight (7.6–166) = 319 mg Fe/kg dry weight (99–2155). When spinach is grown with effective mulch to prevent soil splash, it contains no more than 80 mg Fe/kg DW. Thus 75% of the Fe in commercial spinach is actually soil.

Soil contamination also contributed to misinterpretation of Cu and Co hyperaccumulation by some plants growing on Cu–Co mineralized soils in central Africa. Analysis of soil indicator elements showed that all tissues with high Cu had high soil content. No natural Cu hyperaccumulator has been reported if soil contamination is taken into account (Faucon, Shutcha, & Meerts, 2007). When other elements in a soil are used to estimate the amount of soil in a plant sample, one should separate and analyze the topsoil particle size which becomes part of washed leaf samples. Advances in using other elements to estimate soil has been important in estimating the amount of soil ingested by children through hand-to-mouth play. In estimating soil ingestion based on analyzing several elements in children's feces (based the elements not being absorbed by plants or animals = soil indicator elements), a substantial error resulted in estimates for soil ingestion at Anaconda, MT (Hwang, Bornschein, Grote, Menrath, & Roda, 1997). Calabrese, Stanek, and Barnes (1996) and Stanek, Calabrese, and Barnes (1999) subsequently reviewed the need to take into account particle size that will adhere to hands during children's play which could then be ingested. In US government considerations, only soil particles less than 250 μm are considered important to ingestion by hand-to-mouth play. Similarly, the soil particles analyzed to estimate soil content of plant samples should be the size normally adhering on leaves after soil splash (<250 μm or smaller).

But the biggest errors in claims of phytoextraction has been with Cd (illustrated in Table 2). Soil Cd comprises risks to crops and humans (Chaney, 2015). Although there are some unusual soils with high Cd:Zn ratio mineralization or contamination (e.g., Chaney, 2010, chap. 17; Garrett, Porter, Hunt, & Lalor, 2008), nearly all Zn ores (hence soils contaminated by Zn mining or smelting sources) contain 100–200 times more Zn than Cd. So if a plant (high yielding forage maize is modeled here) tolerates only ~500 mg Zn/kg and accumulated Cd and Zn in approximately the ratio of total Cd and Zn in soil, soil Zn will limit plant growth before significant Cd removal has been achieved. There are now hundreds of papers which suggest some plant species would be useful in Cd phytoextraction because it can accumulate over 100 mg Cd/kg dry plant shoots (part of the definition of Cd hyperaccumulator) when the plant is grown in an acidic Cd salt-spiked soil. But most plant species do not tolerate higher than 500 mg Zn/kg DW leaves, so with the normal ratio of soil Cd to

Table 2 Ability of Maize (*Zea mays* L.) to Phytoextract Zn and Cd at Full Yield or With 50% Yield Reduction Due To Zn Phytotoxicity on Contaminated Soil, or *Noccaea caerulescens*, Either the *Prayon* Ecotype, a Southern France *Ganges* Ecotype or an Improved Cultivar With Higher Yield and 10-Times Higher Cd Accumulation Than the *Prayon* One (Reeves et al., 2001)

		Zn in Crop			
Crop	**Yield (t/ha)**	**(mg/kg)**	**(kg/ha)**	**(% of Soil Zn)**	**Zn in Ash (%)**
Maize (normal soil)	20	50	1.0	0.0025	0.10
Maize-Zn phytotoxicity	10	500	5.0	0.0125	0.50
Noccaea	5	25,000	125	3.12	40
Improved *Noccaea*	10	25,000	250	6.25	40

		Cd in Crop			
Crop	**Yield (t/ha)**	**(mg/kg)**	**(kg/ha)**	**(% of Soil Cd)**	**Cd in Ash (%)**
Maize	20	0.5	0.01	0.025	0.001
Maize	10	5.0	0.05	0.125	0.005
Noccaea Prayon	5	200	1.0	2.5	0.40
Noccaea Ganges	5	2000	10.0	25	4.00
Noccaea cultivar	10	2000	20.0	50	4.00

Annual Cd phytoextraction would be expected to decline as soil Cd was depleted and depends strongly on soil pH management (Wang et al., 2006). For Zn, presume soil has 2000 mg Zn/kg = 4000 kg Zn/(ha-15 cm); for Cd, presume soil has 20 mg Cd/kg = 40 kg Cd/(ha-15 cm).

Zn (1:100) and before Zn phytotoxicity limits growth, the maximum shoot Cd that can be attained in normal non-Zn-hyperaccumulator species would be 5 mg/kg DW. As shown by Brown, Angle, Chaney, and Baker (1995), even the *Prayon* ecotype of *N. caerulescens* cannot accumulate much Cd from fields with normal Cd:Zn ratio before Zn phytotoxicity results. It was the discovery of the southern France ecotypes of *N. caerulescens* with 10-fold higher Cd accumulation than the *Prayon* ecotype (Reeves, Schwartz, Morel, & Edmondson, 2001) that made it possible to achieve significant Cd phytoextraction from common Zn + Cd-contaminated soils.

In the same way, there are now many papers reporting accumulation of Ni by various plant species, but for nearly all species, phytotoxicity is evident before leaves accumulate higher than 100 mg/kg. As noted above, unless a species accumulates higher than 10 g/kg (10,000 mg/kg), that species has little ability to remove enough Ni from a soil to achieve decontamination even over decades of removing the shoot biomass (Table 3). It would be unkind to list papers which suffer this error in testing of phytoextraction of Ni or Cd, but readers need to keep this information in mind when reading the literature on phytoremediation. Dead plants do not remove much element from soils. There are many additional errors from using metal salt-spiked soils to simulate field soils mineralized or contaminated with particular metals, including the soil acidification when soluble metal salts are applied which increases soil metal phytoavailability (e.g., Speir, Kettles, Percival, & Parshotam, 1999).

1.3 Practical Phytoextraction

Many of the researchers who have worked to develop phytoextraction have clarified the potential for this technology to solve environmental problems. Comparison of typical levels of contamination vs. potential annual metal removal by even hyperaccumulator plant species show that annual removals are small compared to amounts in soils (Table 3). Maximizing annual removal in hyperaccumulator plant shoot biomass requires management of soil pH and fertility, weed control, and other general agronomic management practices. Some of the species are so short that harvest is difficult. For example, *N. caerulescens* has a rosette growth pattern and flowering stems are no higher than 30 cm; the majority of the shoot biomass is rosette leaves no more than 10 cm from the soil surface. Hammer and Keller (2003) discuss the difficulty in practical harvest of such a species. The significant issues in the arguments of Ernst (1988, 2000, 2005) were based on the difficulty in harvesting shoot biomass of plants with short stature. In the field, one

Table 3 Potential for Phytoextraction of Elements From Contaminated Soil Using Natural Element Hyperaccumulator Species With No Addition of Chelating Agent to Induce Phytoextraction

Element	Soil Conc. (mg/kg)	Plant Species	Shoot Conc. (g/kg)	Annual Yield (t/ha)	Annual Removal (kg/ha)	Soil Total (kg/(30 cm))	Fraction Removed (%)
As	20	*Pteris vittata*	0.6	5.0	3.0	80	3.8
Cd	20	*Noccaea caerulescens*	2.5	5.0	12.5	80	15.6
Pb	1,000	*Brassica juncea*	0.10	10	1.0	4,000	0.002
Mn	10,000	*Phytolacca acinosa*	10	10	100	40,000	0.025
Ni	2,500	*Alyssum corsicum*	20	20	400	10,000	4.0
Zn	2,000	*Noccaea caerulescens*	30	5.0	150	8,000	1.8

Assumes surface 30 cm of soil is contaminated and requires element remediation.

may be able to observe *N. caerulescens* only when the flowering stems appear near the height of the metal-tolerant grasses which co-occur in natural settings where this species occurs. Another part of Ernst's interpretation of the potential for phytoextraction is based on the density of *N. caerulescens* in Zn-rich field soils. Without the use of fertilizers, yield of *N. caerulescens* is very low and plants even smaller and harder to harvest. *N. caerulescens* would be a small part of the shoot biomass in the field unless monocot herbicides were applied repeatedly. In the view of Simmons et al. (2015), Broadhurst et al. (2015) and Nkrumah et al. (2016), Cd phytoextraction can be effective only with careful management of agronomic practices to maximize harvestable yield of the phytoextraction crop. Application of monocot herbicides can limit competition and promote phytoextraction biomass. Further, if such herbicides are not used, the harvestable biomass is not especially rich in Zn or Cd because of the low density of *N. caerulescens* in the field. Table 3 summarizes more realistic potential annual phytomining for several elements in the field using appropriate element accumulation potentials and agronomic production practices. Clearly the amount of Pb removed per year is trivial despite many papers on this subject. Zn, As and even Ni phytoextraction are limited in terms of soil decontamination, a major point of Ernst (2005) with which we concur. Similarly, Robinson, Anderson, and Dickinson (2015) reviewed phytoextraction and raise important questions which indicate that even given 25 years, the present known accumulator plants and agronomic practices would not remove a high fraction of soil Zn, Cu, Ni or most other elements. But these arguments against the success of phytoextraction do not negate the potential phytoextraction of soil Cd, or phytomining of Ni and perhaps other elements.

Useful Cd phytoextraction might be possible with a high yielding, easy to grow species with unusual Cd accumulation. As noted in Chaney (2015), the most important risk of soil Cd is in areas of subsistence rice production. When rice is grown in soil contaminated by Zn mine wastes, the plants accumulate high amounts of Cd during grain filling, after fields are drained. Rice absorbs Cd on a Mn transporter (NRAMP5), not the Zn transporter (ZIP2) used by other studied species (Ishikawa et al., 2012; Sasaki, Yamaji, Yokosho, & Ma, 2012). When fields are drained at flowering, soil pH and soluble Mn^{2+} decline rapidly and Cd uptake greatly increases (e.g., Arao, Kawasaki, Baba, Mori, & Matsumoto, 2009; Murakami, Nakagawa, Ae, Ito, & Arao, 2009).

Cadmium phytoextraction has seemed to offer important value in environmental remediation. Adverse human health effects of soil Cd have occurred in at least Japan, China and Thailand where rice soils were contaminated with Cd and Zn by mine or smelter emissions (see review in Chaney, 2015). Some other sites in other countries clearly also comprise Cd risk but no government has required or funded remediation of Cd in these locations (e.g., Broadhurst et al., 2015; Chaney et al., 2005; Simmons et al., 2015).

Models for the potential for Zn or Cd phytoextraction to decontaminate soils are shown in Table 2. A crop plant can remove very little Zn compared to the soil pool that requires remediation; even an improved hyperaccumulator grown with good agronomic practices can remove only a small percentage of soil Zn per year. Nearly all Cd-contaminated soils received the Cd with Zn in Zn or Pb–Zn or Cu–Pb–Zn mine wastes or smelter emissions and Zn is on the order of 100- to 200-fold the level of Cd present. Because of this, normal plants (e.g., maize) suffer Zn phytotoxicity (at 400–500 mg Zn/kg dry leaves) when they have accumulated only 5–10 mg Cd/kg dry weight. Even the hyperaccumulator of Cd and Zn, *N. caerulescens* F.K. Mey., ecotype *Prayon* cannot remove useful amounts of Cd (Table 2) because it accumulates Cd and Zn in about the proportion in the soil (e.g., Brown, Chaney, Angle, & Baker, 1994). However, some southern France *N. caerulescens* ecotypes (commonly referred to as Ganges but from many locations in southern France) accumulate up to 10-fold higher levels of Cd than *Prayon* so that these ecotypes can accumulate over 2000 mg Cd/kg DW in the southern France soils where they were collected (Table 2) (Reeves et al., 2001). Because soil pH strongly affects the solubility of both Cd and Zn, lowering pH substantially increases the phytoextraction of Cd by *N. caerulescens* (Wang, Angle, Chaney, Delorme, & Reeves, 2006; Yanai, Zhao, McGrath, & Kosaki, 2006). Wang et al. (2006) found that as high as 40% of soil total Cd could be phytoextracted with one crop of southern France *N. caerulescens* in pots of acidified Zn smelter-contaminated soil. At locations where the Zn/Cd ratio is 10 rather than 200, other plant species may be useful in Cd phytoextraction (e.g., Broadhurst et al., 2015).

N. caerulescens evolved in Europe so growing it for Cd phytoextraction from Asian rice soils was only successful with fungicide applications, and Cd removal benefitted from acidification (Simmons et al., 2015). Alternatively, natural Cd-accumulating species adapted to the Asian rice production environment might offer high yield and moderate Cd accumulation and serve for Cd phytoextraction (Khaokaew & Landrot, 2015). Some rice genotypes can accumulate high levels of Cd in their shoots if soils are managed to be aerobic and acidic (Arao et al., 2009). The unusual natural

Cd-accumulator rice (*Oryza sativa* L.) genotypes with defective HMA3 gene may be useful in Cd phytoextraction from Zn–Cd-contaminated rice soils (Murakami et al., 2009; Tezuka et al., 2010). Plant breeders have even generated a new rice cultivar, Akita 110′ for practical Cd phytoextraction from rice soils (Takahashi et al., 2016). Cutting and disposal of the rice biomass before the grain is mature would be critical to allow use of these Cd-accumulating rice cultivars for practical phytoextraction of soil Cd. Best agronomic management of rice, *Zea mays* or *N. caerulescens* in Cd phytoextraction should involve soil acidification to near pH 5.5 to maximize annual Cd removals without causing Al or Mn phytotoxicity at lower pH (Wang et al., 2006), and then after removal of the needed amount of soil Cd, the soil should be limed to >6.5 postphytoextraction to minimize continuing Cd uptake or potential Zn phytotoxicity to other crop species.

An alternative to using Cd-accumulating rice cultivars to remove Cd from rice soils is the use of new rice mutants which accumulate only low levels of Cd because the Mn/Cd transporter of rice roots was inactivated (Ishikawa et al., 2012). New cultivars have been developed to minimize both Cd and As accumulation in rice grain (Ishikawa et al., 2016).

Another approach for element phytoextraction would use high biomass plant species which are not really hyperaccumulators, but which could remove enough of an element to achieve eventual phytoremediation if the crop had other value. The value for power generation, biogas production or CO_2 abatement could pay for the phytoextraction if the residues rich in Cd were disposed or recycled. It has been suggested that coppiced trees for biomass power generation might be able to remove significant amounts of Cd over time. Genotypes of willow (*Salix* species) and poplar (*Populus* species) have been tested in contaminated fields and achieved slow removal of Cd (Dickinson & Pulford, 2005; Eriksson & Ledin, 1999; Hammer, Kayser, & Keller, 2003; McGrath et al., 2006; Meers, Vandecasteele, Ruttens, Vangronsveld, & Tack, 2007). A team conducted an evaluation of using willow and maize to remove Cd from a smelter-contaminated area in Belgium including economic evaluation for heat or biogas production or considering the CO_2 abatement value. If someone would pay for the decontamination value, several approaches seem useful (Thewys, Witters, Meers, & Vangronsveld, 2010; Witters et al., 2012).

Removal of ^{137}Cs by canola also appears to offer an opportunity to decontaminate ^{137}Cs-contaminated soils. ^{137}Cs is accumulated in the grain, but when oil is produced from the grain and purified by standard industrial practices for canola oil, no ^{137}Cs remains in the oil. If the full value of the canola oil would pay for the crop production, proper disposal of the canola 'oilseed cake' would make a phytoextraction technology. If the crop were harvested as a 'hay' crop, more ^{137}Cs would be removed, but burning high potassium plant biomass is difficult in power generation facilities. The ash would require safe disposal in a radionuclide landfill at some cost, but the overall opportunity does seem to offer a way to remove ^{137}Cs from soils (Chaney, Russell, & Malik, 1998). Similar approaches for phytoextraction of ^{90}Sr or ^{238}U have not been reported because the accumulation by plants is too low. Sas-Nowosielska et al. (2004) and others have challenged the whole concept of phytoextraction suggesting that disposal of the phytoextraction biomass would be expensive or dangerous. We argue that the social value of soil decontamination may be great enough in terms of human health or environmental protection that the cost of proper disposal or recycling of the contaminants is still cost-effective. As discussed above, several elements could be phytoextracted/phytomined for value of the biomass metals as alternative metal ores (Ni, Tl, Zn). In other cases (e.g., As, radionuclides), contaminated biomass may have to be reduced in volume perhaps by composting or biogas production and the residues placed in secure landfills. Landfilling the biomass would be much less expensive than landfilling the entire volume of the contaminated soil, at least for large sites. If biomass is incinerated to produce energy or recover the metals, current exhaust treatment technologies can prevent dispersal of metals in stack emissions quite well. It might even be possible to separate the volatile elements in stack exhaust from other elements to allow cost-effective recycling or disposal of the volatile elements and other use of the decontaminated plant ash (Ljung & Nordin, 1997).

1.4 Impractical Phytoextraction

The section above summarized the cases where phytoextraction could reliably reduce soil metal concentrations. Some cases were cost-effective. The profitable cases are discussed below. There is another subset of proposed phytotechnologies which have not been found to be environmentally acceptable, similar to the case of chelator-induced Pb phytoextraction. Some elements can be accumulated by plants and the plants generate volatile chemical derivatives which could be dispersed into the general environment. For example, Se-accumulating plants produce some $(CH_3)_2Se$ as well as accumulating Se-amino acids in their tissues. Although this could remove Se from a particular area of soil, the dispersal would not be to land areas which needed Se fertilization and such emissions might not be environmentally acceptable. Some plants generate arsine or methyl-arsine compounds with part of the As accumulated from contaminated soils. Dispersal may not be environmentally acceptable. The last example of unacceptable phytovolatilization is Hg. Many plants can reduce Hg^{2+} to Hg^0 which converts the Hg into a volatile form released to the environment. Very creative plant biotechnology was used to produce bioengineered plants which could remove significant amounts of Hg from soils and release them to the atmosphere (Meagher & Heaton, 2005). However, because the major risk of Hg in soil is transfer to the general atmosphere and deposition into areas where sediment organisms generate $(CH_3)_2Hg$ which can be biomagnified in aquatic ecosystems, the potential for contamination of fish makes phytovolatilization of soil or wetland Hg unacceptable. Actually plant uptake of Hg^{2+} is very limited; most Hg in plant shoots arrives there by volatilization from the soil surface rather than root uptake and transport through the xylem (Gustin et al., 2004). Photovolatilization (especially UV light) is the major process which causes soil Hg to be released as Hg^0 to the local atmosphere (Gustin, Biester, & Kim, 2002; Moore & Carpi, 2005). Pedron, Petruzzelli, Barbafieri, and Tassi (2013) showed that induced phytoextraction of Hg could be improved by application of thiocyanate to soils, but phytovolatilization occurred from the plant-absorbed Hg.

1.5 Phytomining

Phytomining is qualitatively different from simple phytoextraction because the goal is to have enough soil element accumulated into the harvestable plant shoots to make an agricultural technology profitable. Table 4 compares annual Ni in shoots of maize or *Alyssum* hypernickelophore species. Even if the soil is acidified to reach 50% yield reduction of maize, the amount of Ni in the shoots has no value. But the natural hyperaccumulator growing wild in a serpentine pasture area, hand harvested separately as a crop, or improved cultivars grown for commercial phytomining offers an economic technology (Bani, Echevarria, Sulçe, & Morel, 2015; Chaney et al., 2007; Van der Ent et al., 2015). Our recent papers on the agronomy of Ni phytomining show the importance of fertilization, soil pH management, etc., on annual Ni harvest (Nkrumah et al., 2016; Van der Ent et al., 2015). Because Ni has enough value, Ni phytomining can be profitable (Table 3). Both our work (Chaney et al., 2007; Li, Chaney, Brewer, Angle, & Nelkin, 2003, Li, Chaney, Brewer, Roseberg, et al., 2003) and the Bani–Morel–Echevarria team (Bani et al., 2015) have shown that the harvestable Ni value is enough higher than production costs that it is profitable agriculture for

Table 4 Importance of Hyperaccumulation in Practical Phytoextraction of Soil Ni

Species	Soil	Yield (dry t/ha)	Ni in the Crop (mg/kg)	(kg/ha)	(% of Soil)	Ash-Ni (%)
Maize	Control	20	1	0.02	0.01	0.002
Maize (50% YD)	Ni-rich	10	100	1	0.01	0.20
Wild *Alyssum* in pasture	Ni-rich	3	10,000	30	0.3	10–15
Wild *Alyssum murale* crop	Ni-rich	10	20,000	200	2.0	20–25
Alyssum murale cultivar crop	Ni-rich	20	25,000	500	5.0	25–30

Estimated Ni phytoextraction by nonaccumulator crop maize (*Zea mays* L.) shoot biomass versus hypernickelophore *Alyssum murale* shoot biomass grown as a phytomining crop; assume control soil contains 25 mg Ni/kg, and the Ni-rich soil contains 2500 mg Ni/kg = 10,000 kg Ni/(ha-30 cm); for line 2, assume soil is acidic enough that Ni causes 50% yield reduction compared to corn grown on similar soil without Ni contamination or with higher pH. Research has shown that unimproved *Alyssum murale* can easily yield 10 t/ha when grown with herbicides and proper fertilization, and selected cultivars can exceed 20 t/ha with appropriate soil and crop management on serpentine soils. Improved cultivars accumulated appreciably higher shoot Ni concentrations (Li, Chaney, Brewer, Roseberg, et al., 2003). Most crop plant species suffer 25% yield reduction when the shoots contain 50–100 mg Ni/kg dry weight (L'Huillier, d'Auzac, Durand, & Michaud-Ferriere, 1996; Kukier & Chaney, 2004).

otherwise infertile and low productivity serpentine soils. It would be a lot more profitable if the value of Ni returned to the $20/kg level from before the recent international economic slowdown.

Researchers have conducted studies to clarify the ways that hyperaccumulator plants accumulate such high levels of metals in shoot biomass. Our team showed that *Alyssum* can extract Ni from many Ni minerals added to nutrient solutions (Centofanti, Siebecker, Chaney, Davis, & Sparks, 2012), but not bunsenite (NiO) a mineral with severely limited kinetics of dissolution. Our recent research illustrated that *Alyssum* roots could obtain high levels of Ni from serpentine subsoil for which the chemical extractability of the Ni is appreciably lower than that of Ni in topsoil (Chaney, Baklanov, Ryan, & Davis, 2017; Nkrumah, Chaney, & Morel, 2017; Paul, Baklanov, & Chaney, 2017).

These studies clearly showed that Ni arrived at rhizodermal cell membranes by diffusion rather than convection. Presuming plants consume about 250 mL of soil solution per gram shoot DW, and the data from Kukier, Peters, Chaney, Angle, and Roseberg (2004) that the Brockman variant gravelly loam soil containing 4700 mg Ni/kg had only 0.047 μg Ni/mL of saturation extract, soil solution convection could supply <0.1%

of the Ni absorbed by the plants (15,000 μg/g DW) (Chaney et al., 2017). Absorbed Ni arrives at the root membrane by diffusion from solid surfaces (organic and mineral) in the soil. So the kinds of minerals which occur in serpentine soils which are the proximal source of plant absorbed Ni are not known specifically (Massoura et al., 2006), but the hyperaccumulator plants absorb Ni from the same 'labile' pool as do crop plants (Shallari, Echevarria, Schwartz, & Morel, 2001). Although some have suspected that hyperaccumulator plants must secrete chelating agents to help dissolve soil metals, there is no evidence to support this hypothesis, nor that localized acidification at hyperaccumulator roots aids in dissolving soil metals. And although Alyssum species very efficiently store Ni in epidermal cell vacuoles (e.g., Broadhurst et al., 2004), other species may retain the absorbed Ni in extracellular space of leaves (Van der Ent & Mulligan, 2015). Although some reported that amino acids such as histidine, or organic acids such as citrate and malate chelated the Ni in xylem fluid of Ni hyperaccumulator Alyssum species, study of steady-state hyperaccumulating plants showed that less than 50% could be chelated by measured ligands (amino and organic acids) (Centofanti et al., 2013). Research on a number of the Ni hyperaccumulators will be needed to clarify the biochemistry of Ni accumulation and transport to and storage in the plant shoots.

Phytomining of elements other than Ni is obviously dependent on the existence of plants which hyperaccumulate high enough concentration in harvestable shoot biomass, and high enough value to this kind of 'bio-ore' to make a business. Table 5 is a list of known hyperaccumulator plant species for a number of elements, the potential concentration and shoot yield, and potential value. This table follows the model of Brooks, Chambers, Nicks, and Robinson (1998) and Brooks, Anderson, Stewart, and Robinson (1999) who compiled a similar table with the then current element values but included Cu and Co phytomining no longer considered valid. Some elements have little or no economic value (As, Cd), while others might have value as trace element fertilizers (Mn, Zn, Ni), and some are clearly economic phytomining opportunities (Ni) or might be economic (Au). Gold phytomining might be profitable if the ores or tailings are already ground and on a plastic liner to collect any leaching cyanide needed to induce gold dissolution and uptake (Anderson et al., 1999). Thallium seems to offer economic return enough to be profitable as phytomining (Brooks et al., 1999).

Selenium phytoextraction might be profitable if 'organic' Se supplements were used instead of inorganic Se salts in cattle diets or human supplements. Bañuelos (2006) showed that Se biofortification can be managed using several

Table 5 Potential Annual Value of Element in Phytomining Crop With Yield of 10 Dry t/ha With High End Concentrations of Elements Attainable With Agronomic Management to Maximize Element Concentrations in Harvestable Shoots When Grown on an Element Mineralized Soil Where the Species Occurs (Follows Approach of Brooks et al., 1998, 1999)

Element	Species	Concentration (kg/t)	Yield (t/ha)	Element in Crop (kg)	Value ($/kg)	Value of Crop ($/ha)
As[1]	*Pteris vittata*	10	10	100	?	?
Au[2]	*Brassica juncea*	0.039	10	0.39	43,800	17,100
Cd[3]	*Noccaea caerulescens*	2.5	5	12.5	1.0	12.5
Mn[4]	*Phytolacca acinosa*	20			2.6	65
Ni[5]	*Alyssum corsicum*	25	20	500	10.6	5,300
Se[6]	*Astragalus pattersoni*	10	10	100	36.9	3,690
Tl[7]	*Biscutella laevigata*	19	10	190	300	57,000
Zn[8]	*Noccaea caerulescens*	30	5	150	2.3	345

Some elements (Se, Mn, Ni and Zn) in phytomining crops might have higher $ value if marketed as 'Organic' nutrient fertilizers or feed supplements (Bañuelos, 2006; Wood et al., 2006). Co, Cu and Pb omitted due to lack of useful hyperaccumulator species. Useful phytoextraction of Au or Hg requires use of cyanide or thiocyanate addition to soil. Production costs would include fertilizers and pH adjustment amendments to optimize annual element yield in biomass (see also Harris, Naidoo, Nokes, Walker, & Orton, 2009).

[1]Ma et al. (2001); [2]Anderson et al. (2005); [3]Reeves et al. (2001); [4]Xue et al. (2004); [5]Li, Chaney, Brewer, Angle, et al. (2003); [6]Beath et al. (1937); [7]LaCoste et al. (1999); [8]Reeves et al. (2001).

? means unknown.

crop plants (canola, mustard, broccoli) and that the Se-rich crop might be marketed as a Se-biofortified food/crop (broccoli, *Brassica oleracea* var. *italica*) (Bañuelos & Dhillon, 2011). Although scientists were able to increase Se accumulation in *B. juncea* in field tests with a bioengineered *B. juncea* (Bañuelos, Terry, Leduc, Pilon-Smits, & Mackey, 2005), the annual removal remained low because crop plants cannot practically hyperaccumulate Se due to the uptake mechanism in roots. Crop plants have limited Se accumulation from most soils because selenate is accumulated on the sulfate transporter and the soils rich in Se are also rich in sulfate, so selenate accumulation is limited (Bell, Parker, & Page, 1992). In contrast, Se-hyperaccumulator species accumulate Se in the presence of high levels of sulfate (see Parker, Feist, Varvel, Thomason, & Zhang, 2003). Feist and Parker (2001) sampled diverse populations of *Stanylea pinnata* (Prush.) Britton in western United States and reported that this species would be considerably more readily managed for Se phytoextraction than other reported US Se hyperaccumulators such as *Astragalus* species. Se-hyperaccumulator crops might be used to produce livestock feed Se supplements because a high Se crop could be dried and ground and analyzed so that the supplement Se value (Bañuelos & Mayland, 2000) could be listed for marketing. Unfortunately the livestock feed industry has not shown interest is using hyperaccumulator biomass as the commercial 'Organic' Se supplement. Considering how much more evenly the biomass Se could be mixed with a feed than the very small amount of a Se salt, it seems rational to try to commercialize that practice.

Similarly, foliage of swamp black gum (*Nyssa sylvatica* vs. *sylvatica*) might be an effective Co supplement for ruminants. Many soils are low or deficient in Co and chemical Co supplements are supplied. The high accumulation of Co by *Nyssa* species suggested to Robinson, Brooks, and Hedley (1999) that leaves of this species could be added to livestock diets to prevent Co deficiency. This species has demonstrated remarkable bioaccumulation of Co from acidic Coastal Plain soils, accumulating 1000-fold levels in common forage grasses growing near the trees (Alban & Kubota, 1960; Kubota, Lazar, & Beeson, 1960). *Alyssum* species are effective Co hyperaccumulators when soils are low in Ni, but Ni inhibited Co accumulation by *Alyssum* compared to *Nyssa* (Malik, Chaney, Brewer, Li, & Angle, 2000); *Alyssum* is not a Co hyperaccumulator under the definition of Reeves (1992) because Co is usually no higher than 10% of the Ni level in serpentine soils. Interestingly, *Alyssum* accumulates both Ni and Co from soils, translocates both to *Alyssum* shoots, but does not store Co in epidermal cell vacuoles in contrast with Ni which is remarkably stored in vacuoles (Tappero et al., 2007).

1.6 Potential for Phytomining and Phytoproducts

Phytomining depends on (1) the existence of element rich soils, (2) plants which hyperaccumulate the element, (3) agronomic practices to maximize annual yield of metal in biomass, (4) methods to recover the element from the biomass postharvest and (5) markets for the element recovered from biomass (Angle et al., 2001; Bani et al., 2015; Bañuelos, 2006; Chaney et al., 2007). Phytomining of Ni has received the greatest attention because the commercial demand for Ni is large, the land area with Ni-rich natural soils is extensive, and hypernickelophore plants have been tested in the field (Bani et al., 2015; Li, Chaney, Brewer, Roseberg, et al., 2003; Nkrumah et al., 2016). A list of hyperaccumulator species for several elements are presented in Table 5 with the typical concentration of element in mineralized soils, the yield and element concentration observed in field grown plants with good agronomic practices and the fraction of soil total element removed in one annual crop biomass. Because ultramafic soils contain substantial levels of Ni, only a small percentage of total Ni in the soil to 30 cm depth (4%) is removed in one year. Thallium phytomining looks like an economic opportunity although the market for global Tl demand is small (Brooks et al., 1999). And gold phytomining might be profitable based on research by Anderson, Moreno, and Meech (2005) but requirement for a ligand to induce phytoextraction makes this a difficult technology to obtain permits to conduct.

The potential for production of 'Organic' Se-enriched or Se-biofortified foods (Bañuelos, 2006), or 'Organic' Ni fertilizer (Wood, Chaney, & Crawford, 2006) or feed Se-supplement products (Bañuelos & Mayland, 2000) from accumulator plants is another commercial opportunity (Robinson, Bañuelos, Conesa, Evangelou, & Schulin, 2009). Chaney et al. (2007) reported that ash from *Alyssum murale* plants grown on serpentine soils was supplied to an electric arc furnace at Sudbury, Ontario, and Ni metal recovered very effectively. Others have recovered Ni using hydrometallurgy and produced $Ni(NH_4)_2(SO_4)_2$ (e.g., Barbaroux et al., 2012; Zhang et al., 2016) which may have higher commercial value per kg of Ni than does Ni metal (easiest product from biomass ash); the latter paper reported purity of the Ni salt over 99%. Van der Ent et al. (2015) reviewed a range of Ni products which may be produced from Ni-rich phytomining biomass. Others have discussed production of nanoparticles in hyperaccumulator plants or plants fed soluble metal ions. No nanoparticle or catalyst products have yet been marketed from phytomining plant biomass.

Production of an 'Organic' plant biomass Zn supplement from *N. caerulescens* must take into account the usual co-contamination with Cd and unacceptability of Cd in such a supplement. If Zn only contaminated soils (e.g., rubber ash; rubber contains 1.5% Zn and low mg/kg Cd) (Taheri, Khoshgoftarmanesh, Shariatmadari, & Chaney, 2011) were used to grow the *Prayon* ecotype *N. caerulescens*, a Zn-rich low Cd product might be attained. Robinson et al. (2005) evaluated Zn accumulation by willow and poplar grown as stock fodder, and Anderson, Robinson, Clucas, West, and Portmann (2012) reported a sheep feeding experiment with grass hay grown on biosolids amended soil to increase Zn concentration, or willow rich in Zn, and found that the willow forage fed to sheep could prevent Zn deficiency problems.

2. PHYTOSTABILIZATION OF SOIL METALS

The above discussion illustrates that phytoextraction might be effective for some contaminated or mineralized soils but that phytoextraction of Cd and other elements will require payment because the value of element in the harvestable shoot biomass is insufficient to pay for the crop production, harvest and processing/disposal. Nickel and thallium were shown to provide a commercial phytomining potential and others theoretically may be profitable if permitted (gold) (Table 5). Although phytoextraction could be less expensive than removal and replacement of contaminated topsoil, until government orders remediation of elements from contaminated soils for which phytoextraction can be effective, it will not become a common practice. Protecting food and feed from excessive Cd or Se could be achieved by not growing food or feed on Cd-contaminated or Se-contaminated soils. Breeding Cd excluder rice and other crops might be an effective alternative but requires new cultivars (fortunately discovered for rice; Ishikawa et al., 2012). Because phytoextraction requires long periods and is applicable to only a few elements in contaminated soils, an alternative technology was sought.

Phytostabilization is the use of soil amendments and low-accumulating plant species to alleviate the environmental risk from contaminated soils. Phytostabilization has been applied to coal mine soils for over 50 years to alleviate phytotoxicity of Al and Mn in highly acidic coal mine wastes. Excessive soil Pb is an important problem in many nations because young children may ingest soil or housedust rich in Pb and suffer health effects of that Pb (Brown, Chaney, & Hettiarachchi, 2016). With the recognition that Pb phytoextraction is not practical with any plant species known to humans, but that treatment of soil with fertilizers and other soil amendments can greatly reduce both plant uptake and bioavailability of soil Pb to animals and children which might ingest soil, it became clear that Pb phytostabilization could be an effective way to manage moderately Pb-contaminated soils (Chaney & Mahoney, 2014). Ryan et al. (2004) reported on a field experiment at Joplin, Missouri, Unites States, where a number of soil treatments were tested and soils fed to rats, swine or humans. High phosphate addition reduced the absorption of soil Pb by all three animals by about 69% compared to the untreated (but normally fertilized) soil. And the grasses grown on these plots accumulated only low levels of Pb, Cd and Zn because the soils were made alkaline after soil treatments to reduce soil Pb bioavailability (Brown, Chaney, Hallfrisch, Ryan, & Berti, 2004). Other research has shown that this approach of inactivating soil Pb and other elements may both revegetate and protect ecosystems where metal smelting or mining caused local soil contamination. Brown, Sprenger, Maxemchu, and Compton (2005) illustrated revegetation of previously barren mine waste and smelter-contaminated soils rich in Zn, Cd and Pb, including protection of earthworms and small mammals. In such cases it is important to make the amended soil calcareous so that mineralization of N and S in the amendments cannot acidify the soil and allow Zn phytotoxicity to recur. Brown and Chaney (2016) summarized current progress in testing use of soil amendments to restore ecosystems at metal-impacted sites.

Although much progress has been made in identification of element hyperaccumulator plants, and in characterization of the agronomic management practices needed for commercialization of phytomining, none of the proposed technologies have been adopted commercially. Our patents and licensing thereof delayed commercialization of Ni phytomining, and traditional mining and smelting companies are leery of agricultural approaches. Failure of governments to require remediation of Cd-contaminated fields has supported development of Cd-excluder rice cultivars. Development of effective phytostabilization methods which protect the general environment limit adoption of phytoextraction approaches. Based on the science summarized in this paper, we believe that economics will eventually lead to adoption of Ni and perhaps several other element phytomining industry.

REFERENCES

Alban, L. A., & Kubota, J. (1960). A study of extractable soil cobalt in soils of the southeastern United States. *Soil Science Society of America, Proceedings, 24*, 183–185.

Anderson, C., Moreno, F., & Meech, J. (2005). A field demonstration of gold phytoextraction technology. *Minerals Engineering, 18*, 385–392.

Anderson, C. W. N., Brooks, R. R., Chiarucci, A., LaCoste, C. J., Leblanc, M., Robinson, B. H., … Stewart, R. B. (1999). Phytomining for nickel, thallium, and gold. *Journal of Geochemical Exploration, 67*, 407–415.

Anderson, C. W. N., Robinson, B. H., Clucas, L., West, D. M., & Portmann, D. (2012). Zinc-enriched and zinc-biofortified feed as a possible animal remedy in pastoral agriculture: Animal health and environmental benefits. *Journal of Geochemical Exploration, 121*, 30–35.

Angle, J. S., Chaney, R. L., Baker, A. J. M., Li, Y.-M., Reeves, R., Volk, V., … Nelkin, J. P. (2001). Developing commercial phytoextraction technologies: Practical considerations. *South African Journal of Science, 97*, 619–623.

Arao, T., Kawasaki, A., Baba, K., Mori, S., & Matsumoto, S. (2009). Effects of water management on cadmium and arsenic accumulation and dimethylarsinic acid concentrations in Japanese rice. *Environmental Science and Technology, 43*, 9361–9367.

Assunção, A. G. L., Schat, H., & Aarts, M. G. M. (2003). *Thlaspi caerulescens*, an attractive model species to study heavy metal hyperaccumulation in plants. *New Phytologist, 159*, 351–360.

Baker, A. J. M. (1981). Accumulators and excluders Strategies in the response of plants to heavy metals. *Journal of Plant Nutrition, 3*, 643–654.

Baker, A. J. M., & Brooks, R. R. (1989). Terrestrial higher plants which hyperaccumulate metal elements. A review of their distribution, ecology, and phytochemistry. *Biorecovery, 1*, 81–126.

Baker, A. J. M., McGrath, S. P., Reeves, R. D., & Smith, J. A. C. (2000). Metal hyperaccumulator plants: A review of the ecology and physiology of a biological resource for phytoremediation of metal-polluted soils. In N. Terry, & G. S. Bañuelos (Eds.), *Phytoremediation of contaminants in soil and water* (pp. 85–107). Boca Raton, FL: CRC Press.

Bani, A., Echevarria, G., Sulçe, S., & Morel, J. L. (2015). Improving the agronomy of *Alyssum murale* for extensive phytomining: A five-year field study. *International Journal of Phytoremediation, 17*, 117–127.

Bañuelos, G. S. (2006). Phyto-products may be essential for sustainability and implementation of phytoremediation. *Environmental Pollution, 144*, 19–23.

Bañuelos, G. S., & Dhillon, K. S. (2011). Developing a sustainable phytomanagement strategy for excessive selenium in Western United States and India. *International Journal of Phytoremediation, 13*(S1), 208–228.

Bañuelos, G. S., & Mayland, H. F. (2000). Absorption and distribution of selenium in animals consuming canola grown for selenium phytoremediation. *Ecotoxicology and Environmental Safety, 46*, 322–328.

Bañuelos, G. S., Terry, N., Leduc, D. L., Pilon-Smits, E. A. H., & Mackey, B. (2005). Field trial of transgenic Indian mustard plants shows enhanced phytoremediation of selenium-contaminated sediment. *Environmental Science and Technology, 39*, 1771–1777.

Barbaroux, R., Plasari, E., Mercier, G., Simonnot, M. O., Morel, J. L., & Blais, J. F. (2012). New process for nickel ammonium disulfate production from ash of the hyperaccumulating plant *Alyssum murale*. *The Science of the Total Environment, 423*, 111–119.

Baumann, A. (1885). Das Verhalten von Zinksalzen gegen Pflanzen und im Boden. *Landwirtschaftliche Versuchsstation, 31*, 1–53.

Beath, O. A. (1943). Toxic vegetation growing on the Salt Wash sandstone member of the Morrison formation. *American Journal of Botany, 30*, 698–707.

Beath, O. A., Eppsom, H. F., & Gilbert, C. S. (1937). Selenium distribution in and seasonal variation of type vegetation occurring on seleniferous soils. *Journal of the American Pharmaceutical Association, 26*, 394–405.

Bell, P. F., Parker, D. R., & Page, A. L. (1992). Contrasting selenate-sulfate interactions in selenium-accumulating and nonaccumulating plant species. *Soil Science Society of American Journal, 56*, 1818–1824.

Blaylock, M. J., Salt, D. E., Dushenkov, S., Zakharova, O., Gussman, C., Kapulnik, Y., … Raskin, I. (1997). Enhanced accumulation of Pb in Indian mustard by soil-applied chelating agents. *Environmental Science and Technology, 31*, 860–865.

Bradshaw, A. D., Humphries, R. N., & Johnson, M. S. (1978). The value of heavy metal tolerance in the vegetation of metalliferous mine wastes. In G. T. Goodman, & M. J. Chadwick (Eds.), *Environmental management of mineral wastes* (pp. 311–334). The Netherlands: Sijthoff & Noordhoff.

Broadhurst, C. L., Chaney, R. L., Angle, J. S., Maugel, T. K., Erbe, E. E., & Murphy, C. A. (2004). Simultaneous hyperaccumulation of nickel, manganese, and calcium in *Alyssum* leaf trichomes. *Environmental Science and Technology, 38*, 5797–5802.

Broadhurst, C. L., Chaney, R. L., Davis, A. P., Cox, A., Kumar, K., Reeves, R. D., & Green, C. E. (2015). Growth and cadmium phytoextraction by Swiss chard, corn, rice, *Noccaea caerulescens* and *Alyssum murale* in pH adjusted biosolids amended soils. *International Journal of Phytoremediation, 17*, 25–39.

Brooks, R. R. (1998). *Plants that hyperaccumulate heavy metals: Their role in phytoremediation, microbiology, archaeology, mineral exploration and phytomining*. Wallingford, Oxon, UK: CAB International, 380 pp.

Brooks, R. R., Anderson, C., Stewart, R., & Robinson, B. (1999). Phytomining: Growing a crop of a metal. *Biologist, 46*, 201–205.

Brooks, R. R., Chambers, M. F., Nicks, L. J., & Robinson, B. H. (1998). Phytomining. *Trends in Plant Science, 3*, 359–362.

Brooks, R. R., Lee, J., Reeves, R. D., & Jaffré, T. (1977). Detection of nickeliferous rocks by analysis of herbarium specimens of indicator plants. *Journal of Geochemical Exploration, 7*, 49–57.

Brooks, R. R., Trow, M. M., Veillon, J.-M., & Jaffré, T. (1981). Studies on manganese-accumulating *Alyxia* species from New Caledonia. *Taxon, 30*, 420–423.

Brown, S. L., Angle, J. S., Chaney, R. L., & Baker, A. J. M. (1995). Zinc and cadmium uptake by *Thlaspi caerulescens* and *Silene cucubalis* grown on sludge-amended soils in relation to total soil metals and soil pH. *Environmental Science and Technology, 29*, 1581–1585.

Brown, S. L., & Chaney, R. L. (2016). Use of amendments to restore ecosystem function to metal mining impacted sites; tools to evaluate efficacy. *Current Pollution Reports, 2*, 91–102. http://dx.doi.org/10.1007/s40726-016-0029-1 (Online).

Brown, S. L., Chaney, R. L., Angle, J. S., & Baker, A. J. M. (1994). Zinc and cadmium uptake by *Thlaspi caerulescens* and *Silene cucubalis* in relation to soil metals and soil pH. *Journal of Environmental Quality, 23*, 1151–1157.

Brown, S., Chaney, R., Hallfrisch, J., Ryan, J. A., & Berti, W. R. (2004). In situ soil treatments to reduce the phyto- and bioavailability of lead, zinc, and cadmium. *Journal of Environmental Quality, 33*, 522–531.

Brown, S. L., Chaney, R. L., & Hettiarachchi, G. (2016). Lead in urban soils. A real or perceived concern for urban agriculture? *Journal of Environmental Quality, 45*, 26–36.

Brown, S., Sprenger, M., Maxemchu, A., & Compton, H. (2005). Ecosystem function in alluvial tailings after biosolids and lime addition. *Journal of Environmental Quality, 34*, 139–148.

Calabrese, E. J., Stanek, E. J., II, & Barnes, R. (1996). Methodology to estimate the amount and particle size of soil ingested by children: Implications for exposure assessment at waste sites. *Regulatory Toxicology and Pharmacology, 24*, 264–268.

Cannon, H. L. (1960). Botanical prospecting for ore deposits. *Science, 132*, 591–598.

Cannon, H. L. (1962). Development of botanical methods of prospecting for uranium on the Colorado Plateau. *US Geological Survey Bulletin, 1085A*, 1–50.

Cappa, J. J., & Pilon-Smits, E. A. H. (2013). Evolutionary aspects of elemental hyperaccumulation. *Planta, 239*, 267–275.

Cary, E. E., Grunes, D. L., Bohman, V. R., & Sanchirico, C. A. (1986). Titanium determination for correction of plant sample contamination by soil. *Agronomy Journal, 78*, 933–936.

Cary, E. E., Grunes, D. L., Dallyn, S. L., Pearson, G. A., Peck, N. H., & Hulme, R. S. (1994). Plant Fe, Al and Cr concentrations in vegetables as influenced by soil inclusion. *Journal of Food Quality, 17*, 467–476.

Cary, E. E., & Kubota, J. (1990). Chromium concentration in plants: Effects of soil chromium concentration and tissue contamination by soil. *Journal of Agricultural and Food Chemistry, 38*, 108–114.

Centofanti, T., Sayers, Z., Cabello-Conejo, M. I., Kidd, P., Nishizawa, N. K., Kakei, Y., … Chaney, R. L. (2013). Xylem exudate composition and root-to-shoot nickel translocation in *Alyssum* species. *Plant and Soil, 373*, 59–75.

Centofanti, T., Siebecker, M. G., Chaney, R. L., Davis, A. P., & Sparks, D. L. (2012). Hyperaccumulation of nickel by *Alyssum* species is related to solubility of Ni mineral species. *Plant and Soil, 359*, 71–83.

Chaney, R. L. (1983). Plant uptake of inorganic waste constituents. In J. F. Parr, P. B. Marsh, & J. M. Kla (Eds.), *Land treatment of hazardous wastes* (pp. 50–76). Park Ridge, NJ: Noyes Data Corp.

Chaney, R. L. (2010). Cadmium and zinc. In P. Hooda (Ed.), *Trace elements in soils* (pp. 409–439). Oxford, UK: Blackwell Publ.

Chaney, R. L. (2015). How does contamination of rice soils with Cd and Zn cause high incidence of human Cd disease in subsistence rice farmers. *Current Pollution Reports, 1*, 16–22.

Chaney, R. L., Angle, J. S., Broadhurst, C. L., Peters, C. A., Tappero, R. V., & Sparks, D. L. (2007). Improved understanding of hyperaccumulation yields commercial phytoextraction and phytomining technologies. *Journal of Environmental Quality, 36*, 1429–1443.

Chaney, R. L., Angle, J. S., McIntosh, M. S., Reeves, R. D., Li, Y.-M., Brewer, E. P., … Baker, A. J. M. (2005). Using hyperaccumulator plants to phytoextract soil Ni and Cd. *Zeitschrift für Naturforschung, 60C*, 190–198.

Chaney, R. L., Baklanov, I. A., Ryan, T. C., & Davis, A. P. (2017). Effect of soil volume on Ni hyperaccumulation from serpentine soil by *Alyssum corsicum*. *International Journal of Phytoremediation* (in press).

Chaney, R. L., Broadhurst, C. L., & Centofanti, T. (2010). Phytoremediation of soil trace elements. In P. Hooda (Ed.), *Trace elements in soils* (pp. 311–352). Oxford, UK: Blackwell Publ.

Chaney, R. L., Hornick, S. B., & Sikora, L. J. (1981). Review and preliminary studies of industrial land treatment practices. In *Proc. Seventh Annual Research Symposium on land disposal of municipal solid and hazardous waste and resource recovery* (pp. 200–212). EPA-600/9-81-002b.

Chaney, R. L., Li, Y.-M., Angle, J. S., Baker, A. J. M., Reeves, R. D., Brown, S. L., … Chin, M. (2000). Improving metal hyperaccumulator wild plants to develop commercial phytoextraction systems: Approaches and progress. In N. Terry, & G. S. Banuelos (Eds.), *Phytoremediation of contaminated soil and water* (pp. 131–160). Boca Raton, FL: CRC Press.

Chaney, R. L., & Mahoney, M. (2014). Phytostabilization and phytomining: Principles and successes. Paper 104. In *Proc. Life of Mines Conf. (July 15–17, 2014; Brisbane Australia)*. Brisbane, Australia: Aust. Inst. Mining Metallurgy.

Chaney, R. L., Reeves, R. D., Baklanov, I. A., Centofanti, T., Broadhurst, C. L., Baker, A. J. M., … Roseberg, R. J. (2014). Phytoremediation and phytomining: using plants to remediate contaminated or mineralized environments. In R. Rajakaruna, R. S. Boyd, & T. Harris (Eds.), *Plant ecology and evolution in harsh environments* (pp. 365–391). NY: Nova Science Publishers.

Chaney, R. L., Russell, P., & Malik, M. (1998). Phytoextraction and phytostabilization of radionuclides in contaminated soils. In *Proc. 1998 International Radiological Post-emergency Response Issues Conference (Sept. 9–11, 1998, Washington, DC)* (pp. 137–138). US Environ. Protect. Agency. EPA 402-S-98-001.

Dickinson, N. M., Baker, A. J. M., Doronila, A., Laidlaw, S., & Reeves, R. D. (2009). Phytoremediation of inorganics: Realism and synergies. *International Journal of Phytoremediation, 11*, 97–114.

Dickinson, N. M., & Pulford, I. D. (2005). Cadmium phytoextraction using short-rotation coppice *Salix*: The evidence trail. *Environment International, 31*, 609–613.

Eriksson, J., & Ledin, S. (1999). Changes in phytoavailability and concentration of cadmium in soil following long term *Salix* cropping. *Water, Air, and Soil Pollution, 114*, 171–184.

Ernst, W. H. O. (1974). *Schwermetallvegetation der Erde*. Stuttgart: Fischer.

Ernst, W. H. O. (1988). Decontamination of mine sites by plants: An analysis of the efficiency. In *Proc Intern. Conf. Environmental Contamination (Venice)* (pp. 305–310). Edinburgh: CEP Consultants, Ltd.

Ernst, W. H. O. (1996). Bioavailability of heavy metals and decontamination of soils by plants. *Applied Geochemistry, 11*, 163–167.
Ernst, W. H. O. (2000). Commentary: Evolution of metal hyperaccumulation and phytoremediation hype. *New Phytologist, 146*, 357–358.
Ernst, W. H. O. (2005). Phytoextraction of mine wastes – Options and impossibilities. *Chemie der Erde-Geochemistry, 65*(S1), 29–42.
Evangelou, M. W. H., Ebel, M., & Schaeffer, A. (2007). Chelate assisted phytoextraction of heavy metals from soil. Effect, mechanism, toxicity, and fate of chelating agents. *Chemosphere, 68*, 989–1003.
Faucon, M.-P., Shutcha, M. N., & Meerts, P. (2007). Revisiting copper and cobalt concentrations in supposed hyperaccumulators from SC Africa: Influence of washing and metal concentrations in soil. *Plant and Soil, 301*, 29–36.
Feist, L. J., & Parker, D. R. (2001). Ecotypic variation in selenium accumulation among populations of *Stanleya pinnata*. *New Phytologist, 149*, 61–69.
Foy, C. D., Chaney, R. L., & White, M. C. (1978). The physiology of metal toxicity in plants. *Annual Review of Plant Physiology, 29*, 511–566.
Garrett, R. G., Porter, A. R. D., Hunt, P. A., & Lalor, G. C. (2008). The presence of anomalous trace element levels in present day Jamaican soils and the geochemistry of Late-Miocene or Pliocene phosphorites. *Applied Geochemistry, 23*, 822–834.
Gustin, M. S., Biester, H., & Kim, C. S. (2002). Investigation of the light-enhanced emission of mercury from naturally enriched substrates. *Atmospheric Environment, 36*, 3241–3254.
Gustin, M. S., Ericksen, J. A., Schorran, D. E., Johnson, D. W., Lindberg, S. E., & Coleman, J. S. (2004). Application of controlled mesocosms for understanding mercury air-soil-plant exchange. *Environmental Science and Technology, 38*, 6044–6050.
Hammer, D., Kayser, A., & Keller, C. (2003). Phytoextraction of Cd and Zn with *Salix viminalis* in field trials. *Soil Use and Management, 19*, 187–192.
Hammer, D., & Keller, C. (2003). Phytoextraction of Cd and Zn with *Thlaspi caerulescens* in field trials. *Soil Use and Management, 19*, 144–149.
Hanikenne, M., & Nouet, C. (2011). Metal hyperaccumulation and hypertolerance: A model for plant evolutionary genomics. *Current Opinion in Plant Biology, 14*, 252–259.
Harris, A. T., Naidoo, K., Nokes, J., Walker, T., & Orton, F. (2009). Indicative assessment of the feasibility of Ni and Au phytomining in Australia. *Journal of Cleaner Production, 17*, 194–200.
Huang, J. W., Chen, J., Berti, W. R., & Cunningham, S. D. (1997). Phytoremediation of lead-contaminated soils: Role of synthetic chelates in lead phytoextraction. *Environmental Science and Technology, 31*, 800–805.
Huang, J. W., & Cunningham, S. D. (1996). Lead phytoextraction: Species variation in lead uptake and translocation. *New Phytologist, 134*, 75–84.
Hwang, Y.-H., Bornschein, R. L., Grote, J., Menrath, W., & Roda, S. (1997). Environmental arsenic exposure of children around a former copper smelter site. *Environmental Research, 72*, 72–81.
Ishikawa, S., Ishimaru, Y., Igura, M., Kuramata, M., Abe, T., Senoura, T., … Nakanishi, H. (2012). Ion-beam irradiation, gene identification, and marker-assisted breeding in the development of low-cadmium rice. *Proceedings of the National Academy of Sciences of the United States of America, 109*, 19166–19171.
Ishikawa, S., Makino, T., Ito, M., Harada, K., Nakada, H., Nishida, I., … Arao, T. (2016). Low-cadmium rice (*Oryza sativa* L.) cultivar can simultaneously reduce arsenic and cadmium concentrations in rice grains. *Soil Science and Plant Nutrition, 62*. http://dx.doi.org/10.1080/00380768.2016.1144452 (Online).

Jaffré, T., Pillon, Y., Thomine, S., & Merlot, S. (2013). The metal hyperaccumulators from New Caledonia can broaden our understanding of nickel accumulation in plants. *Frontiers in Plant Science, 4*, 279. http://dx.doi.org/10.3389/fpls.2013.00279.

Jaffré, T., & Schmid, M. (1974). Accumulation du nickel par une Rubiacée de Nouvelle Calédonia: Psychotria douarrei (G. Beauvisage) Däniker (in French) *Comptes Rendues de l'Academie des Sciences, Paris 278*, D1727–D1730.

Kersten, W. J., Brooks, R. R., Reeves, R. D., & Jaffré, T. (1979). Nickel uptake by New Caledonian species of *Phyllanthus*. *Taxon, 28*, 529–534.

Khaokaew, S., & Landrot, G. (2015). A field-scale study of cadmium phytoremediation in a contaminated agricultural soil at Mae Sot District, Tak Province, Thailand: (1) Determination of Cd-hyperaccumulating plants. *Chemosphere, 138*, 883–887.

Kochian, L. V., Pineros, M. A., & Hoekenga, O. A. (2005). The physiology, genetics and molecular biology of plant aluminum resistance and toxicity. *Plant and Soil, 274*, 175–195.

Koeppe, D. E. (1981). Lead: Understanding the minimal toxicity of lead in plants. In N. W. Lepp (Ed.), *Effects of trace metals on plant function: Vol. 1. Effect of heavy metal pollution on plants* (pp. 55–76). London: Applied Science Publ.

Koopmans, G. F., Römkens, P. F. A. M., Fokkema, M. J., Song, J., Luo, Y. M., Japenga, J., & Zhao, F. J. (2008). Feasibility of phytoextraction to remediate cadmium and zinc contaminated soils. *Environmental Pollution, 156*, 905–914.

Koopmans, G. F., Römkens, P. F. A. M., Song, J., Temminghoff, E. J. M., & Japenga, J. (2007). Predicting the phytoextraction duration to remediate heavy metal contaminated soils. *Water, Air, and Soil Pollution, 181*, 355–371.

Kubota, J., Lazar, V. A., & Beeson, K. C. (1960). The study of cobalt status of soils in Arkansas and Louisiana using the black gum as the indicator plant. *Soil Science Society of America, Proceedings, 24*, 527–528.

Kukier, U., & Chaney, R. L. (2004). *In situ* remediation of Ni-phytotoxicity for different plant species. *Journal of Plant Nutrition, 27*, 465–495.

Kukier, U., Peters, C. A., Chaney, R. L., Angle, J. S., & Roseberg, R. J. (2004). The effect of pH on metal accumulation in two *Alyssum* species. *Journal of Environmental Quality, 32*, 2090–2102.

Kumar, P. B. A. N., Dushenkov, V., Motto, H., & Raskin, I. (1995). Phytoextraction: The use of plants to remove heavy metals from soils. *Environmental Science and Technology, 29*, 1232–1238.

LaCoste, C., Robinson, B., Brooks, R., Anderson, C., Chiarucci, A., & LeBlanc, M. (1999). The phytoremediation potential of thallium-contaminated soils using *Iberis* and *Biscutella* species. *International Journal of Phytoremediation, 1*, 327–338.

L'Huillier, L., d'Auzac, J., Durand, M., & Michaud-Ferriere, N. (1996). Nickel effects on two maize (*Zea mays*) cultivars: Growth, structure, Ni concentration, and localization. *Canadian Journal of Botany, 74*, 1547–1554.

Li, Y.-M., Chaney, R. L., Brewer, E. P., Angle, J. S., & Nelkin, J. P. (2003). Phytoextraction of nickel and cobalt by hyperaccumulator *Alyssum* species grown on nickel-contaminated soils. *Environmental Science and Technology, 37*, 1463–1468.

Li, Y.-M., Chaney, R. L., Brewer, E., Roseberg, R. J., Angle, J. S., Baker, A. J. M., … Nelkin, J. (2003). Development of a technology for commercial phytoextraction of nickel: Economic and technical considerations. *Plant and Soil, 249*, 107–115.

Ljung, A., & Nordin, A. (1997). Theoretical feasibility for ecological biomass ash recirculation: Chemical equilibrium behavior of nutrient elements and heavy metals during combustion. *Environmental Science and Technology, 31*, 2499–2503.

Lombi, E., Zhao, F. J., Dunham, S. J., & McGrath, S. P. (2001). Phytoremediation of heavy metal-contaminated soils: Natural hyperaccumulation versus chemically enhanced phytoextraction. *Journal of Environmental Quality, 30*, 1919–1926.

Ma, L. Q., Komar, K. M., Tu, C., Zhang, W. H., Cai, Y., & Kennelley, E. D. (2001). A fern that hyperaccumulates arsenic. *Nature, 409*, 579.

Mahieu, S., Soussou, S., Cleyet-Marel, J. C., Brunel, B., Maure, L., Lefebvre, C., & Escarre, J. (2013). Local adaptation of metallicolous and non-metallicolous *Anthyllis vulneraria* populations: their utilization in soil restoration. *Restoration Ecology, 21*, 551–559.

Malik, M., Chaney, R. L., Brewer, E. P., Li, Y.-M., & Angle, J. S. (2000). Phytoextraction of soil cobalt using hyperaccumulator plants. *International Journal of Phytoremediation, 2*, 319–329.

Malyuga, D. P. (1964). *Biogeochemical methods of prospecting*. Moscow: Acad. Sci. Press, 1963; Translated. NY: Consultants Bureau. 205 pp.

Massoura, S. T., Echevarria, G., Becquer, T., Ghanbaja, J., Leclerc-Cessac, E., & Morel, J.-L. (2006). Control of nickel availability by nickel bearing minerals in natural and anthropogenic soils. *Geoderma, 136*, 28–37.

McGrath, S. P., Lombi, E., Gray, C. W., Caille, N., Dunham, S. J., & Zhao, F. J. (2006). Field evaluation of Cd and Zn phytoextraction potential by the hyperaccumulators *Thlaspi caerulescens* and *Arabidopsis halleri*. *Environmental Pollution, 141*, 115–125.

McGrath, S. P., Zhao, F. J., & Lombi, E. (2001). Plant and rhizosphere processes involved in phytoremediation of metal-contaminated soils. *Plant and Soil, 232*, 207–214.

Meagher, R. B., & Heaton, A. C. P. (2005). Strategies for the engineered phytoremediation of toxic element pollution: mercury and arsenic. *Journal of Industrial Microbiology and Biotechnology, 32*, 502–513.

Meers, E., Hopgood, M., Lesage, E., Vervaeke, P., Tack, F. M. G., & Verloo, F. G. (2004). Enhanced phytoextraction: In search of EDTA alternatives. *International Journal of Phytoremediation, 6*, 95–109.

Meers, E., Ruttens, A., Hopgood, M. J., Samson, D., & Tack, F. M. G. (2005). Comparison of EDTA and EDDS as potential soil amendments for enhanced phytoextraction of heavy metals. *Chemosphere, 58*, 1011–1022.

Meers, E., Vandecasteele, B., Ruttens, A., Vangronsveld, J., & Tack, F. M. G. (2007). Potential of five willow species (*Salix* spp.) for phytoextraction of heavy metals. *Environmental and Experimental Botany, 60*, 57–68.

Miller, R. J., & Koeppe, D. E. (1971). Accumulation and physiological effects of lead in corn. *Trace Subst Environ Health, 4*, 186–193.

Minguzzi, C., & Vergnano, O. (1948). Nickel content of the ash of *Alyssum bertolonii* (in Italian) *Atti della Societa Toscana Sciienze Naturali, Pisa, Memorie Serie A, 55*, 49–74 (Chem. Abstr. 44:9003h).

Moore, C., & Carpi, A. (2005). Mechanisms of the emission of mercury from soil: Role of UV radiation. *Journal of Geophysical Research, D:110*, 1–9.

Murakami, M., Nakagawa, F., Ae, N., Ito, N., & Arao, T. (2009). Phytoextraction by rice capable of accumulating Cd at high levels: Reduction of Cd content of rice grain. *Environmental Science and Technology, 43*, 5878–5883.

Nkrumah, P. N., Baker, A. J. M., Chaney, R. L., Erskine, P. D., Echevarria, G., Morel, J. L., & van der Ent, A. (2016). Current status and challenges in developing Ni phytomining: an agronomic perspective. *Plant and Soil*. http://dx.doi.org/10.1007/s11104-016-2859-4.

Nkrumah, P. N., Chaney, R. L., & Morel, J. L. (2017). Agronomy of metal crops used in agromining. In G. Echevarria, A. van der Ent, J. L. Morel, & A. J. M. Baker (Eds.), *Agromining: Farming for metals: Extracting unconventional resources using plants*. London: Springer (in press).

Nowack, B., Schulin, R., & Robinson, B. H. (2006). Critical assessment of chelant-enhanced metal phytoextraction. *Environmental Science and Technology, 40*, 5225–5232.

Parker, D. R., Feist, L. J., Varvel, T. J., Thomason, D. N., & Zhang, Y. (2003). Selenium phytoremediation potential of *Stanleya pinnata*. *Plant and Soil, 249*, 157–165.
Paul, A. L. D., Baklanov, I. A., & Chaney, R. L. (2017). Influence of subsoil and soil volume on the accumulation of nickel by *Alyssum corsicum* grown on a serpentine soil. *International Journal of Phytoremediation* (Submitted).
Pedron, F., Petruzzelli, G., Barbafieri, M., & Tassi, T. (2013). Remediation of a mercury-contaminated industrial soil using bioavailable contaminant stripping. *Pedosphere, 23*, 104–110.
Pilon-Smits, E. (2005). Phytoremediation. *Annual Review of Plant Biology, 56*, 15–39.
Pollard, A. J., Powell, K. D., Harper, F. A., & Smith, J. A. C. (2002). The genetic basis of metal hyperaccumulation in plants. *Critical Reviews in Plant Sciences, 21*, 539–566.
Rascio, N. (1977). Metal accumulation by some plants growing on zinc-mine dumps. *Oikos, 29*, 250–253.
Rascio, N., & Navari-Izzo, F. (2011). Heavy metal hyperaccumulating plants: How and why do they do it? And what makes them so interesting? *Plant Sci, 180*, 169–181.
Reeves, R. D. (1992). The hyperaccumulation of nickel by serpentine plants. In A. J. M. Baker, J. Proctor, & R. D. Reeves (Eds.), *The vegetation of ultramafic (serpentine) soils* (pp. 253–277). Andover, Hampshire, UK: Intercept Ltd.
Reeves, R. D., & Brooks, R. R. (1983). Hyperaccumulation of lead and zinc by two metallophytes from mining areas of Central Europe. *Environmental Pollution, A31*, 277–285.
Reeves, R. D., Schwartz, C., Morel, J. L., & Edmondson, J. (2001). Distribution and metal-accumulating behavior of *Thlaspi caerulescens* and associated metallophytes in France. *International Journal of Phytoremediation, 3*, 145–172.
Robinson, B. H., Anderson, C. W. N., & Dickinson, N. M. (2015). Phytoextraction: Where's the action? *Journal of Geochemical Exploration, 151*, 34–40.
Robinson, B. H., Bañuelos, G., Conesa, H. M., Evangelou, M. W. H., & Schulin, R. (2009). The phytomanagement of trace elements in soil. *Critical Reviews in Plant Sciences, 28*, 240–266.
Robinson, B. H., Brooks, R. R., & Hedley, M. J. (1999). Cobalt and nickel accumulation in *Nyssa* (tupelo) species and its significance for New Zealand agriculture. *New Zealand Journal of Agricultural Research, 42*, 235–240.
Robinson, B., Ferandez, J.-E., Madejòn, P., Marañon, T., Murillo, J. M., Green, S., & Clothier, B. (2003). Phytoextraction: An assessment of biogeochemical and economic viability. *Plant and Soil, 249*, 117–125.
Robinson, B., Mills, T., Green, S., Chancerel, B., Clothier, B., Fung, L., … McIvor, I. (2005). Trace element accumulation by poplars and willows used for stock fodder. *New Zealand Journal of Agricultural Research, 48*, 489–497.
Ryan, J. A., Berti, W. R., Brown, S. L., Casteel, S. W., Chaney, R. L., Doolan, M., … Mosby, D. (2004). Reducing children's risk from soil lead: Summary of a field experiment. *Environmental Science and Technology, 38*, 18A–24A.
Sas-Nowosielska, A., Kucharski, R., Makowski, E., Pogrzeba, M., Kuperberg, J. M., & Kryski, K. (2004). Phytoextraction crop disposal. An unsolved problem. *Environmental Pollution, 128*, 373–379.
Sasaki, A., Yamaji, N., Yokosho, K., & Ma, J. F. (2012). Nramp5 is a major transporter responsible for manganese and cadmium uptake in rice. *The Plant Cell, 24*, 2155–2167.
Schaider, L. A., Parker, D. R., & Sedlak, D. L. (2006). Uptake of EDTA-complexed Pb, Cd and Fe by solution- and sand-cultured *Brassica juncea*. *Plant and Soil, 286*, 377–391.
Scheckel, K. G., Chaney, R. L., Basta, N. T., & Ryan, J. A. (2009). Advances in assessing bioavailability of metal(loid)s in contaminated soils. *Advances in Agronomy, 104*, 1–52.

Schnoor, J. L., Licht, L. A., McCutcheon, S. C., Wolfe, N. L., & Carreira, L. H. (1995). Phytoremediation of organic and nutrient contaminants. *Environmental Science and Technology, 29*, A318–A323.

Shallari, S., Echevarria, G., Schwartz, C., & Morel, J. L. (2001). Availability of nickel in soils for the hyperaccumulator *Alyssum murale* (Waldst. & Kit.). *South African Journal of Science, 97*, 568–570.

Simmons, R. W., Chaney, R. L., Angle, J. S., Kruatrachue, M., Klinphoklap, S., Reeves, R. D., & Bellamy, P. (2015). Towards practical cadmium phytoextraction with *Noccaea caerulescens*. *International Journal of Phytoremediation, 17*, 191–199.

Smith, R. A. H., & Bradshaw, A. D. (1979). The use of metal tolerant populations for the reclamation of metalliferous wastes. *Journal of Applied Ecology, 16*, 595–612.

Sopper, W. E. (1993). *Municipal sludge use in land reclamation*. Boca Raton, FL: Lewis Publ.

Speir, T. W., Kettles, H. A., Percival, H. J., & Parshotam, A. (1999). Is soil acidification the cause of biochemical responses when soils are amended with heavy metal salts? *Soil Biology and Biochemistry, 31*, 1953–1961.

Stanek, E. J., III, Calabrese, E. J., & Barnes, R. M. (1999). Soil ingestion estimates for children in Anaconda using trace element concentrations in different particle size fractions. *Human and Ecological Risk Assessment, 5*, 547–558.

Taheri, S., Khoshgoftarmanesh, A. H., Shariatmadari, H., & Chaney, R. L. (2011). Kinetics of zinc release from ground tire rubber and rubber ash in a calcareous soil as alternatives to Zn fertilizers. *Plant and Soil, 341*, 89–97.

Takahashi, R., Ito, M., Katou, K., Sato, K., Nakagawa, S., Tezuka, K., … Kawamoto, T. (2016). Breeding and characterization of the rice (*Oryza sativa* L.) line 'Akita 110' for cadmium phytoremediation. *Soil Science and Plant Nutrition, 62*, 373–378. http://dx.doi.org/10.1080/00380768.2016.1150791.

Tang, Y.-T., Deng, T.-H.-B., Wu, Q.-H., Wang, S.-Z., Qiu, R.-L., Wei, Z.-B., … Morel, J.-L. (2012). Designing cropping systems for metal-contaminated sites: A review. *Pedosphere, 22*, 470–488.

Tappero, R. V., Peltier, E., Gräfe, M., Heidel, K., Ginder-Vogel, M., Livi, K. J. T., … Sparks, D. L. (2007). Hyperaccumulator *Alyssum murale* relies on a different metal storage mechanism for cobalt than for nickel. *New Phytologist, 175*, 641–654.

Tezuka, K., Miyadate, H., Katou, K., Kodama, I., Matsumoto, S., Kawamoto, T., … Akagi, H. (2010). A single recessive gene controls cadmium translocation in the cadmium hyperaccumulating rice cultivar Cho-Ko-Koku. *Theoretical and Applied Genetics, 120*, 1175–1182.

Thewys, T., Witters, N., Meers, E., & Vangronsveld, J. (2010). Economic viability of phytoremediation of a cadmium contaminated agricultural area using energy maize. Part II: Economics of anaerobic digestion of metal contaminated maize in Belgium. *International Journal of Phytoremediation, 12*, 663–679.

Van der Ent, A., Baker, A. J. M., van Balgooy, M. M. J., & Tjoa, A. (2013). Ultramafic nickel laterites in Indonesia (Sulawesi, Halmahera): Mining, nickel hyperaccumulators and opportunities for phytomining. *Journal of Geochemical Exploration, 128*, 72–79.

Van der Ent, A., Baker, A. J. M., Reeves, R. D., Chaney, R. L., Anderson, C. W. N., Meech, J. A., … Mulligan, D. R. (2015). Agromining: Farming for metals in the future? *Environmental Science and Technology, 49*, 4773–4780.

Van der Ent, A., Baker, A. J. M., Reeves, R. D., Pollard, A. J., & Schat, H. (2012). Hyperaccumulators of metal and metalloid trace elements: Facts and fiction. *Plant and Soil, 362*, 319–334.

Van der Ent, A., & Mulligan, D. (2015). Multi-element concentrations in plant parts and fluids of Malaysian nickel hyperaccumulator plants and some economic and ecological considerations. *Journal of Chemical Ecology, 41*, 396–408.

Van Nevel, L., Mertens, J., Oorts, K., & Verheyen, K. (2007). Phytoextraction of metals: How far from practice. *Environmental Pollution, 150*, 34–40.

Vangronsveld, J., Herzig, R., Weyens, N., Boulet, J., Adriaensen, K., Ruttens, A., ... Mench, M. (2009). Phytoremediation of contaminated soils and groundwater: Lessons from the field. *Environmental Science and Pollution Research International, 16*, 765–794.

Vassil, A. D., Kapulnik, Y., Raskin, I., & Salt, D. E. (1998). The role of EDTA in lead transport and accumulation by Indian mustard. *Plant Physiology, 117*, 447–453.

Vergnano, O., & Hunter, J. G. (1952). Nickel and cobalt toxicities in oat plants. *Annals of Botany, 17*, 317–329.

Wang, A. S., Angle, J. S., Chaney, R. L., Delorme, T. A., & Reeves, R. D. (2006). Soil pH effects on uptake of Cd and Zn by *Thlaspi caerulescens*. *Plant and Soil, 281*, 325–337.

Witters, N., Mendelsohn, R. O., Van Slycken, S., Weyens, N., Schreurs, E., Meers, E., ... Vangronsveld, J. (2012). Phytoremediation, a sustainable remediation technology? Conclusions from a case study. I: Energy production and carbon dioxide abatement. *Biomass Bioenergy, 39*, 454–469.

Wójcik, M., Gonnelli, C., Selvi, F., Dresler, S., Rostański, A., Vangronsveld, J., & Schat, H. (2017). Metallophytes of serpentine and calamine soils – their unique ecophysiology and potential for phytoremediation. In A. Cuypers, & J. Vangronsveld (Eds.), *Advances in botanical research* (Vol. 83, pp 1–42).

Wood, B. W., Chaney, R. L., & Crawford, M. (2006). Correcting micronutrient deficiency using metal hyper accumulators: *Alyssum* biomass as a natural product for nickel deficiency correction. *HortScience, 41*, 1231–1234.

Wu, J., Hsu, F. C., & Cunningham, S. C. (1999). Chelate-assisted Pb phytoextraction: Pb availability, uptake, and translocation constraints. *Environmental Science and Technology, 33*, 1898–1904.

Xue, S. G., Chen, Y. X., Reeves, R. D., Baker, A. J. M., Lin, Q., & Fernando, D. R. (2004). Manganese uptake and accumulation by the hyperaccumulator plant *Phytolacca acinosa* Roxb. (Phytolaccaceae). *Environmental Pollution, 131*, 393–399.

Yanai, J., Zhao, F.-J., McGrath, S. P., & Kosaki, T. (2006). Effect of soil characteristics on Cd uptake by the hyperaccumulator *Thlaspi caerulescens*. *Environmental Pollution, 139*, 167–175.

Zhang, X., Laubie, B., Houzelot, V., Plasari, E., Echevarria, G., & Simonnot, M.-O. (2016). Increasing purity of ammonium nickel sulfate hexahydrate and production sustainability in a nickel phytomining process. *Chemical Engineering Research and Design, 106*, 26–32.

FURTHER READING

Brooks, R. R., McCleave, J. A., & Schofield, E. K. (1977). Cobalt and nickel uptake by the *Nyssaceae*. *Taxon, 26*, 197–201.

Verbruggen, N., Hermans, C., & Schat, H. (2009). Molecular mechanisms of metal hyperaccumulation in plants. *New Phytologist, 181*, 759–776.

Wang, Y., Yu, K.-F., Poysa, V., Shi, C., & Zhou, Y.-H. (2012). A single point mutation in GmHMA3 affects cadmium (Cd) translocation and accumulation in soybean seeds. *Molecular Plant, 5*, 1154–1156.

CHAPTER SIX

Molecular and Cellular Aspects of Contaminant Toxicity in Plants: The Importance of Sulphur and Associated Signalling Pathways

Sophie Hendrix*,[1], Peter Schröder[§], Els Keunen*, Christian Huber[§] and Ann Cuypers*

*Hasselt University, Diepenbeek, Belgium
[§]Helmholtz Zentrum München, Neuherberg, Germany
[1]Corresponding author: E-mail: sophie.hendrix@uhasselt.be

Contents

Abstract

Environmental contamination with metals and organic compounds poses a serious threat to human health. Investigating plant responses to these contaminants at the molecular and cellular level is crucial to optimize phytoremediation strategies to clean up contaminated soils. Two key players in plant stress responses are the

Advances in Botanical Research, Volume 83
ISSN 0065-2296
http://dx.doi.org/10.1016/bs.abr.2016.12.007

sulphur-containing amino acids cysteine and methionine. Cysteine is an important constituent of the metal-chelating metallothioneins and is also the precursor for glutathione and subsequent phytochelatin synthesis. During stress conditions, glutathione is involved in (1) metal chelation, (2) xenobiotic detoxification and (3) antioxidative defence. The activated form of methionine, S-adenosylmethionine, is involved in the synthesis of ethylene and polyamines, both playing important roles in signal transduction. This review provides an overview of sulphur uptake and assimilation and its conversion into basic metabolites essential for detoxification and signal transduction during metal and organic contaminant exposure in plants. Furthermore, the cross talk between these pathways and their relation to the contaminant-induced oxidative challenge are discussed.

1. GENERAL INTRODUCTION

Since the industrial revolution, but especially over the past few decades, abiotic contaminants – mainly from agricultural and industrial origin – have emerged into the environment and are recognized to pose a serious threat to the environment and human health (Ibrahim, Hayyan, AlSaadi, Hayyan, & Ibrahim, 2016; Noguera-Oviedo & Aga, 2016). These contaminants can be categorized into different groups based on their physicochemical properties, but their persistence (e.g., metals) and degradability (e.g., organics) are determining factors for their presence in the environment. As a result, different strategies are needed to remediate the environment. Microbe-assisted phytoremediation is a promising technology to be implemented in the cleanup of metal-polluted soils and organics in soil and air (chapter: Phytoremediation and Phytomining: Status and Promise by Chaney & Baklanov, 2017; Chirakkara, Cameselle, & Reddy, 2016; chapters: Mycorrhiza-Assisted Phytoremediation by Coninx, Martinova, & Rineau, 2017; Bio- and Phytoremediation of Pesticide Contaminated Environments: A Review by Eevers, White, Vangronsveld, & Weyens, 2017; Plants in Air Phytoremediation by Gawronski, Gawronska, Lomnicki, Sæbo, & Vangronsveld, 2017; Gerhardt, Huang, Glick, & Greenberg, 2009; chapter: Potential Role of Plant-Associated Bacteria in Plant Metal Uptake and Implications in Phytotechnologies by Kidd, Alvarez-Lopez, Becerra-Castro, Cabello-Conejo, & Prieto-Fernández, 2017; Ma, Oliveira, Freitas, & Zhang, 2016; Vangronsveld et al., 2009). However, to exploit this technology, it is important to unravel the underlying molecular and cellular responses of plants to minimize phytotoxic responses.

Besides growth reduction and retardation, morphological and physiological responses as a result of organics or metal stress have been extensively described in plants (Verkleij, Golan-Goldhirsh, Antosiewisz,

Schwitzguébel, & Schröder, 2009). Multiple studies have focused on the underlying molecular mechanisms of these stress responses. To cope with toxic compounds, plants rely on different detoxification mechanisms such as (1) biotransformation, (2) conjugation and (3) sequestration, all of which should be highly coordinated to prevent cellular damage. In addition to the detoxification of excess contaminants, other factors such as environmental conditions (e.g., temperature, humidity) determine the stress intensity perceived by the plant (Reichenauer & Germida, 2008). Pivotal modulators of plant metabolism upon multiple environmental challenges are reactive oxygen species (ROS), which are involved in oxidative damage as well as signalling. In response to the alteration of ROS levels, metabolic adjustments lead to a newly established cellular homeostasis essential for plant performance (Cuypers et al., 2016; Foyer & Noctor, 2005).

When considering cellular detoxification and regulation mechanisms, the sulphur-containing amino acids, cysteine and methionine, play an important role. Cysteine is a major constituent of the metal chelators metallothioneins (MTs), glutathione (GSH) and phytochelatins (PCs) (Anjum, Gill, & Gill, 2014). MTs are gene-encoded peptides synthesized by ribosomes, whereas GSH and PCs are metabolically synthesized (Anjum et al., 2015). In addition to its function in metal chelation, GSH is often consumed in conjugation reactions of organic compounds before their sequestration into the vacuole (Dixon, Skipsey, & Edwards, 2010). Besides its role in detoxification, GSH is a major cellular antioxidant in plants (Noctor et al., 2012). As such, it is clear that detoxification and regulation are closely interconnected in plant stress acclimation. Methionine, the other sulphur-containing amino acid, forms the basis of S-adenosylmethionine (SAM), a central component in plant metabolism. It is also the initial biosynthetic compound in the production of the stress hormone ethylene (Sauter, Moffatt, Saechao, Hell, & Wirtz, 2013). Besides ROS, ethylene is an important regulator of stress perception and is known to be involved in multiple responses induced by abiotic stress (Keunen, Schellingen, Vangronsveld, & Cuypers, 2016). In consequence, when plants are exposed to organic contaminants or metals, sulphur is an essential element in (1) detoxification pathways and (2) the regulation of stress responses. Therefore, this review focusses on sulphur uptake and assimilation and its conversion into basic metabolites essential for detoxification and regulation mechanisms during organic contaminant and metal exposure in plants. These processes are strongly interconnected and are discussed in view of the oxidative challenge and signal transduction during stress conditions.

2. SULPHUR UPTAKE AND ASSIMILATION

Sulphur is an essential macronutrient for all organisms, serving different metabolic functions. It is incorporated into the amino acids cysteine and methionine, controlling the structure and biological activity of many proteins (Davidian & Kopriva, 2010; Gotor et al., 2015). Furthermore, it is an important component of several coenzymes and prosthetic groups and is involved in plant responses to both biotic and abiotic stress factors (Davidian & Kopriva, 2010; Gigolashvili & Kopriva, 2014; Romero et al., 2014). Sulphur present as inorganic sulphate (SO_4^{2-}) in soils is uptaken by plants. To enter metabolic pathways, sulphate is assimilated in plants through a pathway consisting of (1) sulphate uptake, (2) sulphate activation, (3) sulphate reduction and (4) synthesis of cysteine (Fig. 1) (Droux, 2004; Ravilious & Jez, 2012).

Sulphate is uptaken by plants via the action of proton/sulphate cotransport systems (Gigolashvili & Kopriva, 2014; Takahashi, Kopriva, Giordano, Saito, & Hell, 2011). The plant family of sulphate transporters (SULTRs), composed of 12–16 genes depending on the plant species, is subdivided into four functional groups based on their sequence, biochemical characteristics and physiological function (Takahashi et al., 2011). The SULTR family is best characterized in the model organism *Arabidopsis thaliana* (Gigolashvili & Kopriva, 2014). The most important transporters involved in sulphate uptake in roots are the high-affinity SULTR1;1 and SULTR1;2 proteins (group 1). However, as the expression of the *SULTR1;1* gene in roots is much lower than that of *SULTR1;2*, the latter is considered the most important transporter involved in plant sulphate uptake during normal sulphate supply (Gigolashvili & Kopriva, 2014; Rouached et al., 2008). During sulphate starvation, however, *SULTR1;1* expression is strongly upregulated, a response that is likely due to the presence of a sulphur-responsive (SURE) *cis* element in its promoter (Maruyama-Nakashita et al., 2005). Another known SULTR is SULTR1;3, which is localized in phloem companion cells and involved in sulphate transport from source to sink (Yoshimoto, Inoue, Saito, Yamaya, & Takahashi, 2003). Furthermore, the low-affinity group 2 SULTRs facilitate long-distance sulphate transport (Gigolashvili & Kopriva, 2014; Takahashi et al., 2000). The activity of SULTR2;1 is modulated by SULTR3;5, which does not transport sulphate itself (Kataoka, Hayashi, Yamaya, & Takahashi, 2004). In contrast to SULTR3;5, other group 3 SULTRs function in sulphate import into plastids (Cao et al., 2013; Gigolashvili & Kopriva,

2014), as the enzymes necessary for sulphate reduction are exclusively present in these organelles (Davidian & Kopriva, 2010). Transporters belonging to group 4 are localized in the tonoplast, releasing sulphate from vacuoles (Kataoka, Watanabe-Takahashi, et al., 2004; Takahashi et al., 2011).

After its uptake in plant cells, sulphate is activated in an adenylation reaction catalyzed by ATP sulphurylase (ATPS). The resulting adenosine 5′-phosphosulphate (APS) is an important branching point in the sulphate assimilation pathway (Fig. 1). It can be phosphorylated by the action of APS kinase (APK), leading to the formation of 3′-phosphoadenosine 5′-phosphosulphate (PAPS), an important activated sulphate donor for many sulphation reactions in plant secondary metabolism. In the primary sulphate assimilation pathway, however, APS reductase (APR) reduces APS to sulphite (SO_3^{2-}), which is further reduced to sulphide (S^{2-}) by the ferredoxin-dependent sulphite reductase (SiR). Sulphide is subsequently used for the synthesis of cysteine (Droux, 2004; Takahashi et al., 2011). Whereas ATPS is also present in the cytosol, the reductive steps of the sulphate assimilation pathway only take place in plastids (Davidian & Kopriva, 2010).

The final step in the assimilation pathway of reduced sulphate is the synthesis of cysteine (Fig. 1). This sulphur-containing amino acid is subsequently incorporated into different compounds including MTs, GSH and PCs, important in metal chelation, detoxification of xenobiotics and antioxidative defence (Anjum et al., 2014). Furthermore, cysteine is an important building block for the synthesis of methionine, the second sulphur-containing amino acid (Wirtz & Droux, 2005). In the first and rate-limiting step of cysteine biosynthesis, O-acetylserine (OAS) is synthesized from L-serine and acetyl coenzyme A by serine acetyl transferase (SAT) (Wirtz & Droux, 2005; Wirtz & Hell, 2006). Subsequently, cysteine is formed by substitution of the acetate of OAS with sulphide in a β-replacement reaction catalyzed by OAS (thiol) lyase (OASTL). Both SAT and OASTL are present in the cytosol, plastids and mitochondria (Anjum et al., 2014). These enzymes were demonstrated to associate, forming the hetero-oligomeric cysteine synthase complex. This association with OASTL is necessary for the stability and activity of SAT, which is otherwise subjected to feedback inhibition by cysteine. In contrast, OASTL activity is silenced when bound to SAT, suggesting that this enzyme only functions as a chaperone-like subunit in the cysteine synthase complex (Wirtz et al., 2010; Wirtz & Hell, 2006). However, as cellular OASTL activity is 100–300 times higher than SAT activity, free OASTL enzymes

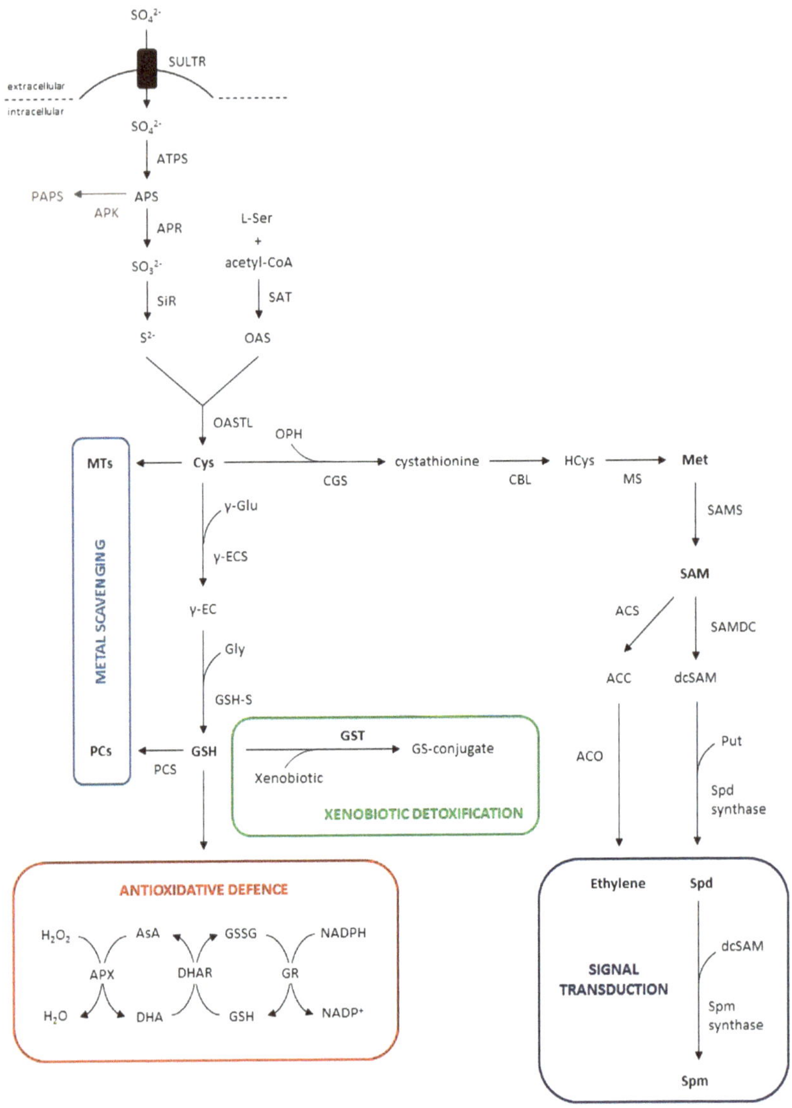

Figure 1 ***Biosynthesis pathway of sulphur-derived cellular compounds and their involvement in metal scavenging, xenobiotic detoxification, antioxidative defence and signal transduction.*** *γ-EC*, γ-glutamylcysteine; *γ-ECS*, γ-EC synthetase; *γ-Glu*, γ-glutamate; *ACC*, 1-aminocyclopropane-1-carboxylic acid; *acetyl-CoA*, acetyl coenzyme A; *ACO*, ACC oxidase; *ACS*, ACC synthase; *APK*, APS kinase; *APR*, APS reductase; *APS*, adenosine 5′-phosphosulphate; *APX*, ascorbate peroxidase; *AsA*, antioxidative ascorbate; *ATPS*, ATP sulphurylase; *CBL*, cystathionine β-lyase; *CGS*, cystathionine γ-synthase; *Cys*, cysteine; *dcSAM*, decarboxylated SAM; *DHA*, dehydroascorbate; *DHAR*, DHA reductase; *Gly*, glycine; *GR*, glutathione reductase; *GS-conjugate*, glutathione S-conjugate; *GSH*,

are present in cells and catalyze the formation of cysteine from sulphide and OAS released by the complex (Droux, 2004; Saito, 2000). It must be noted that several metabolites regulate the stability of the cysteine synthase complex. During sulphur-deprived conditions, for example, increased OAS levels stimulate dissociation of the complex, thereby inhibiting SAT activity and further OAS accumulation. In contrast, the presence of sulphide counteracts the dissociation of SAT and OASTL, resulting in the maintenance of SAT activity as long as sulphide availability is not a limiting factor (Droux, Ruffet, Douce, & Job, 1998; Takahashi et al., 2011; Wirtz & Hell, 2006).

2.1 Responses to Metal Stress

In addition to its regulation by internal stimuli, the sulphate assimilation pathway is affected by environmental factors including metal exposure (Ernst, Krauss, Verkleij, & Wesenberg, 2008; Na & Salt, 2011). Indeed, several metals were shown to influence the expression of genes encoding both high- and low-affinity SULTRs, thereby affecting sulphate uptake and assimilation. Nocito et al. (2006), for example, demonstrated that exposure to different concentrations of cadmium (Cd), copper (Cu) and zinc (Zn) increased the expression of *ST1;1*, a high-affinity SULTR in *Zea mays*. As a result, sulphate uptake capacity in roots was significantly enhanced during metal exposure (Nocito et al., 2006). Furthermore, Zn was shown to enhance root *SULTR1;1*, *SULTR1;2*, *SULTR4;1* and *SULTR4;2* expression and sulphate uptake in *Brassica pekinensis* (Stuiver et al., 2014). Similarly, Dixit et al. (2016) demonstrated an upregulation of *SULTR1;1*, *SULTR1;2 SULTR2;2* and *SULTR4;1* expression in roots of *Oryza sativa* exposed to the metalloid arsenic (As), resulting in increased sulphur accumulation and translocation. As a consequence, cysteine, GSH and PC levels were higher in both roots and shoots (Dixit et al., 2016). However, exposure to chromium (Cr) clearly decreased the expression

glutathione; *GSH-S*, glutathione synthetase; *GSSG*, glutathione disulphide; *GST*, glutathione transferase; *H_2O*, water; *H_2O_2*, hydrogen peroxide; *HCys*, homocysteine; *L-Ser*, L-serine; *Met*, methionine; *MS*, methionine synthase; *MT*, metallothionein; *OAS*, O-acetylserine; *OASTL*, OAS (thiol) lyase; *OPH*, O-homophosphoserine; *PAPS*, 3′-phosphoadenosine 5′-phosphosulphate; *PC*, phytochelatin; *PCS*, phytochelatin synthase; *Put*, putrescine; *S^{2-}*, sulphide; *SAM*, S-adenosylmethionine; *SAMDC*, SAM decarboxylase; *SAMS*, SAM synthetase; *SAT*, serine acetyl transferase; *SiR*, sulphite reductase; *SO_3^{2-}*, sulphite; *SO_4^{2-}*, sulphate; *Spd*, spermidine; *Spm*, spermine; *SULTR*, sulphate transporter.

levels of *ST1;1* and significantly suppressed sulphate uptake in *Z. mays* (Schiavon et al., 2007). Cr was also shown to reduce sulphate uptake and negatively affect the transcription of several SULTRs, including *SULTR1;3*, *SULTR2;1*, *SULTR3;2*, *SULTR3;5* and *SULTR4;1*, in roots of *Brassica juncea* after 24 h of exposure, whereas expression levels of *SULTR4;2* were significantly increased (Schiavon et al., 2012).

In addition to their effect on sulphate uptake, metals also influence transcript levels, protein abundance and activity of enzymes catalyzing different reactions in the sulphate assimilation pathway. For example, an increased activity of the sulphate-activating enzyme ATPS was reported in leaves of *A. thaliana* exposed to 50 μM Cd for 3 or 5 days (Bashir, Ahmad, Bagheri, Nauman, & Qureshi, 2013). Similar effects were demonstrated in *Sedum alfredii*, with Cd increasing root and shoot *ATPS* transcript levels in a concentration-dependent manner. Moreover, the expression of the *SAT* gene, encoding the first enzyme in cysteine biosynthesis, was also significantly upregulated by Cd exposure in the roots (Liang et al., 2014). In addition to Cd, Cu was shown to affect sulphate assimilation, as demonstrated by a strongly increased *ATPS* expression in roots of both *A. thaliana* and *Arabidopsis halleri* (Weber, Trampczynska, & Clemens, 2006). Furthermore, Cu-induced increases in the protein abundance of APR, SAT and OASTL in *O. sativa* roots were reported by Song et al. (2013). By affecting enzymes involved in sulphate assimilation, several metal(loid)s including As (Dixit et al., 2016; Talukdar & Talukdar, 2014), Cd (Bashir et al., 2013; Liang et al., 2014; Nocito et al., 2006), Cr (Schiavon et al., 2012; Schiavon, Pilon-Smits, Wirtz, Hell, & Malagoli, 2008), Cu (Gajewska, Glowacki, Mazur, & Sklodowska, 2013), nickel (Ni) (Gajewska et al., 2013), lead (Pb) (Mandavian, Ghaderian, & Schat, 2016) and Zn (Stuiver et al., 2014) were shown to alter cysteine levels in a wide range of plant species.

However, it is important to note that plant responses to metal exposure depend on sulphate supply, as metal-induced effects observed during sulphate-deprived conditions often differ from those observed during normal sulphate availability. Results of a study by Bashir et al. (2015) indicate that under conditions of sulphate starvation, *B. juncea* displayed higher Cd-induced oxidative stress levels and an increased Cd sensitivity. Furthermore, sulphur addition was reported to increase Cd uptake in roots of *Brassica chinensis*, while inhibiting root-to-shoot translocation of the metal. Although the addition of sulphate to the growth medium increased plant Cd uptake, it also relieved the Cd-induced inhibition of root and shoot growth and significantly reduced malondialdehyde (MDA) and superoxide ($O_2^{\bullet -}$) levels in both organs (Liang et al., 2016). Decreases in *Corchorus olitorius* dry weight induced by aluminium (Al), Cd, Cu and Pb were alleviated by the addition of either K_2SO_4 or cysteine to the growth medium (Mazen, 2004). These results are of particular interest in the context of phytoremediation, as this soil remediation strategy strongly benefits from plants exhibiting a high metal uptake capacity and increased metal tolerance.

The involvement of sulphate uptake and assimilation in plant responses to metal stress is further supported by the observation that modification of genes involved in sulphate assimilation alters plant tolerance to a broad array of metals. Indeed, overexpression of *SAT* was shown to increase OAS, cysteine and GSH levels in *A. thaliana*, thereby increasing its resistance to Ni-induced growth inhibition (Freeman et al., 2004). Furthermore, overexpression of different *OASTL* isoforms increased Cd tolerance of both *A. thaliana* and *Nicotiana tabacum* (Dominguez-Solis, Gutierrez-Alcala, Romero, & Gotor, 2001; Dominguez-Solis et al., 2004; Harada, Choi, Tsuchisaka, Obata, & Sano, 2001; Kawashima et al., 2004; Ning, Zhang, Yao, & Yu, 2010). Interestingly, transgenic *Brassica napus* plants overexpressing *miR395* also showed diminished Cd-induced oxidative stress levels and were more tolerant to Cd. This miRNA is induced by sulphate starvation and controls the expression levels of *SULTR2;1* and three *ATPS* isoforms. Expression of *miRNA395* is also increased during Cd exposure, highlighting the similarity between plant responses to both stress factors (Gielen, Remans, Vangronsveld, & Cuypers, 2016; Zhang, Song, Shu, Zhang, & Yang, 2013).

The importance of sulphate assimilation in plant responses to metal stress is supported by the observation that transcript levels and activity of enzymes involved in sulphate uptake often differ between metal hyperaccumulators and their nonaccumulator counterparts. For example, Freeman et al. (2004) reported strong positive correlations between OAS, cysteine and GSH levels and shoot Ni concentrations in various *Thlaspi* hyperaccumulators and nonaccumulators. Indeed, the metal-accumulating species *Thlaspi goesingense* displayed increased SAT activity and OAS, cysteine and GSH levels in comparison to the genetically related nonaccumulator *A. thaliana* under control conditions. When exposed to Ni, this resulted in an increased resistance of the former species against Ni-induced oxidative stress, as indicated by considerably lower lipid peroxidation levels (Freeman et al., 2004). Metal-tolerant plants of the same species often exhibit increased sulphate assimilation capacities when compared to their metal-sensitive counterparts. For example, an As-tolerant *Lens culinaris* cultivar showed strongly increased *SAT* and *OASTL* transcript levels and activities and significantly enhanced cysteine and GSH levels when compared to an As-sensitive cultivar under As exposure conditions. Interestingly, As differently affected transcript levels of group 1 SULTRs between both cultivars, with *SULTR1;1* and *SULTR1;2* upregulation in the tolerant cultivar and downregulation in the sensitive cultivar after 24 h of exposure (Talukdar & Talukdar, 2014). Similarly, Song et al. (2013) reported more pronounced Cu-induced increases in SAT and OASTL protein abundance in a Cu-tolerant *O. sativa* variety as compared to a Cu-sensitive variety. Furthermore, the induction of ATPS and OASTL activities by Cd exposure in a tolerant *B. chinensis* cultivar was less pronounced or even absent in its sensitive counterpart (Liang et al., 2016). In conclusion, it is obvious that sulphate uptake and assimilation play a prominent role in plant responses to metal stress. This can be explained by the crucial role of cysteine-containing compounds in metal chelation and detoxification, as discussed in Section 3. Therefore, metallophytes could be useful in phytoremediation technologies (chapter: Metallophytes of Serpentine and Calamine Soils – Their Unique Ecophysiology and Potential for Phytoremediation by Wójcik et al., 2017), which might be further enhanced by plant-associated bacteria and fungi (chapter: The Bacterial and Fungal Microbiota of Hyperaccumulator Plants: Small Organisms, Large Influence by Thijs, Langill, & Vangronsveld, 2017).

2.2 Responses to Organic Contaminants

Although our knowledge on the effects of organic environmental contaminants on sulphate assimilation is scarce, the effects of herbicides and safeners – used to selectively protect crops from herbicide damage – have been frequently described (Abu-Qare & Duncan, 2002; Hirase & Molin, 2003). In 1985, the Lamoureux Laboratory described that pretreatment of *Z. mays* with low levels of the chloroacetamide herbicide 2-chloro-N,N-di-2-propenylacetamide protected them from later exposure to higher concentrations of the herbicide by increasing GSH synthesis (Ezra, Rusness, Lamoureux, & Stephenson, 1985). Similarly, Adams, Blee, and Casida (1983) showed that dichloroacetamide herbicide antidotes enhanced sulphate metabolism in *Z. mays* roots. Both dichlormid and benoxacor induced ATPS, whereas flurazole acted on OASTL (Hirase & Molin, 2003). Such an effect was also observed a few years later for *Sorghum bicolor* by Gronwald, Fuerst, Eberlein, and Egli (1987). In 1990, evidence for a direct interaction of N,N-diallyl-2,2-dichloroacetamide and 4-dichloroacetyl-3,4-dihydro-3-methyl-2H-1,4-benzo-oxazine (CGA 154 281) with sulphate assimilation and GSH concentrations was presented in *Z. mays* (Farago & Brunold,

1990). Similar to the safener CGA 154 281, cysteine formation and GSH synthesis were also induced by the safener 1-dichloroacetyl-hexahydro-3,3,8-δα-trimethyl-pyrrolo-[1,2-α]-pyrimidine-6-(2H)-one-(dicyclonone) (BAS 145 138) in *Z. mays* (Kocsy et al., 2001). These specific effects of safeners on sulphate assimilation have been attributed to their influence on oxidative stress levels, but are not completely unravelled yet. Recently, the effect of three herbicide safeners (mefenpyr-diethyl, fenchlorazole-ethyl and dichlormid) on the content of sulphur-containing metabolites in Fe-deficient barley was investigated (Bartucca et al., 2016). All three safeners effectively induced ATPS activity, but the effect on OASTL activity was dependent on concentration as well as exposure time. An initial reduction of OASTL activity was followed by a strong induction. The authors speculated that safeners initially induce some membrane damage and the generation of hydrogen peroxide (H_2O_2) in an oxidative burst, which leads to the activation of defence genes. Different from herbicidal metabolic action injuring plants, the inertness of safeners to plants does not cause toxicity. Hence, the initial decrease in OASTL activity after treatment would be regarded as an unspecific response of the plant to stress. Thereafter, the safening action prevails and the chemicals activate defence responses inducing OASTL activity, leading to enhanced GSH availability (Bartucca et al., 2016).

3. THIOLS PLAY AN IMPORTANT ROLE IN DETOXIFICATION

3.1 Metallothioneins

MTs are cysteine-rich low-molecular-weight (<10 kDa) metal-chelating proteins in many organisms (Fig. 1) (Anjum et al., 2015; Freisinger, 2008). Plant MTs are divided into four groups based on the number and arrangement of their cysteine residues (Cobbett & Goldsbrough, 2002). Type 1–3 MTs contain two cysteine-rich domains connected by a cysteine-poor linker region, the length of which depends on the specific MT type and the plant species (Leszczyszyn, Imam, & Blindauer, 2013). The cysteine-rich α- and β-domains are involved in metal binding. The α-domain is located at the C-terminus of the protein and contains six cysteine residues arranged according to the consensus sequence CxCxxxCxCxxCxC, where x represents any other amino acid besides cysteine. The β-domain is present at the N-terminus and has a more variable amino acid sequence. It generally contains six cysteine residues in MT1 proteins, eight in MT2 proteins and four in MT3 proteins. Type 4 MTs – also referred to as E_c proteins – can be distinguished from other MT classes based on the presence of three cysteine-rich domains separated by two linker regions. Furthermore, they are characterized by two highly conserved histidine residues in the central cysteine-rich domain (Freisinger, 2008, 2011). In contrast to PCs, which are also important metal chelators in plant cells (see Section 3.2.1), MTs are encoded by genes and thus are products of mRNA translation (Anjum et al., 2015). Although almost all MT genes contain an intron near the N-terminal cysteine-rich domain, it is interesting to note that the exact position of this intron varies according to the specific type of MT encoded (Cobbett & Goldsbrough, 2002).

Different types of MTs display distinct spatiotemporal expression patterns. In general, type 1 MTs are more strongly expressed in roots than in shoots, whereas the opposite holds true for type 2 MTs (Guo, Bundithya, & Goldsbrough, 2003). Although type 3 MTs are also present in leaves, they are mainly expressed in ripening, fleshy fruits (Clendennen & May, 1997; Moyle, Fairbairn, Ripi, Crowe, & Botella, 2005). In contrast, MTs of type 4 are exclusively localized in developing seeds, implying a role in metal storage and accumulation in seeds (Guo et al., 2003). As proposed by Ren et al. (2012), these MTs could provide a means to store Zn required for seed germination. Furthermore, different MT genes are strongly upregulated in ageing leaves, suggesting their involvement in leaf senescence. During this process, MTs possibly serve a dual function. First, they could protect cells against metal toxicity arising from the breakdown of different metal-containing cellular components. Second, they could also be involved in translocating the released metals to nonsenescing plant tissues (Guo et al., 2003; Leszczyszyn et al., 2013).

Although many questions remain with regard to the different functions of plant MTs, their role in metal chelation is well established. MTs have the ability to bind both mono- and divalent metal ions in typical metal-thiolate clusters, which are characterized by high thermodynamic and low kinetic stability. As a consequence, metals are tightly bound by MTs, but part of the metal ions can be readily relocated to other proteins (Hassinen, Tervahauta, Schat, & Karenlampi, 2011). MTs chelate both essential and nonessential metals, indicating their involvement in both nutrient homeostasis and metal detoxification. The metals Zn, Cu and Cd are bound to MTs with the highest affinity (Blindauer & Leszczyszyn, 2010). The importance of MTs in detoxifying excess metals is highlighted by their induction during metal exposure, as discussed in Section 3.1.1. However, many other biotic and abiotic stress factors including pathogen attack (Dauch & Jabaji-Hare, 2006), wounding (Razem & Bernards, 2002), light (Chen et al., 2003), drought (Li et al., 2016) and low temperature (Zhu et al., 2009) also induce the expression of *MT* genes, suggesting additional roles besides metal chelation. Transcriptional responses of MTs to different stress factors are possibly mediated by the presence of upstream regulatory elements in the promoter regions of genes encoding these proteins. Metal-, antioxidant-, wounding- and stress-responsive elements were identified in the promoters of type 1, 2 and 3 MTs. Elements responsive to different phytohormones including ethylene, methyl jasmonate, gibberellic acid and salicylic acid were also found in MT promoters, suggesting hormonal regulation of MT expression (Leszczyszyn et al., 2013).

Interestingly, MTs also exhibit antioxidant properties. This is illustrated in *N. tabacum*, where ectopic expression of a type 1 MT from *O. sativa* clearly decreased H_2O_2 accumulation during salinity stress (Kumar et al., 2012). Similar results were obtained by Xue et al. (2009), who reported increased stress tolerance and decreased H_2O_2 levels in transgenic *N. tabacum* plants overexpressing the *Gossypium hirsutum MT3a* gene as compared to the wild type (WT) exposed to cold, salt and drought stress. Furthermore, 5,5′-dithiobis-2-nitrobenzoic acid (DTNB) oxidized MT3a in vitro, suggesting the ROS scavenging capacity of this MT (Xue et al., 2009). H_2O_2 levels were significantly higher in an *A. thaliana* T-DNA insertion line lacking functional MT2a as compared to the WT subjected to drought and cold stress (Zhu et al., 2009). Additional evidence supporting an antioxidative function for MTs is provided by the observation that exposure to ROS or ROS-inducing stress factors (e.g., paraquat) increased the expression of several *MT* genes in many plant species including *Quercus suber* (Mir et al., 2004), *G. hirsutum* (Xue et al., 2009), *Ipomoea batatas* (Kim, Jeong, Ahn, Lee, & Kwak, 2014) and *O. sativa* (J. Liu et al., 2015). The antioxidative function of MTs can be explained by the fact that the cysteine residues responsible for metal scavenging also have the capability to reduce ROS. During metal exposure, MTs could also exert an indirect antioxidative function by chelating redox-active metals such as Cu, thereby preventing ROS formation as a consequence of the Fenton and Haber–Weiss reactions (Hassinen et al., 2011; Leszczyszyn et al., 2013).

3.1.1 Responses to Metal Stress

As mentioned earlier, metal exposure induces MT expression in numerous plant species. Exposure to Cu, for example, induced the expression of *MT1a*, *MT2a*, *MT2b* and *MT3* in different organs of *A. thaliana*. This increased transcription was especially pronounced in leaf trichomes, which are often reported to accumulate large amounts of metal ions (Guo et al., 2003). Similarly, Cd exposure significantly induced the expression of *MT2a* in *A. thaliana* roots and leaves (Jozefczak et al., 2014). In another study, 24 h exposure to Cd, Cu or Zn was shown to induce the expression of *MT1* in *Cajanus cajan* (Sekhar, Priyanka, Reddy, & Rao, 2011). These metals also increased *MT3* transcription levels in leaves of *Porteresia coarctata* (Usha, Keeran, Harikrishnan, Kavitha, & Parida, 2011). In addition to Cd, Cu and Zn, which are most often associated with MT binding, other metals were shown to induce MT expression as well. Exposure to Pb, for example, strongly induced *MT2a* and *MT2b* expression in shoots of *Hirschfeldia incana* (Auguy et al., 2016). Similarly, *MT1* and *MT2* expression levels were elevated in As-exposed *O. sativa* (Nath et al., 2014). Furthermore, Fe (Ahn et al., 2012), Mn (Ahn et al., 2012; Zhao et al., 2012), mercury (Hg) (Venkatachalam, Srivastava, Raghothama, & Sahi, 2009) and Cr (Gautam et al., 2012) were shown to induce the expression of different *MT* genes in a broad range of plant species. The induction of MT expression in metal-exposed plants suggests a role for these proteins in metal stress defence responses.

A positive role for MTs in metal tolerance is also suggested by several studies reporting differences in *MT* expression between metal-sensitive and metal-tolerant plants. Indeed, *MT3* transcript levels were approximately 2.5-fold higher in the metal hyperaccumulator *Thlaspi caerulescens* than in the nonaccumulator *A. thaliana* grown under control conditions. The cysteine positions in the MT amino acid sequence of both species were different, predicting a smaller metal binding cavity in *A. thaliana* than in *T. caerulescens* (Roosens, Bernard, Leplae, & Verbruggen, 2004). Interestingly, *MT2a*, *MT2b* and *MT3* expression levels were also 1.5- to 4-fold higher in the metal hyperaccumulator *A. halleri* than in *A. thaliana* (Chiang, Lo, & Yeh, 2006). In addition, *MT2b* expression was considerably higher in two independently evolved Cu-tolerant *Silene paradoxa* populations than in a Cu-sensitive population, both under control conditions and after Cu exposure. This effect is probably related to the presence of multiple *MT2b* gene copies in both Cu-tolerant populations (Mengoni et al., 2003). Hassinen et al. (2009) reported that *MT2* and *MT3* were more highly expressed in metallicolous *T. caerulescens* accessions than in a nonmetallicolous accession. However, Zn accumulation and *MT* transcript levels did not cosegregate, implying that MTs are not the major determinants of Zn accumulation in these plants (Hassinen et al., 2009).

A role for MTs in plant defence against metal stress is also supported by the fact that plants overexpressing different *MT* genes often display enhanced metal tolerance. Overexpression of a putative MT from *Colocasia esculenta* (*CeMT*), for example, reduced the negative effects of Cd, Cu and Zn exposure on root growth in *N. tabacum* seedlings. Furthermore, it significantly increased metal accumulation in these plants. The positive function of this MT during metal stress could be related not only to its metal-chelating function but also to its antioxidant properties. This is confirmed by significantly reduced H_2O_2 and lipid peroxidation levels after 24 h of metal exposure in the *CeMT*-overexpressing plants as compared to their WT counterparts (Kim, Jung, Kim, & Bae, 2013). Similarly, overexpression of *MT2* from *S. alfredii* resulted in increased Cd and Zn tolerance and accumulation in *N. tabacum*. Again, this response was accompanied by significantly decreased H_2O_2 levels in metal-exposed transgenic tobacco plants as compared to the WT. When exposed to Cd and Zn, the *MT2*-overexpressing plants also displayed increased superoxide dismutase (SOD), catalase (CAT) and peroxidase (POD) activities in comparison to WT plants (J. Zhang et al., 2014). Overexpression of MTs derived from yeast and mammals can also increase metal tolerance and/or accumulation in plants (Daghan, Arslan, Uygur, & Koleli, 2013; Ruiz, Alvarez, Torres, Roman, & Daniell, 2011; Vrbova et al., 2013) and is proposed to be an interesting phytoremediation strategy.

Although MTs are important players in plant metal stress responses, it should be noted that their responses depend on many factors, including the plant species, tissue and specific metal considered. Schiller et al. (2014), for example, demonstrated that leaf expression levels of *MT1a* were upregulated by Cu and Zn exposure and downregulated by Cd exposure in *Hordeum vulgare*. Similarly, transcript levels of *MT2c* and *MT3* increased in response to Cu, whereas they decreased in leaves of Cd-exposed plants. In contrast, expression levels of these *MTs* remained unchanged in response to Zn. Furthermore – as mentioned earlier – expression levels of different MTs were strongly tissue-dependent, with *MT1b* and *MT4* transcripts exclusively detected in roots and grains, respectively (Schiller et al., 2014). Taken together, these data point towards distinct physiological functions for different MT isoforms. The first step towards unravelling these functions involves the analysis of MTs at the protein level. This remains a challenge, however, as plant MTs are highly prone to proteolysis, hampering the application of proteomics approaches (Blindauer & Leszczyszyn, 2010).

3.2 Glutathione

GSH is the most abundant nonprotein thiol in almost all aerobic species, occurring at intracellular concentrations of 0.5–10 mM. It is a tripeptide consisting of γ-glutamate, cysteine and glycine and is synthesized in two ATP-dependent steps (Fig. 1). First, a peptide bond is formed between γ-glutamate and cysteine, producing γ-glutamylcysteine (γ-EC). This reaction is catalyzed by γ-EC synthetase (γ-ECS; GSH1), also known as glutamate–cysteine ligase (GCL), and constitutes the rate-limiting step in GSH biosynthesis. Subsequently, GSH is formed by the addition of glycine to γ-EC in a reaction catalyzed by glutathione synthetase (GSH-S; GSH2) (Noctor et al., 2012). GSH1 is exclusively localized in chloroplasts, whereas GSH2 is also present in the cytosol, with cytosolic GSH2 activity strongly exceeding that in chloroplasts (Wachter, Wolf, Steininger, Bogs, & Rausch, 2005). Transport of γ-EC into the cytosol, necessary for cytosolic GSH production, is mediated by chloroquine-resistance transporter-like transporters (CLTs) in the plastid envelope (Maughan et al., 2010). Once synthesized, GSH is distributed to different organelles by the action of plastidial CLTs and transporters in mitochondrial and nuclear membranes (Zechmann, 2014). Mitochondria, chloroplasts and the cytosol have the highest GSH abundance in plant cells (Noctor et al., 2012). In addition to its intracellular transport, GSH is also transported between cells and is one of the major reduced sulphur forms translocated in the phloem (Noctor et al., 2012).

An important determinant of GSH biosynthesis is GSH1 activity, which is regulated at three levels. First, GSH1 activity is subject to feedback inhibition by GSH itself. Under stress conditions, GSH is consumed and the feedback inhibition is alleviated, thereby increasing GSH synthesis. In addition, GSH1 is transcriptionally regulated via the interaction of a redox-sensitive repressor-binding protein with the 5′-untranslated region of its encoding gene (Noctor, Gomez, Vanacker, & Foyer, 2002). Furthermore, GSH1 is posttranscriptionally modulated by the cellular redox state. Under oxidizing conditions, it functions as a homodimeric enzyme with two intermolecular disulphide bonds. Reducing conditions, however, disrupt one of these disulphide bonds, thereby altering the dimer interface and shifting the enzyme to its less active monomeric form (Hothorn et al., 2006). Whether this mechanism is responsible for GSH1 feedback inhibition by GSH is currently unknown (Jozefczak, Remans, Vangronsveld, & Cuypers, 2012). It is interesting to note that GSH biosynthesis is tightly coupled to sulphate assimilation, as stress-induced GSH accumulation is often accompanied by increased sulphate uptake and transcriptional upregulation of sulphate assimilation genes, including *APR* and *SAT* (Noctor et al., 2012; Queval et al., 2009; Smith, Kendall, Keys, Turner, & Lea, 1985).

In plants, GSH is involved in many processes including cell cycle regulation and defence against biotic and abiotic stress factors (Noctor et al., 2012). The essential role of GSH in plant physiology is underlined by the observation that plants lacking GSH due to a knockout mutation in either *GSH1* or *GSH2* display a lethal phenotype (Cairns, Pasternak, Wachter, Cobbett, & Meyer, 2006; Lim, Meyer, & Cobbett, 2011). In contrast, plants with lower GSH levels are viable but show an altered phenotype as compared to WT plants. For example, the *root meristemless 1-1* (*rml1-1*) mutant contains less than 5% of WT GSH levels and thereby fails to develop a root apical meristem (Cheng, Seeley, & Sung, 1995; Vernoux et al., 2000). Other GSH-deficient mutants such as *cadmium-sensitive 2-1* (*cad2-1*), *regulator of ascorbate peroxidase 1-1* (*rax1-1*) and *phytoalexin-deficient 2-1* (*pad2-1*) display GSH levels between 25% and 50% of WT levels. Even though these plants develop normally, they are more sensitive to environmental stress factors, emphasizing the crucial role of GSH during stress (Ball et al., 2004; Howden, Andersen, Goldsbrough, & Cobbett, 1995; Parisy et al., 2007).

In this review, we will focus on the most important functions of GSH in stress responses induced by metals and organic contaminants. These include (1) metal chelation by PCs, (2) detoxification of xenobiotics by glutathione transferases (GSTs) and (3) antioxidative defence (Fig. 1).

3.2.1 Glutathione in Chelation: Phytochelatins

An important role of GSH during stress relies on its function as a precursor of PCs. These molecules are important metal scavengers in plant cells and are characterized by a typical $(\gamma\text{-Glu-Cys})_n$-Gly structure, with n ranging from 2 to 11 (Zagorchev, Seal, Kranner, & Odjakova, 2013). PCs are synthesized by PC synthase (PCS) in a two-step reaction. In the first step, the cysteine–glycine peptide bond of a donor GSH molecule is cleaved, yielding γ-EC and glycine. Subsequently the γ-EC unit is transferred to an acceptor molecule – either another GSH molecule or an oligomeric PC peptide – in a transpeptidation reaction. Even though its encoding gene is constitutively expressed, PCS activity is dependent on the presence of metals. Although Cd is the most important inducer of PCS activity, other metal(loid)s including As, Cu, Hg, Ni, Pb and Zn were also reported to increase PC synthesis (Anjum et al., 2014). The metal-induced activation of PCS is related to the presence of four highly conserved cysteine residues in its N-terminal domain, which also contains the active site. The C-terminal domain, in contrast, is rich in cysteine residues that possibly bind metal ions to transfer them to the N-terminal active site (Vestergaard et al., 2008). Interestingly, Cd–PC complexes were demonstrated to be translocated into the vacuole, possibly reducing Cd-induced damage to other cellular compartments (Cobbett & Goldsbrough, 2002). In *A. thaliana*, this vacuolar translocation is likely mediated by the ATP-binding cassette (ABC) C3 (ABCC3) transporter, which is transcriptionally induced by Cd exposure (Brunetti et al., 2015).

As mentioned earlier, PC synthesis is induced by several metals. The importance of PCs in metal-induced plant responses is further supported by the fact that the PC-deficient *A. thaliana* mutants *cad1-3* and *cad1-6* – carrying a mutation in the *PCS1* gene – display an enhanced sensitivity towards several metals such as Cd, Pb and Zn (Fischer, Kuhnlenz, Thieme, Schmidt, & Clemens, 2014; Tennstedt, Peisker, Bottcher, Trampczynska, & Clemens, 2009). Furthermore, differences in the extent of PC synthesis between metal-sensitive and metal-tolerant varieties are often observed. Indeed, a Cd-tolerant *Triticum aestivum* cultivar displayed stronger Cd-induced increases in GSH levels and *PCS* expression in comparison to a Cd-sensitive cultivar (Kumari, Parmar, & Sharma, 2015). Similarly, *PCS1* expression and total PC levels were significantly increased by As exposure in an As-tolerant *O. sativa* cultivar, whereas they remained unaffected in its As-sensitive counterpart. As-induced increases in GSH, cysteine and methionine levels were also more pronounced in the tolerant cultivar than in the sensitive cultivar (Begum et al., 2016).

As PCs are efficient metal chelators, transgenic plants with increased PC levels are often able to accumulate high metal concentrations. For example, *N. tabacum* plants overexpressing *PCS1* displayed an enhanced accumulation of Cd, Cu, Ni and Zn as compared to WT plants. These plants even accumulated higher metal concentrations than the metal hyperaccumulator *T. caerulescens* (Martínez et al., 2006). The use of transgenic plants with increased PC production could therefore offer a promising strategy for the phytoremediation of metal-polluted soils. However, it should be noted that overproduction of PCs is not always beneficial for plants, as it can cause a strong decrease in cellular GSH levels. Since GSH also serves important antioxidative functions (see Section 3.2.3), GSH depletion possibly results in oxidative stress and cellular damage. This was shown by Jozefczak et al. (2014), who reported a rapid increase in PC levels in roots and leaves of *A. thaliana* plants exposed to 5 and 10 μM Cd. In the roots, this response coincided with a fast (2 h) reduction in GSH levels. Despite the activation of alternative antioxidative defence systems after 24 h, decreased biomass and increased lipid peroxidation were observed in Cd-exposed plants (Jozefczak et al., 2014). These results indicate that plant metal tolerance is related not only to the production of PCs but also to the ability to prevent the resulting GSH depletion. Therefore, the use of transgenic plants overexpressing both *PCS* and GSH biosynthesis genes could be more efficient to remediate metal-contaminated areas (Seth et al., 2012).

3.2.2 Glutathione in Detoxification: Glutathione Transferases

Another important function of GSH is related to its involvement in detoxification reactions mediated by GSTs (Fig. 1). These enzymes catalyze the conjugation of xenobiotics to GSH by addition or substitution reactions to reduce their toxicity and increase their water solubility (Coleman, Blake-Kalff, & Davies, 1997; Dixon et al., 2010). As GSTs possess broad and overlapping substrate specificities, they are able to detoxify a chemically diverse array of compounds. Xenobiotics susceptible to GSH conjugation include electrophilic herbicides, several drugs and organic contaminants (Cole, Cummins, Hatton, Dixon, & Edwards, 1997; Frova, 2003). Generally, these compounds are characterized by the presence of two carbon atoms coupled by a double bond, adjacent to an electron-withdrawing group (Frova, 2003). Given the reactivity of these molecules, conjugation to GSH occurs spontaneously but is considerably speeded up by GST activity (Coleman et al., 1997). In addition to their role in GSH conjugation reactions, GSTs also serve a broad range of other catalytic as well as noncatalytic functions (Frova, 2003). Although the molecular structure and catalytic mechanisms of GSTs have been studied by several research groups, their natural functions remains largely obscure. An overview of the natural GST functions elucidated so far is provided in Table 1. The first functions ascribed to GSTs were all related to conjugation (e.g., of xenobiotics), whereas more sophisticated functions such as the synthesis of glucosinolates were only discovered more recently (Table 1).

Even though most GSTs are soluble proteins present in the cytosol, they have also been detected in several organelles including chloroplasts, mitochondria, peroxisomes, the vacuole and nucleus.

Table 1 Natural Functions of Glutathione Transferases (GSTs) in Plants

Postulated Natural Role of GSTs	References
Conjugation of endogenous metabolites	Diesperger and Sandermann (1979), Edwards and Dixon (1991), and Dean, Devarenne, Lee, and Orlofsky (1995)
Conjugation of DNA degradation products	Morgenstern, Depierre, and Jornvall, (1985)
Conjugation of phytohormones (auxin, ethylene, gibberellic acid)	Lamoureux and Frear (1987), Meyer, Goldsbrough, and Woodson (1991), Takahashi and Nagata (1992), and Zettl, Schell, & Palme (1994)
Detoxification of lipid hydroperoxides	Mannervik and Danielson (1988), Marrs (1996), and Sommer and Boger (1999)
Transport of (thio-)phenols, chlorophyllin and anthocyans	Singh and Shaw (1988) and Martinoia, Grill, Tommasini, Kreuz, and Amrhein (1993)
Regulation of GSH pool	Lamoureux and Rusness (1989)
Detoxification of fungal toxins and pathogen defence	Dudler, Hertig, Rebmann, Bull, and Mauch (1991) and Mauch and Dudler (1993)
Increase of drought tolerance	Dhindsa (1991)
Antioxidative protective protein	Levine, Tenhaken, Dixon, and Lamb (1994)
Conjugation and transport of phytoalexins and similar compounds	Li, Alfenito, Rea, Walbot, and Dixon (1997) and Marrs, Alfenito, Lloyd, and Walbot (1995)
Regulation of UV-induced genes	Loyall, Uchida, Braun, Furuya, and Frohnmeyer (2000)
Transporter proteins for secondary metabolites and their unstable intermediates	Dixon et al., 2010
Synthesis of glucosinolates; conjugation, transport and storage of reactive oxylipins, phenolics and flavonoids	Dixon et al. (2010) and Dixon and Edwards (2010)
Isomerization and peroxidation	Cummins, Dixon, Freitag-Pohl, Skipsey, and Edwards (2011)
Deglutathionation of GS-conjugates	Lallement et al. (2014)

Microsomal GSTs belonging to the 'membrane-associated proteins in eicosanoid and glutathione metabolism' (MAPEG) family have also been characterized (Oztetik, 2008). For example, microsomal GSTs with herbicide-detoxifying activities have been detected in *Pachyrhizus erosus*

(Belford, Dorfler, Stampfl, & Schroder, 2004) and in lower plants (Pflugmacher, Schroder, & Sandermann, 2000). It is important to note that cytosolic GSTs are very abundant, constituting up to 2% of all soluble proteins in plants (Scalla & Roulet, 2002).

To date, plant GSTs are subdivided into several classes according to a classification system proposed by Droog (1997) and subsequently refined by Edwards, Dixon, and Walbot (2000). Using this system, GSTs are classified based on gene organization (number and position of introns), amino acid similarity and conservation of specific residues in the protein. The phi (GSTF) and tau (GSTU) classes constitute the largest classes of plant GSTs and are plant specific. Interestingly, they are also predominantly active in xenobiotic detoxification (Dixon et al., 2010; Frova, 2006). Indeed, phi class GSTs have been cocrystallized with a range of herbicide and other xenobiotic conjugates (Prade, Huber, & Bieseler, 1998). In contrast, the zeta (GSTZ) and theta (GSTT) families are much smaller and are related to mammalian GSTs. Whereas GSTs belonging to these classes display GSH-conjugating activities, other classes including the lambda GSTs (GSTL) and dehydroascorbate reductases (DHARs) have no known ability to conjugate or detoxify synthetic compounds. Instead, they seem to have a redox-related role (Dixon, Davis, & Edwards, 2002).

Structurally, most cytosolic GSTs occur as dimers consisting of two subunits (approximately 25 kDa each), each encoded by one gene (Oztetik, 2008). Although homodimers and heterodimers can be formed, dimerization has only been observed between GSTs from the same class (Armstrong, 1997). An exception are the DHARs and lambda GSTs, which function as monomers (Dixon et al., 2010). Each GST subunit is characterized by a common overall structure with a well-defined N-terminal GSH-binding domain (G-site) and a less specific C-terminal domain binding the hydrophobic cosubstrate (H-site). Most GST classes, including the phi, tau, zeta and theta class, contain a serine in the active site responsible for the formation and stabilization of the reactive thiolate anion from GSH, which is the target for the nucleophilic attack of an electrophilic substrate (Dixon et al., 2010). In DHARs and lambda GSTs, however, the serine is replaced by a cysteine that forms a mixed disulphide with GSH. The catalytic cysteine performs a nucleophilic attack on GSH-conjugated substrates. Under these conditions, the catalytic cysteine of the GST becomes glutathionylated, thereby deconjugating the substrate that is subsequently released. Regeneration of the resulting glutathionylated GST forms requires a GSH molecule, forming GSSG as another end product. The reduced GSTs can be used in another catalytic cycle and GSSG is reduced back to GSH by glutathione reductase (GR) at the expense of NADPH (Lallement, Brouwer, Keech, Hecker, & Rouhier, 2014).

In general, phi and tau GST-catalyzed substitution reactions change the parent compound to an extent that the toxicity of the xenobiotic is significantly lowered (Schroder & Collins, 2002). Along with the resulting increase in polarity, GSH addition is a very effective means of detoxification, especially when the electrophilic centre of the target molecule is a leaving group. In some situations, however, the formation of glutathione S (GS)-conjugates could be disadvantageous to plants when they accumulate in high amounts. In *Picea abies*, for example, it is clearly proven that GS-conjugates accumulating in the cytosol inhibit the stress-reducing activity of GR (Schroder & Pflugmacher, 1996). Furthermore, when detoxification reactions withdraw GSH from the cellular pool, GSH depletion due to exaggerated utilization in conjugation reactions or inhibition of GSH supply could become a problem, particularly when the xenobiotic conjugates still remain reactive and/or are not efficiently stabilized (Brazier-Hicks et al., 2008). However, GS-conjugate formation has generally been accepted to be beneficial to plants (Edwards et al., 2000; Schröder, 2006).

Furthermore, conjugation of substrates with GSH allows them to be transported into the vacuole by ABC transporters (Mohanasundaram et al., 2015; Theodoulou, 2000). This removal of GS-conjugates from the cytosol is of ample importance, as they can inhibit the activities of GSTs as well as other GSH-dependent enzymes (cfr. *supra*) (Coleman et al., 1997). Once present in the vacuole, GS-conjugates can undergo further metabolism, as discussed by Coleman et al. (1997).

Interestingly, GSTs are induced by a broad range of biotic and abiotic stress conditions, including exposure to organic contaminants. The relevance of this induction is mostly related to direct detoxification of xenobiotics, but can also be explained by the fact that certain GSTs detoxify products of lipid peroxidation and oxidative DNA damage. This characteristic of GSTs also possibly underlies their involvement in plant responses to metal stress (Frova, 2003).

3.2.2.1 Responses to Stress Conditions

As mentioned earlier, GSTs are important players in plant responses to organic compounds. In this context, especially herbicides are known to affect plant GST levels and activities. For example, glyphosate significantly increased GST activities in *Pisum sativum* roots and leaves (Miteva, Ivanov,

& Alexieva, 2010). Similarly, exposure to the chloroacetamide herbicide metazachlor induced GST activity in *B. napus* leaves (Vercampt, Koleva, Vassilev, Vangronsveld, & Cuypers, 2016). The role of GSTs in herbicide detoxification is further supported by the enhanced tolerance to the diphenyl ether herbicide fluorodifen of *N. tabacum* plants overexpressing a *Citrus sinensis* tau class GST. Interestingly, salt and drought tolerance were also increased in the overexpressor plants. However, this effect was not due to scavenging of oxidative stress by-products, suggesting additional GST-mediated mechanisms provoking tolerance to these stress conditions (Lo Cicero, Madesis, Tsaftaris, & Lo Piero, 2015). Even though GSTs play an important role in detoxifying herbicides, their activities are not always induced in herbicide-exposed plants. The photosynthetic herbicide isoproturon reduced GST activity in *Z. mays* leaves, indicating that certain herbicides function as GST inhibitors (Alla, Hassan, El-Bastawisy, 2008).

Herbicide safeners often exert their function by increasing the rate of GSH conjugation in several crop species (Abu-Qare & Duncan, 2002; Davies & Caseley, 1999). For example, the safener benoxacor induced the activity of a tau class GST in *Festuca arundinacea* (Del Buono, Scarponi, & Espen, 2007). Furthermore, the safener cloquintocet-mexyl was shown to increase the protein abundance of several GSTs belonging to the phi, tau and lambda classes in the roots and coleoptile of *Triticum tauschii* seedlings (Zhang, Xu, Lambert, & Riechers, 2007). However, it should be noted that safeners could also increase the GSH conjugation rate in weeds competing with the target crop. Mefenpyr-diethyl and fenchlorazole-ethyl safeners, for example, were shown to increase the protein abundance of certain phi and lambda class GSTs in the weed *Alopecurus myosuroides* treated with the graminicidal herbicide fenoxaprop-ethyl (Cummins, Bryant, & Edwards, 2009). In addition, in *A. thaliana* seedlings, the gene encoding GSTU26 was transcriptionally induced by exposure to the chloroacetanilide herbicides alachlor and metolachlor and the safener benoxacor. In contrast, expression levels of *GSTF9* were not affected under these conditions, indicating that specific GST isoforms are responsive to specific safeners and/or herbicides (Nutricati, Miceli, Blando, & De Bellis, 2006). Similarly, different safeners were shown to transcriptionally induce several phi and theta class GSTs in *A. thaliana*. Nonetheless, this response did not protect the plants from herbicide-induced damage, suggesting that other players besides GSTs are responsible for the differences in safener-induced herbicide tolerance between plant species (DeRidder, Dixon, Beussman, Edwards, & Goldsbrough, 2002).

Furthermore, it should be noted that GSTs can bioactivate instead of detoxify certain organic compounds. Two examples of toxic GS-conjugates in plants refer to GST-mediated isomerization involved in herbicide bioactivation of isourazoles and thiadiazolidines to toxic urazoles and triazolidines, respectively (Edwards et al., 2000).

In addition to herbicides and safeners, metal exposure is also known to affect GSTs in plants. As mentioned earlier, this observation is likely related to the involvement of GSTs in detoxifying products resulting from oxidative damage to cellular macromolecules. Exposure to Cd, Cu, cobalt (Co), Hg, Ni and Zn significantly increased GST activity in root tips of *H. vulgare*. This response was also induced by salt, cold, drought and H_2O_2 exposure (Halušková, Valentovičová, Huttová, Mistrík, & Tamás, 2009). Furthermore, Cd was shown to induce GST activity in the roots and leaves of *Ricinus communis*. This Cd-induced increase in GST activity was also observed in *B. napus* roots. As Cd exposure significantly increased MDA concentrations in both plant species, the induction of GST activity could be targeted towards detoxifying lipid peroxidation products (Bauddh, Kumar, Srivastava, Singh, & Tripathi, 2016). Kumar et al. (2013) reported that gene expression levels of *GSTL1* and *GSTL2* were increased by As, Cd and Pb exposure in *O. sativa* seedlings, whereas they were not affected by Cr. *GSTL3* expression was induced by As and Pb, but not by Cd. Taken together, these data indicate that metal-induced transcriptional responses of GSTs depend on the metal and GST isoform under study (Kumar et al., 2013; Lyubenova & Schroder, 2011).

3.2.3 Glutathione in Antioxidative Defence

GSH, as the most abundant essential nonenzymatic antioxidant in plant cells, contributes to antioxidative defence in several ways. First, it can directly react with singlet oxygen ($^1O_2$), H_2O_2 and hydroxyl radicals ($^{\bullet}OH$) (Gill & Tuteja, 2010b). In addition, GSH plays an important role in the antioxidative ascorbate (AsA)-GSH cycle, also referred to as the Foyer-Halliwell-Asada pathway. In this cycle, H_2O_2 is reduced to H_2O by the action of ascorbate peroxidase (APX), simultaneously oxidizing AsA to dehydroascorbate (DHA). The latter molecule is again reduced to AsA by the action of DHAR, using GSH as an electron donor (Fig. 1) (Foyer & Noctor, 2011). GSH also contributes to antioxidative defence as an electron donor for glutathione peroxidase (GPX), catalyzing the reduction of H_2O_2, organic peroxides and lipid peroxides (Anjum et al., 2012; Gill & Tuteja, 2010b). In addition, it reduces glutaredoxins (GRXs), which are involved in several processes including the regeneration of antioxidative enzymes (Rouhier, 2010).

In all the reactions described earlier, GSH donates an electron – derived from the cysteine residue of its thiol group – to another molecule and thereby becomes reactive. It then readily reacts with another reactive GSH molecule, forming glutathione disulphide (GSSG). Subsequently, GSSG can be reduced to GSH by the action of the NADPH-dependent GR enzyme (Fig. 1). Plants possess two GR-encoding genes: (1) *GR1*, present in the cytosol and (2) *GR2*, dually targeted to plastids and mitochondria (Noctor et al., 2012). As GR is constitutively active, more than 90% of the cellular GSH pool is in its reduced form under control conditions. During stress conditions, however, GSH demand can exceed GR activity, thereby decreasing the GSH/GSSG ratio (Jozefczak et al., 2012). As the most abundant redox couple in plant cells, GSH/GSSG plays a crucial role in redox regulation (Noctor et al., 2012). However, the relevance of the GSH/GSSG redox potential as the driving force of biological processes has to be critically discussed (Flohe, 2013).

Plant GSH levels and redox state are often affected by metal(loid) exposure (Jozefczak et al., 2012). For example, Dixit et al. (2016) demonstrated that exposure to 25 and 50 μM as significantly increased GSH and GSSG levels in both roots and shoots of *O. sativa* plants. The observed oxidation of GSH was related to the activation of the AsA-GSH cycle, as indicated by significantly increased APX activities in both organs. However, the GSH/GSSG ratio remained unaffected after As exposure. This is probably due to the observed increases in the activities of GR and enzymes involved in sulphate assimilation and GSH biosynthesis, such as SAT and GSH1. It is interesting to note that under all conditions, GSH levels were positively correlated with sulphur concentrations in the growth medium, whereas GSSG levels decreased with increasing sulphur levels. These data indicate that antioxidative defence in As-exposed plants is more efficient under high-sulphur conditions (Dixit et al., 2016). Furthermore, the GSH redox state was significantly shifted towards a more reduced state in leaves of *A. thaliana* exposed to Zn concentrations ranging from 100 to 500 μM, despite a significant decrease in GR activity. However, the increased GSH/GSSG ratio could be explained by an increased GSH synthesis, as Zn significantly increased total GSH equivalents and free GSH levels. In contrast, Zn slightly decreased the GSH/GSSG ratio in roots, possibly as a result of the observed decrease in root GR activity. PC synthesis was strongly induced, thereby decreasing free GSH levels available for antioxidative defence reactions in roots of Zn-exposed plants (Remans et al., 2012). Cuypers et al. (2011) reported that both Cd and Cu exposure shifted the GSH redox state in *A. thaliana* roots towards a more oxidized state, which was much more pronounced in Cu-exposed plants. In contrast, the GSH/GSSG ratio was not significantly affected by either metal in the leaves (Cuypers et al., 2011). Nocito et al. (2006) reported that Cd exposure significantly decreased total GSH levels in roots of *Z. mays*. Reduced GSH levels strongly diminished, whereas GSSG concentrations significantly increased in roots of Cd-exposed plants, thereby shifting the cellular redox balance towards a more oxidized state. Zn exposure, in contrast, did not alter the GSH/GSSG ratio, as it increased both GSH and GSSG levels to the same extent. Again, exposure to Cu caused the most dramatic effects, as it decreased the GSH/GSSG ratio by more than 20-fold (Nocito et al., 2006). These data indicate that metal-induced effects related to the GSH/GSSG ratio depend on many factors including the chemical properties of the metal, its concentration, the exposure duration and the plant species and organ considered.

As mentioned earlier, the role of GSH in plants exposed to organic contaminants is mainly related to its involvement in GST-mediated xenobiotic detoxification. In addition, GSH possibly functions in antioxidative defence reactions triggered by organic compounds such as herbicides. Exposure to metazachlor for 14 days increased the activities of APX and GR in leaves of *B. napus*, suggesting activation of the AsA-GSH cycle. This assumption was further supported by the observation that the GSH redox state was significantly shifted towards its oxidized form. Activities of other antioxidative enzymes, including SOD and CAT, were also induced by exposure to metazachlor (Vercampt et al., 2016). Similarly, exposure to the recommended field dose of metribuzin increased GSSG levels and slightly decreased the GSH/GSSG ratio in leaves of *Z. mays*, possibly as a result of the observed decrease in GR activity. In contrast, treatment with the recommended field dose of pretilachlor increased GR activity and the GSH/GSSG ratio. These results possibly explain the fact that the induction of H_2O_2 levels and lipid peroxidation was more pronounced in plants exposed to metribuzin than in those exposed to pretilachlor. Taken together, these data indicate that herbicide-induced antioxidative plant responses depend on the specific herbicide applied (Alla, Badawi, Hassan, El-Bastawisy, & Badran, 2008).

4. THIOLS PLAY AN IMPORTANT ROLE IN CELLULAR REGULATION

4.1 Methionine and S-adenosylmethionine Synthesis

In addition to its function as a building block of proteins and, in particular, MTs, GSH and PCs, cysteine also plays an important role in the biosynthesis of the second sulphur-containing amino acid methionine. Like lysine and threonine, methionine belongs to the family of aspartate-derived amino acids (Hesse et al., 2001). It is composed of (1) a sulphur atom derived from cysteine, (2) the nitrogen/carbon skeleton from a phosphorylated serine and (3) a folate-derived methyl group (Wirtz & Droux, 2005). The synthesis of methionine comprises three steps (Fig. 1). In the first step, cystathionine γ-synthase (CGS) catalyzes the formation of the thioether cystathionine from its substrates cysteine and O-homophosphoserine (OPH). Subsequently, cystathionine β-lyase (CBL) cleaves cystathionine to homocysteine in a β-cleavage reaction. These two steps exclusively take place in plastids and are together referred to as the transsulphuration pathway (Droux, 2004; Wirtz & Droux, 2005). In the third and final step of methionine synthesis, methionine is formed by the methylation of homocysteine, which is catalyzed by a vitamin B12-independent methionine synthase (MS), using methyltetrahydrofolate as the methyl donor (Fig. 1) (Takahashi et al., 2011). This enzyme is present in both plastids and the cytosol. The plastidial MS isoform is involved in only de novo methionine synthesis, whereas the cytosolic isoform also mediates the recycling of homocysteine resulting from the hydrolysis of S-adenosylhomocysteine (Gigolashvili & Kopriva, 2014; Wirtz & Droux, 2005).

More than 80% of the synthesized methionine is used for the production of SAM (Ravanel, Gakiere, Job, & Douce, 1998). This metabolite is produced from methionine and ATP in a cytosolic reaction catalyzed by SAM synthetase (SAMS) (Fig. 1). SAM serves as the methyl donor in a broad array of methylation reactions catalyzed by methyltransferases. Furthermore, SAM is also the precursor for the synthesis of ethylene, polyamines (PAs), nicotianamine, phytosiderophores and biotin (Roje, 2006).

The regulation of methionine and SAM synthesis mainly takes place at the level of the CGS enzyme. Although CGS is not subject to feedback inhibition by methionine or its metabolites at the activity level, it is posttranscriptionally regulated by SAM (Amir, 2010). When SAM is present, it induces a temporary arrest of the translation elongation process of the methionine overaccumulation 1 (MTO1) domain in the CGS mRNA, causing degradation of the mRNA upstream of the stalled ribosome. As a result, 5′-truncated RNA species are produced, causing the decay of the transcript (Amir, 2010; Onouchi et al., 2005). The importance of the tight regulation of CGS is highlighted by the observation that plants with increased or decreased CGS levels display severe morphological phenotypes (Amir, 2010).

Control of methionine production also occurs at the level of OPH, which is the last common intermediate for methionine and threonine biosynthesis. Although both CGS and threonine synthase (TS) use OPH as a substrate, the affinity of TS for OPH strongly exceeds that of CGS. In addition, SAM enhances the activity and substrate affinity of TS. Therefore, OPH mainly flows to the threonine synthesis pathway when methionine and SAM levels are sufficiently high. When SAM levels decline, however, TS activity decreases and OPH supply to the methionine synthesis pathway increases (Takahashi et al., 2011).

In addition to cysteine, methionine levels can be affected in metal-exposed plants. Exposure to 120 μM Ni was demonstrated to increase methionine concentrations in roots and shoots of *Matricaria chamomilla* (Kovacik, Klejdus, Hedbavny, & Backor, 2009). Furthermore, exposure to different sublethal Cd concentrations for 24 and 72 h significantly increased methionine concentrations in both roots and leaves of *A. thaliana* (Keunen, Florez-Sarasa, et al., 2016). MS protein abundance was significantly increased by Cd exposure in *T. aestivum* leaves. This response was accompanied by an increased level of 1-aminocyclopropane-1-carboxylate (ACC) synthase (ACS), an intermediate in the ethylene biosynthesis pathway (Fig 1). Interestingly, similar effects were induced by exposure to trichlorobenzene, suggesting that both metals and organic contaminants can affect methionine and ethylene biosynthesis pathways in plants (Ge et al., 2009). In contrast, Barkla, Vera-Estrella, Miranda-Vergara, and Pantoja (2014) reported that Zn decreased MS abundance in leaves of *A. thaliana*. Serine hydroxymethyltransferase levels were also reduced in Zn-exposed plants. This enzyme catalyzes the conversion of glycine to serine, thereby yielding methyl units that are channelled into the methionine biosynthesis pathway. The Zn-induced decrease of this enzyme possibly results in a decreased methionine concentration. The authors hypothesize that this mechanism protects Zn-stressed cells by attenuating the initiation of protein translation, thereby preventing the synthesis of misfolded proteins (Barkla et al., 2014). A Zn-induced decrease in methionine levels was also observed in shoots of *N. tabacum* (Pavlíková et al., 2014). Exposure to As, however, increased methionine concentrations in *O. sativa*. This response was more pronounced in a genotype accumulating higher As levels. A similar response was observed for cysteine and other stress-responsive amino acids such as proline, glycine and glutamate. Clear correlations were reported among the levels of these amino acids, the extent of lipid peroxidation and the activities of antioxidative enzymes, suggesting that their levels are increased as a defence mechanism against As-induced oxidative stress (Dave et al., 2013).

Further evidence supporting a role for methionine in plant responses to metal stress is again derived from comparing metal-tolerant and metal-sensitive species. Begum et al. (2016) reported that As exposure significantly increased methionine levels in a tolerant *O. sativa* genotype, whereas they were not affected in an As-sensitive genotype. Similar results were demonstrated for cysteine and GSH, underlining the importance of sulphur metabolism in plant metal tolerance (Begum et al., 2016). Similarly, Liang et al. (2014) demonstrated that Cd increased root methionine levels in a hyperaccumulating *S. alfredii* ecotype, whereas this response was absent in its nonhyperaccumulating counterpart. Furthermore, *SAMS* expression levels were only significantly enhanced by Cd exposure in roots and shoots of the hyperaccumulator plants. Taken together, these data point towards a role for SAM in mediating Cd tolerance in *S. alfredii* (Liang et al., 2014). MS and SAMS protein abundance were significantly increased by Cu exposure in roots of a metallicolous population of *Agrostis capillaris*, whereas they were not affected in a nonmetallicolous population, further supporting a role for methionine and SAM in plant metal tolerance (Hego et al., 2014).

The effects of metals on methionine and SAM biosynthesis are relatively well studied, whereas information regarding the influence of organic contaminants on these processes is scarce and therefore constitutes an interesting topic for future research. In this regard, evidence has been presented that SAM-methyltransferase is involved in the detoxification of the herbicide fluorodifen in *Picea*. The fluorodifen metabolite 2-nitro-4-trifluoromethyl-thiophenol was rapidly converted to the corresponding 2-nitro-4-trifluoromethyl-thioanisole, a volatile end product of the degradation pathway in needles of *P. abies*, *Picea glauca* and *Picea pungens* (Lamoureux, Rusness, & Schroder, 1993). A similar pathway was demonstrated for the metabolism of pentachloronitrobenzene in *Allium cepa* (Lamoureux & Rusness, 1980).

4.2 Ethylene: Hormonal Regulation

As mentioned earlier, thiols are important precursors in a large number of cellular processes and regulatory networks. SAM, the activated form of methionine, is a key player in the synthesis of ethylene and PAs. As a precursor of ethylene, SAM is directly linked to hormonal signalling.

In the ethylene biosynthesis pathway, SAM is converted to ACC by the ACS enzyme belonging to the class of pyridoxal phosphate (PLP)-dependent enzymes utilizing vitamin B6 as a cofactor (Fig. 1). This reaction involves the release of 5′-methylthioadenosine (MTA), which is recycled back to methionine in the so-called Yang cycle (Murr & Yang, 1975). The ACS enzyme is located in the cytosol and encoded by a multigene family of 12 members in *Arabidopsis*, of which 8 encode functional proteins (Van de Poel & Van Der Straeten, 2014). Expression of different members of this large gene family is tissue-dependent and single isoforms are specifically involved in distinct physiological or developmental tasks (Tsuchisaka & Theologis, 2004). Furthermore, complex interactions are known to occur between different ACS members (Tsuchisaka et al., 2009).

The second step in ethylene synthesis consists of the oxidation of ACC to ethylene via ACC oxidase (ACO) in the presence of oxygen (Fig. 1) (Kende, 1993). As a member of the superfamily of dioxygenases, ACO requires Fe^{2+} as a cofactor and bicarbonate as an activator (Zhang, Ren, Clifton, & Schofield, 2004). Information on its cellular localization remains unclear: it could be localized in either the cytosol (Chung, Chou, Kuang, Charng, & Yang, 2002; Hudgins, Ralph, Franceschi, & Bohlmann, 2006) or the plasma membrane (Ramassamy, Olmos, Bouzayen, Pech, & Latche, 1998; Rombaldi et al., 1994). Even though ACS is considered the rate-limiting enzyme of ethylene biosynthesis in plants, there are reports indicating that ACO limits ethylene synthesis under certain conditions, e.g., postclimacteric fruit ripening in tomato (Van de Poel & Van Der Straeten, 2014). In plants, ethylene synthesis is regulated by a variety of internal and external signals, often with ACS as the main target (for a review, see Van de Poel and Van Der Straeten (2014) and Keunen, Schellingen, et al. (2016)).

Many reports have shown enhanced ethylene production upon metal exposure in plants, with the extent depending on the metal and its applied concentration. From all inorganic ions, Cd probably causes the strongest induction of ethylene production in plants (Keunen, Schellingen, et al., 2016). While this was shown for several species under short time exposure conditions, longer Cd exposure decreased ethylene concentrations in

A. thaliana (Carrio-Segui, Garcia-Molina, Sanz, & Penarrubia, 2015). Summing up literature data, an intimate relationship exists between metal stress and ethylene production in plants. More and more reports suggest an implication of ethylene signalling in plant adaptation and tolerance to metal stress (Keunen, Schellingen, et al., 2016; Thao et al., 2015). As discussed in Section 5, extensive cross talk is observed between ethylene and other key players in the plant metal stress response, such as ROS and GSH.

In contrast with the vast amount of data available on the involvement of ethylene in plant metal stress responses, knowledge on its role in plants exposed to organic contaminants is limited. Nonetheless, Kummerová, Zezulka, Váňová, and Fišerová (2012) have demonstrated a significant stimulation of ethylene release in germinating *Z. mays* and *P. sativum* seeds during exposure to the polycyclic aromatic hydrocarbon fluoranthene. Seed germination was significantly inhibited by this compound in both species (Kummerová et al., 2012).

4.3 Polyamines: Hormonelike Signalling Compounds

Next to ethylene synthesis, SAM is also a precursor for the synthesis of PAs, representing a group of low-molecular-weight polycationic amines ubiquitous in all living organisms (J.-H. Liu, Wang, Wu, Gong, & Moriguchi, 2015). In plants, the positively charged PAs are either bound to negatively charged molecules or conjugated to small molecules and proteins, but they also occur as free forms (Walters, 2003). PAs are involved in a large number of cellular functions and regulatory processes, including the stabilization of proteins and other biomolecules, the regulation of cell division, growth and differentiation or senescence as well as general adaptive stress responses (Bouchereau, Aziz, Larher, & Martin-Tanguy, 1999; Groppa & Benavides, 2008; Walters, 2003).

Putrescine (Put), spermidine (Spd) and spermine (Spm) are the three major PAs and are therefore most studied in plants. They can be present in free, soluble conjugated and insoluble bound forms. In plants, PA synthesis is well studied and based on the two precursors L-ornithine and L-arginine from which Put is generated by the catalytic actions of ornithine decarboxylase (ODC) and arginine decarboxylase (ADC). From Put, both Spd and Spm are formed, with a strong connection to SAM (Fig. 1) (Sauter et al., 2013). In the first step, SAM is decarboxylated to decarboxylated SAM (dcSAM) by SAM decarboxylase (SAMDC). This metabolite then provides the aminopropyl group for the conversion of Put into Spd by Spd synthase.

Spd is then further converted into Spm by Spm synthase, again using dcSAM as an aminopropyl donor (Fig. 1) (J.-H. Liu et al., 2015). From this reaction, MTA is released and directed back into the SAM cycle where SAM is synthesized from methionine by SAMS (Sauter et al., 2013). Apart from de novo synthesis of PAs, it is worth mentioning that catabolism of these molecules takes place, involving both Cu-containing diamine oxidases (CuAOs) and FAD-dependent polyamine oxidases (PAOs) (Cona, Rea, Angelini, Federico, & Tavladoraki, 2006; J.-H. Liu et al., 2015).

The relationship between abiotic stress and altered (mostly elevated) PA levels is well established and has been known for a long time. Nevertheless, their actual function still remains unclear. Although PAs could protect cells against abiotic stress, for example by their antioxidative properties, they could also cause cell damage by the production of H_2O_2 via their catabolism (Minocha, Majumdar, & Minocha, 2014).

As indicated in Fig. 1, SAM represents a direct link between ethylene and PA synthesis, allowing plants to directly switch between both pathways. Interestingly, PAs and ethylene display counteracting functions in plants. Ethylene is known to induce senescence, whereas PAs play the opposite role by decelerating chlorophyll loss and cell membrane degradation and slowing down senescence-induced increases in protease and RNase activity (Pandey, Ranade, Nagar, & Kumar, 2000). This is because of their ability to neutralize acids, their antioxidative properties and their potential to stabilize cell walls and membranes (Gill & Tuteja, 2010a).

Different reports indicate a role for PAs in plant metal stress responses. For example, exposure to Cu and Cd led to an increase in ADC and ODC activity in *T. aestivum* leaves. This effect was stronger under Cd exposure, where Put levels were elevated up to threefold as compared to control levels. Spd concentration was not affected by any of the metals, whereas Spm levels were significantly reduced. Interestingly, externally applied Spm reduced the formation of H_2O_2 and led to a reduction in lipid peroxidation levels as well. A potential antioxidative function of Spm could be concluded from these results, although the actual mechanism of protection is still unclear (Groppa, Tomaro, & Benavides, 2007).

Although it is well established that in plants many organic pollutants cause the formation of ROS, there is only little information available on how the PA metabolism is affected by these xenobiotic compounds. Nevertheless, Burritt (2008) could link phenanthrene exposure to the synthesis of PAs. Phenanthrene is a polycyclic and highly toxic aromatic hydrocarbon that is often found in aquatic environments. Whereas data regarding its effect on aquatic macrophytes are scarce, polycyclic aromatic hydrocarbons often induce oxidative stress in plants. The involvement of PAs in the protection of plants against oxidative stress could be shown for *Riccia fluitans* exposed to phenanthrene. Concentrations up to 500 μM led to an induction of oxidative stress to which the plants responded with an increase in PA synthesis linked to elevated ADC and SAMDC activity. Chemical inhibition of these enzymes caused an inhibition of plant recovery, whereas externally applied PAs could reduce the negative effects caused by phenanthrene exposure (Burritt, 2008). Studies suggest PAs as priming agents before all sorts of abiotic stress, including metal exposure (Savvides, Ali, Tester, & Fotopoulos, 2016). This opens the window to more extensive research on the involvement of PAs in plant responses to metals as well as organic pollutants.

5. INTERACTION BETWEEN DETOXIFICATION AND SIGNAL TRANSDUCTION PATHWAYS

As discussed earlier, sulphur-containing metabolites play an important role in plant responses to metal stress. Compounds directly derived from cysteine (e.g., GSH, PCs and MTs) are involved in metal chelation and ROS detoxification, whereas molecules derived from methionine and its primary metabolite SAM (e.g., ethylene and PAs) mainly function in cellular signal transduction. However, different components of the detoxification and signal transduction pathways interact with each other under both physiological and stress conditions, including metal exposure (Fig. 2).

For example, cross talk exists between ethylene and other players involved in plant responses to metal stress, as reviewed by Keunen, Schellingen, et al. (2016). Several studies demonstrate a role for ethylene in the oxidative burst induced by a broad range of stress conditions. Cao, Jiang, and Zhang (2006) reported that paraquat-induced increases in $O_2^{\bullet -}$ and H_2O_2 levels were less pronounced in an *ethylene insensitive 2-1* (*ein2-1*) mutant as compared to those in WT *A. thaliana* plants. Therefore, the extent of lipid peroxidation was significantly lower in the mutant (Cao et al., 2006). Similar responses were described in *ein2-1* mutants exposed to Al (Y.Y. Zhang, He, Zhao, Huang, & Hao, 2014). In addition, blocking of ethylene biosynthesis and/or signalling resulted in a reduced extent of H_2O_2 production in camptothecin-exposed *Lycopersicon esculentum* suspension cells (de Jong, Yakimova, Kapchina, & Woltering, 2002) and

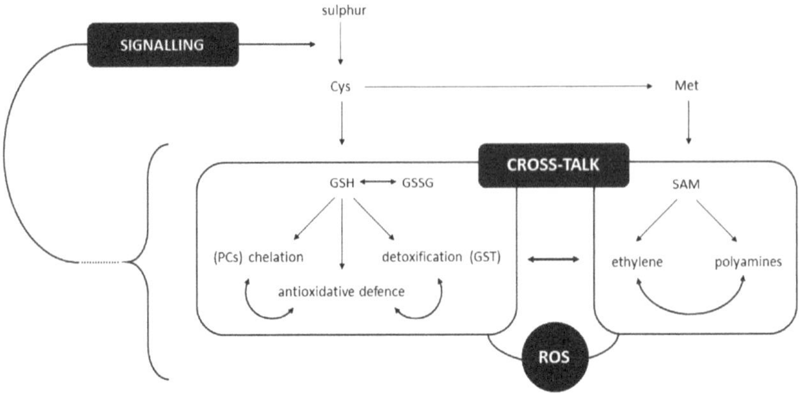

Figure 2 ***Cross talk between detoxification and signal transduction pathways in plants.*** Significant cross talk exists between glutathione-mediated detoxification mechanisms and ethylene- and polyamine-related signal transduction pathways in plants, both under physiological and metal exposure conditions. Reactive oxygen species are put forward as central players in this interaction. Furthermore, several components involved in detoxification and signal transduction pathways are known to affect sulphur uptake and assimilation. *Cys*, cysteine; *GSH*, glutathione; *GSSG*, glutathione disulphide; *GST*, glutathione transferase; *Met*, methionine; *PC*, phytochelatin; *ROS*, reactive oxygen species; *SAM*, S-adenosylmethionine.

Cd-exposed *Phaseolus coccineus* roots (Maksymiec, 2011). Furthermore, inhibition of ethylene biosynthesis limited Cd-induced $O_2^{\bullet -}$ accumulation in root hair tips of *B. napus*, suggesting that ethylene is an upstream regulator of $O_2^{\bullet -}$ generation (Sun & Guo, 2013). The effect of ethylene on $O_2^{\bullet -}$ production is possibly mediated by its interaction with ROS-producing NADPH oxidases, also referred to as respiratory burst oxidase homologues (RBOHs). This is confirmed by the observation that H_2O_2 production induced by the ethylene-releasing compound ethephon in *I. batatas* was limited by treatment with the NADPH inhibitor diphenyleneiodonium (Chen, Huang, Huang, Chow, & Lin, 2013). This relationship between ethylene and NADPH oxidases is also supported under metal exposure conditions, as Hg-induced increases in apoplastic H_2O_2 accumulation and NADPH oxidase activity in apical root segments of *Medicago sativa* were reduced by treatment with an ethylene signalling inhibitor (Montero-Palmero, Martin-Barranco, Escobar, & Hernandez, 2014). Keunen et al. (2015) reported that the Cd-induced *RBOHC* upregulation in several *A. thaliana* ethylene biosynthesis and signal transduction mutants was less pronounced than that in WT plants.

In addition to its effects on ROS production, ethylene is also linked to several players of the antioxidative defence system. It interacts with both enzymatic antioxidants, such as SOD and CAT, and nonenzymatic antioxidants, such as AsA and α-tocopherol (Keunen, Schellingen, et al., 2016). In the context of this review, the interaction between ethylene and GSH is of particular interest. The cross talk between these two compounds is probably related to the link of ethylene biosynthesis and signalling with sulphate assimilation (Fig. 2), as reviewed by Iqbal et al. (2013) and Wawrzynska, Moniuszko, and Sirko (2015). Indeed, sulphur nutrition has been shown to modulate plant responses to a diverse array of stress factors by increasing ethylene production (Wawrzynska et al., 2015). Nazar, Khan, Iqbal, Masood, and Khan (2014), for example, demonstrated that excess sulphur alleviated the effects of salt stress in *B. juncea*. However, this response was absent when ethylene synthesis was inhibited, suggesting a role for ethylene in the sulphur-induced salt stress alleviation (Nazar et al., 2014). Several genes involved in ethylene signalling were strongly upregulated in sulphate-treated *Vitis vinifera* (Giraud, Ivanova, Gordon, Whelan, & Considine, 2012). In addition to sulphur excess, sulphur limitation was also shown to increase ethylene concentrations and the expression of ethylene-related genes in *N. tabacum* plants (Lewandowska et al., 2010; Moniuszko et al., 2013).

Interestingly, the interplay between ethylene and sulphate assimilation also functions in the other direction, with ethylene levels affecting several proteins of the sulphate assimilation pathway (Fig. 2). It was shown that ethephon significantly induced ATPS activity and sulphur content in *B. juncea* (Iqbal, Khan, Nazar, & da Silva, 2012). Furthermore, treatment of *A. thaliana* with 200 μM ACC was shown to induce APR at the transcriptional and activity level (Koprivova, North, & Kopriva, 2008). As GSH synthesis depends on sulphur availability for the production of its precursor cysteine, ethylene possibly modulates the sulphate assimilation pathway to meet the increasing demand for GSH during stress conditions (Keunen, Schellingen, et al., 2016).

The cross talk between ethylene and GSH is further supported by the observation that application of exogenous GSH in *A. thaliana* increased both gene expression levels and protein abundance of ACS2 and ERF2, involved in ethylene biosynthesis and signal transduction, respectively (Sinha et al., 2015). The ethephon-induced increased ROS production in *I. batatas* was attenuated when plants were treated with exogenous GSH (Chen et al., 2013). Expression levels of ethylene-related genes are often affected in transgenic plants with altered GSH levels (Keunen, Schellingen, et al., 2016). Schnaubelt et al. (2015), for example, reported several *ERF* genes to be significantly upregulated in the severely GSH-deficient *rml1-1* *A. thaliana* mutant as compared to WT plants. In contrast, *ERF2* expression was significantly downregulated in the *A. thaliana cad2-1* mutant, which is characterized by a milder GSH deficiency (Han, Mhamdi, Chaouch, & Noctor, 2013). Furthermore, transgenic tobacco plants with an enhanced GSH content displayed a significant upregulation of genes involved in ethylene biosynthesis and signal transduction (Ghanta et al., 2014). Similarly, Datta et al. (2015) reported that transgenic *A. thaliana* plants with elevated GSH levels displayed increased ACS2, ACS6 and ACO1 transcription levels and protein abundance as compared to WT plants, while the opposite response was observed in the GSH-deficient *pad2-1* mutant. The authors demonstrated that the GSH-induced upregulation of *ACS2* and *ACS6* was mediated by a WRKY33-related mechanism, whereas the increased *ACO1* expression was due to an increased stability of its encoding mRNA. Interestingly, application of exogenous GSH also increased the tolerance of WT *A. thaliana* plants to necrotrophic infection and salt stress, while it did not affect the stress tolerance of *ein2-1* mutants deficient in ethylene signalling. These data suggest that GSH induces plant tolerance to different stress factors by an ethylene-dependent mechanism (Datta et al., 2015).

Interestingly, the relationship between ethylene and GSH is also supported during metal exposure (Keunen, Schellingen, et al., 2016). Indeed, ethephon application was shown to increase GSH concentrations in *B. juncea* exposed to Cd (Masood, Iqbal, & Khan, 2012), Ni and Zn (Khan & Khan, 2014). Schellingen et al. (2015) reported that the Cd-induced upregulation of genes involved in GSH biosynthesis was significantly weaker in an *acs2-1/6-1* knockout mutant than in WT *A. thaliana* plants, suggesting that ethylene biosynthesis is essential for the induction of efficient GSH-dependent defence responses during Cd exposure (Schellingen et al., 2015). Similarly, the crucial role of EIN2 in Pb tolerance in *A. thaliana* was also demonstrated to be partially related to its stimulating effect on GSH levels (Cao et al., 2009). The interaction between ethylene and GSH during metal exposure also functions in the other direction, with GSH levels affecting ethylene signalling. Hasan et al. (2016) reported that the Cd-induced upregulation of *ERF1* and *ERF2* was more pronounced in plants supplied with 5 mM GSH, whereas it was weaker in plants treated with the GSH biosynthesis inhibitor buthionine sulphoximine.

In addition to ethylene, PAs are also subject to cross talk with the antioxidative defence system during stress conditions. Addition of Spd to *Solanum lycopersicum* exposed to chilling stress, for example, strongly enhanced the chilling-induced increase in GSH and AsA levels. Furthermore, it significantly increased the GSH/GSSG and AsA/DHA ratios, possibly as a result of the observed induction of enzymes involved in the AsA-GSH cycle at the transcriptional and/or activity level. While the enzymatic activities of SOD, POD and CAT were decreased by chilling stress, this effect was reversed by adding Spd, with activities even increasing those measured in control plants. These data possibly explain the observed reduction of chilling-induced $O_2^{\bullet-}$ and H_2O_2 production by addition of exogenous Spd (Diao, Song, & Qi, 2015). Similarly, positive effects of PA treatment on the antioxidative defence system were also reported in salt-exposed *O. sativa* (Jain, Vart, Verma, & Malhotra, 2015) and *Cucumis sativus* (Shu, Yuan, Guo, Sun, & Yuan, 2013) and *Glycine max*. subjected to osmotic stress (Radhakrishnan & Lee, 2013).

PAs were also shown to positively affect several components of the antioxidative defence system in metal-exposed plants. Spd treatment was shown to further enhance the Cr-induced increase in GSH levels in *Rhapus sativus*. Furthermore, Spd significantly alleviated Cr-induced lipid peroxidation and H_2O_2 concentrations. In contrast, it counteracted the increase in CAT activity caused by Cr exposure. Overall, exogenous Spd increased plant tolerance to Cr, as indicated by its positive effects on root and shoot length and fresh weight of Cr-exposed plants (Choudhary, Kanwar, Bhardwaj, Yu, & Tran, 2012). Similarly, Rady and Hemida (2015) reported that addition of either Spd or Spm counteracted the negative effects of Cd exposure on several growth parameters of *T. aestivum* plants. This effect was likely related to the fact that PA application attenuated Cd-induced effects on leaf GSH levels and SOD, CAT and APX activity. As a result, Cd-induced lipid peroxidation and H_2O_2 production were reduced in leaves of PA-treated plants. However, these data should be interpreted with caution, as the application of Spd and Spm also significantly reduced shoot Cd concentrations potentially explaining the observed attenuation of metal stress (Rady & Hemida, 2015).

Taken together, available data indicate that significant cross talk exists between cysteine-related detoxification and SAM-related signal transduction pathways under different stress conditions including exposure to metals and possibly also organic contaminants. This interaction should not be surprising, as the sulphur atom of cysteine is used to synthesize methionine and its primary metabolite SAM, thereby linking both pathways. However, further research is needed to fully elucidate the mechanisms connecting both pathways (Fig. 2).

6. CONCLUDING REMARKS AND FUTURE PERSPECTIVES

The sulphur-containing amino acids cysteine and methionine play crucial roles in plant responses to metals and organic contaminants through their incorporation into the primary metabolites GSH and SAM. GSH is primarily involved in chelation and detoxification mechanisms, whereas SAM mainly contributes to signal transduction reactions mediated by ethylene and PAs. Significant cross talk exists between these detoxification and signal transduction pathways, which is explained by the fact that cysteine is a precursor for both GSH and methionine synthesis. Furthermore, ROS appear to play an important role in mediating the interaction between both pathways. The effects of metal exposure on plant detoxification and signal transduction pathways are relatively well described, whereas information regarding the influence of organic contaminants is rather scarce and mostly limited to herbicides and safeners, indicating a need for further research. Elucidating plant responses to metals and organic compounds can significantly contribute to optimizing phytoremediation strategies for the cleanup of contaminated soils. In this strategy, special attention should be given to sulphur availability in the environment, as it affects sulphate assimilation and downstream detoxification and regulation pathways that are important factors controlling metal uptake and translocation as well as phytotoxic responses. With view to the importance of sulphur nutrition and the internal regulation of sulphur-dependent metabolic pathways for plant health and performance, future research on breeding and practical application of plants should focus on ways to establish stable and adaptive sulphur metabolism in crops. This could be achieved by molecular approaches, fostering plant–microbe interactions or the use of sustainable agrochemical amendments.

ACKNOWLEDGEMENTS

Funding: This work was supported by Hasselt University by a PhD grant for SH (BOF14DOC04). Additional funding came from the Research Foundation Flanders (FWO) by a postdoctoral grant for EK.

REFERENCES

Abu-Qare, A. W., & Duncan, H. J. (2002). Herbicide safeners: Uses, limitations, metabolism, and mechanisms of action. *Chemosphere, 48*, 965–974.

Adams, C. A., Blee, E., & Casida, J. E. (1983). Dichloroacetamide herbicide antidotes enhance sulfate metabolism in corn roots. *Pesticide Biochemistry and Physiology, 19*, 350–360.

Ahn, Y. O., Kim, S. H., Lee, J., Kim, H., Lee, H.-S., & Kwak, S.-S. (2012). Three *Brassica rapa* metallothionein genes are differentially regulated under various stress conditions. *Molecular Biology Reports, 39*, 2059–2067.

Alla, M. M. N., Badawi, A. H. M., Hassan, N. M., El-Bastawisy, Z. M., & Badran, E. G. (2008). Herbicide tolerance in maize is related to increased levels of glutathione and glutathione-associated enzymes. *Acta Physiologiae Plantarum, 30*, 371–379.

Alla, M. M. N., Hassan, N. M., & El-Bastawisy, Z. M. (2008). Changes in antioxidants and kinetics of glutathione-S-transferase of maize in response to isoproturon treatment. *Plant Biosystems, 142*, 5–16.

Amir, R. (2010). Current understanding of the factors regulating methionine content in vegetative tissues of higher plants. *Amino Acids, 39*, 917–931.

Anjum, N. A., Ahmad, I., Mohmood, I., Pacheco, M., Duarte, A. C., Pereira, E., … Prasad, M. N. V. (2012). Modulation of glutathione and its related enzymes in plants' responses to toxic metals and metalloids – A review. *Environmental and Experimental Botany, 75*, 307–324.

Anjum, N. A., Gill, S. S., & Gill, R. (2014). Cysteine – Jack of all glutathione-based plant stress defence trades. In N. A. Anjum, S. S. Gill, & R. Gill (Eds.), *Plant adaptation to environmental change – Significance of amino acids and their derivatives* (pp. 35–52). London: CAB International.

Anjum, N. A., Hasanuzzaman, M., Hossein, M. A., Thangavel, P., Roychoudhury, A., Gill, S. S., … Ahmad, I. (2015). Jacks of metal/metalloid chelation trade in plants – An overview. *Frontiers in Plant Science, 6*, 1–17.

Armstrong, R. N. (1997). Structure, catalytic mechanism, and evolution of the glutathione transferases. *Chemical Research in Toxicology, 10*, 2–18.

Auguy, F., Mouna, F., Moulin, P., El Mzibri, M., Abdelaziz, S., Filali-Maltouf, A., … Doumas, P. (2016). Transcriptome changes in *Hirschfeldia incana* in response to lead exposure. *Frontiers in Plant Science, 6*, 1–13.

Ball, L., Accotto, G. P., Bechtold, U., Creissen, G., Funck, D., Jimenez, A., … Mullineaux, P. M. (2004). Evidence for a direct link between glutathione biosynthesis and stress defense gene expression in *Arabidopsis*. *The Plant Cell, 16*, 2448–2462.

Barkla, B. J., Vera-Estrella, R., Miranda-Vergara, M. C., & Pantoja, O. (2014). Quantitative proteomics of heavy metal exposure in *Arabidopsis thaliana* reveals alterations in one-carbon metabolism enzymes upon exposure to zinc. *Journal of Proteomics, 111*, 128–138.

Bartucca, M. L., Celletti, S., Astolfi, S., Mimmo, T., Cesco, S., Panfili, I., & Del Buono, D. (2016). Effect of three safeners on sulfur assimilation and iron deficiency response in barley (*Hordeum vulgare*) plants. *Pest Management Science.*

Bashir, H., Ahmad, J., Bagheri, R., Nauman, M., & Qureshi, M. I. (2013). Limited sulfur resource forces *Arabidopsis thaliana* to shift towards non-sulfur tolerance under cadmium stress. *Environmental and Experimental Botany, 94*, 19–32.

Bashir, H., Ibrahim, M. M., Bagheri, R., Ahmad, J., Arif, I. A., Baig, M. A., & Qureshi, M. I. (2015). Influence of sulfur and cadmium on antioxidants, phytochelatins and growth in Indian mustard. *AoB Plants, 7*, 1–13.

Bauddh, K., Kumar, A., Srivastava, S., Singh, R. P., & Tripathi, R. D. (2016). A study on the effect of cadmium on the antioxidative defense system and alteration in different functional groups in castor bean and Indian mustard. *Archives of Agronomy and Soil Science, 62*, 877–891.

Begum, M. C., Islam, M. S., Islam, M., Amin, R., Parvez, M. S., & Kabir, A. H. (2016). Biochemical and molecular responses underlying differential arsenic tolerance in rice (*Oryza sativa* L.). *Plant Physiology and Biochemistry, 104*, 266–277.

Belford, E. J., Dorfler, U., Stampfl, A., & Schroder, P. (2004). Microsomal detoxification enzymes in yam bean [*Pachyrhizus erosus* (L.) urban]. *Zeitschrift fur Naturforschung. C, 59*, 693–700.

Blindauer, C. A., & Leszczyszyn, O. I. (2010). Metallothioneins: Unparalleled diversity in structures and functions for metal ion homeostasis and more. *Natural Product Reports, 27*, 720–741.

Bouchereau, A., Aziz, A., Larher, F., & Martin-Tanguy, J. (1999). Polyamines and environmental challenges: Recent development. *Plant Science, 140*, 103–125.

Brazier-Hicks, M., Evans, K. M., Cunningham, O. D., Hodgson, D. R. W., Steel, P. G., & Edwards, R. (2008). Catabolism of glutathione conjugates in *Arabidopsis thaliana* – Role in metabolic reactivation of the herbicide safener fenclorim. *Journal of Biological Chemistry, 283*, 21102–21112.

Brunetti, P., Zanella, L., De Paolis, A., Di Litta, D., Cecchetti, V., Falasca, G., … Cardarelli, M. (2015). Cadmium-inducible expression of the ABC-type transporter AtABCC3 increases phytochelatin-mediated cadmium tolerance in *Arabidopsis*. *Journal of Experimental Botany, 66*, 3815–3829.

Burritt, D. J. (2008). The polycyclic aromatic hydrocarbon phenanthrene causes oxidative stress and alters polyamine metabolism in the aquatic liverwort *Riccia fluitans* L. *Plant Cell and Environment, 31*, 1416–1431.

Cairns, N. G., Pasternak, M., Wachter, A., Cobbett, C. S., & Meyer, A. J. (2006). Maturation of *Arabidopsis* seeds is dependent on glutathione biosynthesis within the embryo. *Plant Physiology, 141*, 446–455.

Cao, S. Q., Chen, Z. Y., Liu, G. Q., Jiang, L., Yuan, H. B., Ren, G., … Ma, X. (2009). The *Arabidopsis* Ethylene-Insensitive 2 gene is required for lead resistance. *Plant Physiology and Biochemistry, 47*, 308–312.

Cao, S. Q., Jiang, S. T., & Zhang, R. X. (2006). Evidence for a role of Ethylene-Insensitive 2 gene in the regulation of the oxidative stress response in *Arabidopsis*. *Acta Physiologiae Plantarum, 28*, 417–425.

Cao, M. J., Wang, Z., Wirtz, M., Hell, R., Oliver, D. J., & Xiang, C. B. (2013). SULTR3; 1 is a chloroplast-localized sulfate transporter in *Arabidopsis thaliana*. *Plant Journal, 73*, 607–616.

Carrio-Segui, A., Garcia-Molina, A., Sanz, A., & Penarrubia, L. (2015). Defective copper transport in the *copt5* mutant affects cadmium tolerance. *Plant & Cell Physiology, 56*, 442–454.

Chaney, R., & Baklanov, I. (2017). Phytoremediation and phytomining: Status and promise. In A. Cuypers, & J. Vangronsveld (Eds.), *Phytoremediation* (Vol. 83, pp. 189–222).

Chen, H. J., Hou, W. C., Yang, C. Y., Huang, D. J., Liu, J. S., & Lin, Y. H. (2003). Molecular cloning of two metallothionein-like protein genes with differential expression patterns from sweet potato (*Ipomoea batatas*) leaves. *Journal of Plant Physiology, 160*, 547–555.

Chen, H. J., Huang, C. S., Huang, G. J., Chow, T. J., & Lin, Y. H. (2013). NADPH oxidase inhibitor diphenyleneiodonium and reduced glutathione mitigate ethephon-mediated leaf senescence, H_2O_2 elevation and senescence-associated gene expression in sweet potato (*Ipomoea batatas*). *Journal of Plant Physiology, 170*, 1471–1483.

Cheng, J. C., Seeley, K. A., & Sung, Z. R. (1995). *RML1* and *RML2*, *Arabidopsis* genes required for cell proliferation at the root tip. *Plant Physiology, 107*, 365–376.

Chiang, H. C., Lo, J. C., & Yeh, K. C. (2006). Genes associated with heavy metal tolerance and accumulation in Zn/Cd hyperaccumulator *Arabidopsis halleri*: A genomic survey with cDNA microarray. *Environmental Science & Technology, 40*, 6792–6798.

Chirakkara, R. A., Cameselle, C., & Reddy, K. R. (2016). Assessing the applicability of phytoremediation of soils with mixed organic and heavy metal contaminants. *Reviews in Environmental Science and Bio/Technology, 15*, 299–326.

Choudhary, S. P., Kanwar, M., Bhardwaj, R., Yu, J. Q., & Tran, L. S. P. (2012). Chromium stress mitigation by polyamine-brassinosteroid application involves phytohormonal and physiological strategies in *Raphanus sativus* L. *PLoS One, 7*, 11.

Chung, M. C., Chou, S. J., Kuang, L. Y., Charng, Y. Y., & Yang, S. F. (2002). Subcellular localization of 1-aminocyclopropane-1-carboxylic acid oxidase in apple fruit. *Plant & Cell Physiology, 43*, 549–554.

Clendennen, S. K., & May, G. D. (1997). Differential gene expression in ripening banana fruit. *Plant Physiology, 115*, 463–469.

Cobbett, C., & Goldsbrough, P. (2002). Phytochelatins and metallothioneins: Roles in heavy metal detoxification and homeostasis. *Annual Review of Plant Biology, 53*, 159–182.

Cole, D. J., Cummins, I., Hatton, P. J., Dixon, D. P., & Edwards, R. (1997). Glutathione transferases in crops and major weeds. In K. K. Hatzios (Ed.), *Regulation of enzymatic systems detoxifying xenobiotics in plants* (Vol. 37, pp. 139–154). Dordrecht: Kluwer Academic Publishers.

Coleman, J., Blake-Kalff, M., & Davies, E. (1997). Detoxification of xenobiotics by plants: Chemical modification and vacuolar compartmentation. *Trends in Plant Science, 2*, 144–151.

Cona, A., Rea, G., Angelini, R., Federico, R., & Tavladoraki, P. (2006). Functions of amine oxidases in plant development and defence. *Trends in Plant Science, 11*, 80–88.

Coninx, L., Martinova, V., & Rineau, F. (2017). Mycorrhiza-assisted phytoremediation. In A. Cuypers, & J. Vangronsveld (Eds.), *Phytoremediation* (Vol. 83, pp. 127–188).

Cummins, I., Bryant, D. N., & Edwards, R. (2009). Safener responsiveness and multiple herbicide resistance in the weed black-grass (*Alopecurus myosuroides*). *Plant Biotechnology Journal, 7*, 807–820.

Cummins, I., Dixon, D. P., Freitag-Pohl, S., Skipsey, M., & Edwards, R. (2011). Multiple roles for plant glutathione transferases in xenobiotic detoxification. *Drug Metabolism Reviews, 43*, 266–280.

Cuypers, A., Hendrix, S., dos Reis, R. A., De Smet, S., Deckers, J., Gielen, H., … Keunen, E. (2016). Hydrogen peroxide, signaling in disguise during metal phytotoxicity. *Frontiers in Plant Science, 7*, 1–25.

Cuypers, A., Smeets, K., Ruytinx, J., Opdenakker, K., Keunen, E., Remans, T., … Vangronsveld, J. (2011). The cellular redox state as a modulator in cadmium and copper responses in *Arabidopsis thaliana* seedlings. *Journal of Plant Physiology, 168*, 309–316.

Daghan, H., Arslan, M., Uygur, V., & Koleli, N. (2013). Transformation of tobacco with *ScMTII* gene-enhanced cadmium and zinc accumulation. *Clean-Soil Air Water, 41*, 503–509.

Datta, R., Kumar, D., Sultana, A., Hazra, S., Bhattacharyya, D., & Chattopadhyay, S. (2015). Glutathione regulates 1-aminocyclopropane-1-carboxylate synthase transcription via WRKY33 and 1-aminocyclopropane-1-carboxylate oxidase by modulating messenger RNA stability to induce ethylene synthesis during stress. *Plant Physiology, 169*, 2963–2981.

Dauch, A. L., & Jabaji-Hare, S. H. (2006). Metallothionein and bZIP transcription factor genes from velvetleaf and their differential expression following *Colletotrichum coccodes* infection. *Phytopathology, 96*, 1116–1123.

Dave, R., Tripathi, R. D., Dwivedi, S., Tripathi, P., Dixit, G., Sharma, Y. K., … Chakrabarty, D. (2013). Arsenate and arsenite exposure modulate antioxidants and amino acids in contrasting arsenic accumulating rice (*Oryza sativa* L.) genotypes. *Journal of Hazardous Materials, 262*, 1123–1131.

Davidian, J. C., & Kopriva, S. (2010). Regulation of sulfate uptake and assimilation – The same or not the same? *Molecular Plant, 3*, 314–325.

Davies, J., & Caseley, J. C. (1999). Herbicide safeners: A review. *Pesticide Science, 55*, 1043–1058.

Dean, J. V., Devarenne, T. P., Lee, I. S., & Orlofsky, L. E. (1995). Properties of a maize glutathione S-transferase that conjugates coumaric acid and other phenylpropanoids. *Plant Physiology, 108*, 985–994.

Del Buono, D., Scarponi, L., & Espen, L. (2007). Glutathione S-transferases in *Festuca arundinacea*: Identification, characterization and inducibility by safener benoxacor. *Phytochemistry, 68*, 2614–2624.

DeRidder, B. P., Dixon, D. P., Beussman, D. J., Edwards, R., & Goldsbrough, P. B. (2002). Induction of glutathione S-transferases in *Arabidopsis* by herbicide safeners. *Plant Physiology, 130*, 1497–1505.

Dhindsa, R. S. (1991). Drought stress, enzymes of glutathione metabolism, oxidation injury, and protein synthesis in *Tortula ruralis*. *Plant Physiology, 95*, 648–651.

Diao, Q. N., Song, Y. J., & Qi, H. Y. (2015). Exogenous spermidine enhances chilling tolerance of tomato (*Solanum lycopersicum* L.) seedlings via involvement in polyamines metabolism and physiological parameter levels. *Acta Physiologiae Plantarum, 37*, 15.

Diesperger, H., & Sandermann, H. (1979). Soluble and microsomal glutathione S-transferase activities in pea seedlings (*Pisum sativum* L.). *Planta, 146*, 643–648.

Dixit, G., Singh, A. P., Kumar, A., Mishra, S., Dwivedi, S., Kumar, S., … Tripathi, R. D. (2016). Reduced arsenic accumulation in rice (*Oryza sativa* L.) shoot involves sulfur mediated improved thiol metabolism, antioxidant system and altered arsenic transporters. *Plant Physiology and Biochemistry, 99*, 86–96.

Dixon, D. P., Davis, B. G., & Edwards, R. (2002). Functional divergence in the glutathione transferase superfamily in plants – Identification of two classes with putative functions in redox homeostasis in *Arabidopsis thaliana*. *Journal of Biological Chemistry, 277*, 30859–30869.

Dixon, D. P., & Edwards, R. (2010). Glutathione transferases. *The Arabidopsis Book, 8*, e0131.

Dixon, D. P., Skipsey, M., & Edwards, R. (2010). Roles for glutathione transferases in plant secondary metabolism. *Phytochemistry, 71*, 338–350.

Dominguez-Solis, J. R., Gutierrez-Alcala, G., Romero, L. C., & Gotor, C. (2001). The cytosolic O-acetylserine(thiol)lyase gene is regulated by heavy metals and can function in cadmium tolerance. *Journal of Biological Chemistry, 276*, 9297–9302.

Dominguez-Solis, J. R., Lopez-Martin, M. C., Ager, F. J., Ynsa, M. D., Romero, L. C., & Gotor, C. (2004). Increased cysteine availability is essential for cadmium tolerance and accumulation in *Arabidopsis thaliana*. *Plant Biotechnology Journal, 2*, 469–476.

Droog, F. (1997). Plant glutathione S-transferases, a tale of theta and tau. *Journal of Plant Growth Regulation, 16*, 95–107.

Droux, M. (2004). Sulfur assimilation and the role of sulfur in plant metabolism: A survey. *Photosynthesis Research, 79*, 331–348.

Droux, M., Ruffet, M. L., Douce, R., & Job, D. (1998). Interactions between serine acetyltransferase and O-acetylserine (thiol) lyase in higher plants – Structural and kinetic properties of the free and bound enzymes. *European Journal of Biochemistry, 255*, 235–245.

Dudler, R., Hertig, C., Rebmann, G., Bull, J., & Mauch, F. (1991). A pathogen-induced wheat gene encodes a protein homologous to glutathione-S-transferases. *Molecular Plant-Microbe Interactions, 4*, 14–18.

Edwards, R., Dixon, D. P., & Walbot, V. (2000). Plant glutathione S-transferases: Enzymes with multiple functions in sickness and in health. *Trends in Plant Science, 5*, 193–198.

Edwards, R., & Dixon, R. A. (1991). Glutathione S-cinnamoyl transferases in plants. *Phytochemistry, 30*, 79–84.

Eevers, N., White, J. C., Vangronsveld, J., & Weyens, N. (2017). Bio- and phytoremediation of pesticide contaminated environments: A review. In A. Cuypers, & J. Vangronsveld (Eds.), *Phytoremediation* (Vol. 83, pp. 277–318).

Ernst, W. H. O., Krauss, G. J., Verkleij, J. A. C., & Wesenberg, D. (2008). Interaction of heavy metals with the sulphur metabolism in angiosperms from an ecological point of view. *Plant Cell and Environment, 31*, 123–143.

Ezra, G., Rusness, D. G., Lamoureux, G. L., & Stephenson, G. R. (1985). The effect of CDAA (N,N,-diallyl-2-chloroacetamide) pretreatments on subsequent CDAA injury to corn (*Zea mays* L.). *Pesticide Biochemistry and Physiology, 23*, 108–115.

Farago, S., & Brunold, C. (1990). Regulation of assimilatory sulfate reduction by herbicide safeners in *Zea mays* L. *Plant Physiology, 94*, 1808–1812.

Fischer, S., Kuhnlenz, T., Thieme, M., Schmidt, H., & Clemens, S. (2014). Analysis of plant Pb tolerance at realistic submicromolar concentrations demonstrates the role of phytochelatin synthesis for Pb detoxification. *Environmental Science & Technology, 48*, 7552–7559.

Flohe, L. (2013). The fairytale of the GSSG/GSH redox potential. *Biochimica et Biophysica Acta–General Subjects, 1830*, 3139–3142.

Foyer, C. H., & Noctor, G. (2005). Oxidant and antioxidant signalling in plants: A re-evaluation of the concept of oxidative stress in a physiological context. *Plant Cell and Environment, 28*, 1056–1071.

Foyer, C. H., & Noctor, G. (2011). Ascorbate and glutathione: The heart of the redox hub. *Plant Physiology, 155*, 2–18.

Freeman, J. L., Persans, M. W., Nieman, K., Albrecht, C., Peer, W., Pickering, I. J., & Salt, D. E. (2004). Increased glutathione biosynthesis plays a role in nickel tolerance in *Thlaspi* nickel hyperaccumulators. *The Plant Cell, 16*, 2176–2191.

Freisinger, E. (2008). Plant MTs – Long neglected members of the metallothionein superfamily. *Dalton Transactions*, 6663–6675.

Freisinger, E. (2011). Structural features specific to plant metallothioneins. *Journal of Biological Inorganic Chemistry, 16*, 1035–1045.

Frova, C. (2003). The plant glutathione transferase gene family: Genomic structure, functions, expression and evolution. *Physiologia Plantarum, 119*, 469–479.

Frova, C. (2006). Glutathione transferases in the genomics era: new insights and perspectives. *Biomolecular Engineering, 23*, 149–169.

Gajewska, E., Glowacki, R., Mazur, J., & Sklodowska, M. (2013). Differential response of wheat roots to Cu, Ni and Cd treatment: Oxidative stress and defense reactions. *Plant Growth Regulation, 71*, 13–20.

Gautam, N., Verma, P. K., Verma, S., Tripathi, R. D., Trivedi, P. K., Adhikari, B., & Chakrabarty, D. (2012). Genome-wide identification of rice class I metallothionein gene: Tissue expression patterns and induction in response to heavy metal stress. *Functional & Integrative Genomics, 12*, 635–647.

Gawronski, S. W., Gawronska, H., Lomnicki, S., Sæbo, A., & Vangronsveld, J. (2017). Plants in air phytoremediation. In A. Cuypers, & J. Vangronsveld (Eds.), *Phytoremediation* (Vol. 83, pp. 319–346).

Ge, C., Ding, Y., Wang, Z., Wan, D., Wang, Y., Shang, Q., & Luo, S. (2009). Responses of wheat seedlings to cadmium, mercury and trichlorobenzene stresses. *Journal of Environmental Sciences, 21*, 806–813.

Gerhardt, K. E., Huang, X.-D., Glick, B. R., & Greenberg, B. M. (2009). Phytoremediation and rhizoremediation of organic soil contaminants: Potential and challenges. *Plant Science, 176*, 20–30.

Ghanta, S., Datta, R., Bhattacharyya, D., Sinha, R., Kumar, D., Hazra, S., … Chattopadhyay, S. (2014). Multistep involvement of glutathione with salicylic acid and ethylene to combat environmental stress. *Journal of Plant Physiology, 171*, 940–950.

Gielen, H., Remans, T., Vangronsveld, J., & Cuypers, A. (2016). Toxicity responses of Cu and Cd: The involvement of miRNAs and the transcription factor SPL7. *BMC Plant Biology, 16*, 16.

Gigolashvili, T., & Kopriva, S. (2014). Transporters in plant sulfur metabolism. *Frontiers in Plant Science, 5*, 1–16.

Gill, S. S., & Tuteja, N. (2010a). Polyamines and abiotic stress tolerance in plants. *Plant Signaling & Behavior, 5*, 26–33.

Gill, S. S., & Tuteja, N. (2010b). Reactive oxygen species and antioxidant machinery in abiotic stress tolerance in crop plants. *Plant Physiology and Biochemistry, 48*, 909–930.

Giraud, E., Ivanova, A., Gordon, C. S., Whelan, J., & Considine, M. J. (2012). Sulphur dioxide evokes a large scale reprogramming of the grape berry transcriptome associated with oxidative signalling and biotic defence responses. *Plant Cell and Environment, 35*, 405–417.

Gotor, C., Laureano-Marin, A. M., Moreno, I., Aroca, A., Garcia, I., & Romero, L. C. (2015). Signaling in the plant cytosol: Cysteine or sulfide? *Amino Acids, 47*, 2155–2164.

Gronwald, J. W., Fuerst, E. P., Eberlein, C. V., & Egli, M. A. (1987). Effect of herbicide antidotes on glutathione content and glutathione S-transferase activity of *Sorghum* shoots. *Pesticide Biochemistry and Physiology, 29*, 66–76.

Groppa, M. D., & Benavides, M. P. (2008). Polyamines and abiotic stress: Recent advances. *Amino Acids, 34*, 35–45.

Groppa, M. D., Tomaro, M. L., & Benavides, M. P. (2007). Polyamines and heavy metal stress: The antioxidant behavior of spermine in cadmium- and copper-treated wheat leaves. *Biometals, 20*, 185–195.

Guo, W. J., Bundithya, W., & Goldsbrough, P. B. (2003). Characterization of the *Arabidopsis* metallothionein gene family: Tissue-specific expression and induction during senescence and in response to copper. *New Phytologist, 159*, 369–381.

Halušková, L. U., Valentovičová, K., Huttová, J., Mistrík, I., & Tamás, L. (2009). Effect of abiotic stresses on glutathione peroxidase and glutathione S-transferase activity in barley root tips. *Plant Physiology and Biochemistry, 47*, 1069–1074.

Han, Y., Mhamdi, A., Chaouch, S., & Noctor, G. (2013). Regulation of basal and oxidative stress-triggered jasmonic acid-related gene expression by glutathione. *Plant Cell and Environment, 36*, 1135–1146.

Harada, E., Choi, Y. E., Tsuchisaka, A., Obata, H., & Sano, H. (2001). Transgenic tobacco plants expressing a rice cysteine synthase gene are tolerant to toxic levels of cadmium. *Journal of Plant Physiology, 158*, 655–661.

Hasan, M. K., Liu, C., Wang, F., Ahammed, G. J., Zhou, J., Xu, M.-X., … Xia, X. J. (2016). Glutathione-mediated regulation of nitric oxide, S-nitrosothiol and redox homeostasis confers cadmium tolerance by inducing transcription factors and stress response genes in tomato. *Chemosphere, 161*, 536–545.

Hassinen, V. H., Tervahauta, A. I., Schat, H., & Karenlampi, S. O. (2011). Plant metallothioneins - metal chelators with ROS scavenging activity? *Plant Biology, 13*, 225–232.

Hassinen, V. H., Tuomainen, M., Peraniemi, S., Schat, H., Karenlampi, S. O., & Tervahauta, A. I. (2009). Metallothioneins 2 and 3 contribute to the metal-adapted phenotype but are not directly linked to Zn accumulation in the metal hyperaccumulator, *Thlaspi caerulescens*. *Journal of Experimental Botany, 60*, 187–196.

Hego, E., Bes, C. M., Bedon, F., Palagi, P. M., Chaumeil, P., Barre, A., … Plomion, C. (2014). Differential accumulation of soluble proteins in roots of metallicolous and nonmetallicolous populations of *Agrostis capillaris* L. exposed to Cu. *Proteomics, 14*, 1746–1758.

Hesse, H., Kreft, O., Maimann, S., Zeh, M., Willmitzer, L., & Höfgen, R. (2001). Approaches towards understanding methionine biosynthesis in higher plants. *Amino Acids, 20*, 281–289.

Hirase, K., & Molin, W. T. (2003). Sulfur assimilation in plants and weed control: Potential targets for novel herbicides and action sites of certain safeners. *Weed Biology and Management, 3*, 147–157.

Hothorn, M., Wachter, A., Gromes, R., Stuwe, T., Rausch, T., & Scheffzek, K. (2006). Structural basis for the redox control of plant glutamate cysteine ligase. *Journal of Biological Chemistry, 281*, 27557–27565.

Howden, R., Andersen, C. R., Goldsbrough, P. B., & Cobbett, C. S. (1995). A cadmium-sensitive, glutathione-deficient mutant of *Arabidopsis thaliana*. *Plant Physiology, 107*, 1067–1073.

Hudgins, J. W., Ralph, S. G., Franceschi, V. R., & Bohlmann, J. (2006). Ethylene in induced conifer defense: cDNA cloning, protein expression, and cellular and subcellular localization of 1-aminocyclopropane-1-carboxylate oxidase in resin duct and phenolic parenchyma cells. *Planta, 224*, 865–877.

Ibrahim, R. K., Hayyan, M., AlSaadi, M. A., Hayyan, A., & Ibrahim, S. (2016). Environmental application of nanotechnology: Air, soil, and water. *Environmental Science and Pollution Research, 23*, 13754–13788.

Iqbal, N., Khan, N. A., Nazar, R., & da Silva, J. A. T. (2012). Ethylene-stimulated photosynthesis results from increased nitrogen and sulfur assimilation in mustard types that differ in photosynthetic capacity. *Environmental and Experimental Botany, 78*, 84–90.

Iqbal, N., Masood, A., Khan, M. I., Asgher, M., Fatma, M., & Khan, N. A. (2013). Cross-talk between sulfur assimilation and ethylene signaling in plants. *Plant Signaling & Behavior, 8*, e22478.

Jain, V., Vart, S., Verma, E., & Malhotra, S. P. (2015). Spermine reduces salinity-induced oxidative damage by enhancing antioxidative system and decreasing lipid peroxidation in rice seedlings. *Journal of Plant Biochemistry and Biotechnology, 24*, 316–323.

de Jong, A. J., Yakimova, E. T., Kapchina, V. M., & Woltering, E. J. (2002). A critical role for ethylene in hydrogen peroxide release during programmed cell death in tomato suspension cells. *Planta, 214*, 537–545.

Jozefczak, M., Keunen, E., Schat, H., Bliek, M., Hernandez, L. E., Carleer, R., … Cuypers, A. (2014). Differential response of *Arabidopsis* leaves and roots to cadmium: Glutathione-related chelating capacity *vs* antioxidant capacity. *Plant Physiology and Biochemistry, 83*, 1–9.

Jozefczak, M., Remans, T., Vangronsveld, J., & Cuypers, A. (2012). Glutathione is a key player in metal-induced oxidative stress defenses. *International Journal of Molecular Sciences, 13*, 3145–3175.

Kataoka, T., Hayashi, N., Yamaya, T., & Takahashi, H. (2004). Root-to-shoot transport of sulfate in *Arabidopsis*. Evidence for the role of SULTR3;5 as a component of low-affinity sulfate transport system in the root vasculature. *Plant Physiology, 136*, 4198–4204.

Kataoka, T., Watanabe-Takahashi, A., Hayashi, N., Ohnishi, M., Mimura, T., Buchner, P., … Takahashi, H. (2004). Vacuolar sulfate transporters are essential determinants controlling internal distribution of sulfate in *Arabidopsis*. *The Plant Cell, 16*, 2693–2704.

Kawashima, C. G., Noji, M., Nakamura, M., Ogra, Y., Suzuki, K. T., & Saito, K. (2004). Heavy metal tolerance of transgenic tobacco plants over-expressing cysteine synthase. *Biotechnology Letters, 26*, 153–157.

Kende, H. (1993). Ethylene biosynthesis. *Annual Review of Plant Physiology and Plant Molecular Biology, 44*, 283–307.

Keunen, E., Florez-Sarasa, I., Obata, T., Jozefczak, M., Remans, T., Vangronsveld, J., … Cuypers, A. (2016). Metabolic responses of *Arabidopsis thaliana* roots and leaves to sublethal cadmium exposure are differentially influenced by ALTERNATIVE OXIDASE1a. *Environmental and Experimental Botany, 124*, 64–78.

Keunen, E., Schellingen, K., Van Der Straeten, D., Remans, T., Colpaert, J., Vangronsveld, J., & Cuypers, A. (2015). ALTERNATIVE OXIDASE1a modulates the oxidative challenge during moderate Cd exposure in *Arabidopsis thaliana* leaves. *Journal of Experimental Botany, 66*, 2967–2977.

Keunen, E., Schellingen, K., Vangronsveld, J., & Cuypers, A. (2016). Ethylene and metal stress: Small molecule, big impact. *Frontiers in Plant Science, 7*, 1–18.

Khan, M. I. R., & Khan, N. A. (2014). Ethylene reverses photosynthetic inhibition by nickel and zinc in mustard through changes in PS II activity, photosynthetic nitrogen use efficiency, and antioxidant metabolism. *Protoplasma, 251*, 1007–1019.

Kidd, P. S., Alvarez-Lopez, V., Becerra-Castro, C., Cabello-Conejo, M., & Prieto-Fernández, Á. (2017). Potential role of plant-associated bacteria in plant metal uptake and implications in phytotechnologies. In A. Cuypers, & J. Vangronsveld (Eds.), *Phytoremediation* (Vol. 83, pp. 87–126).

Kim, S. H., Jeong, J. C., Ahn, Y. O., Lee, H. S., & Kwak, S. S. (2014). Differential responses of three sweetpotato metallothionein genes to abiotic stress and heavy metals. *Molecular Biology Reports, 41*, 6957–6966.

Kim, Y. O., Jung, S., Kim, K., & Bae, H. J. (2013). Role of pCeMT, a putative metallothionein from *Colocasia esculenta*, in response to metal stress. *Plant Physiology and Biochemistry, 64*, 25–32.

Kocsy, G., von Ballmoos, P., Ruegsegger, A., Szalai, G., Galiba, G., & Brunold, C. (2001). Increasing the glutathione content in a chilling-sensitive maize genotype using safeners increased protection against chilling-induced injury. *Plant Physiology, 127*, 1147–1156.

Koprivova, A., North, K. A., & Kopriva, S. (2008). Complex signaling network in regulation of adenosine 5′-phosphosulfate reductase by salt stress in *Arabidopsis* roots. *Plant Physiology, 146*, 1408–1420.

Kovacik, J., Klejdus, B., Hedbavny, J., & Backor, M. (2009). Nickel uptake and its effect on some nutrient levels, amino acid contents and oxidative status in *Matricaria chamomilla* plants. *Water, Air, & Soil Pollution, 202*, 199–209.

Kumar, G., Kushwaha, H. R., Panjabi-Sabharwal, V., Kumari, S., Joshi, R., Karan, R., … Pareek, A. (2012). Clustered metallothionein genes are co-regulated in rice and ectopic expression of *OsMT1e-P* confers multiple abiotic stress tolerance in tobacco via ROS scavenging. *BMC Plant Biology, 12*, 1–15.

Kumar, S., Asif, M. H., Chakrabarty, D., Tripathi, R. D., Dubey, R. S., & Trivedi, P. K. (2013). Differential expression of rice lambda class GST gene family members during plant growth, development, and in response to stress conditions. *Plant Molecular Biology Reporter, 31*, 569–580.

Kumari, N., Parmar, P., & Sharma, V. (2015). Differential gene expression in two contrasting wheat cultivars under cadmium stress. *Biologia Plantarum, 59*, 701–707.

Kummerová, M., Zezulka, S., Váňová, L., & Fišerová, H. (2012). Effect of organic pollutant treatment on the growth of pea and maize seedlings. *Central European Journal of Biology, 7*, 159–166.

Lallement, P. A., Brouwer, B., Keech, O., Hecker, A., & Rouhier, N. (2014). The still mysterious roles of cysteine-containing glutathione transferases in plants. *Frontiers in Pharmacology, 5*, 1–22.

Lamoureux, G. L., & Frear, D. S. (1987). Current problems, trends and developments in pesticide metabolism in plants. In R. Greenhalgh, & T. R. Roberts (Eds.), *Pesticide science and biotechnology* (pp. 455–463). Oxford: Blackwell Scientific.

Lamoureux, G. L., & Rusness, D. G. (1980). In vitro metabolism of pentachloronitrobenzene to pentachloromethylthiobenzene by onion: Characterization of glutathione S-transferase, cysteine C-S lyase, and S-adenosylmethionine methyl transferase activities. *Pesticide Biochemistry and Physiology, 14*, 50–61.

Lamoureux, G. L., & Rusness, D. G. (1989). The role of glutathione and glutathione S-transferases in pesticide metabolism, selectivity and mode of action in plants and insects. In D. Dolphin, R. Poulson, & O. Avramovic (Eds.), *Glutathione: Chemical, biochemical, and medical aspects* (Vol. IIIB, pp. 153–196). New York: Wiley & Sons.

Lamoureux, G. L., Rusness, D. G., & Schroder, P. (1993). Metabolism of a diphenylether herbicide to a volatile thioanisole and a polar sulfonic-acid metabolite in spruce (*Picea*). *Pesticide Biochemistry and Physiology, 47*, 8–20.

Leszczyszyn, O. I., Imam, H. T., & Blindauer, C. A. (2013). Diversity and distribution of plant metallothioneins: A review of structure, properties and functions. *Metallomics, 5*, 1146–1169.

Levine, A., Tenhaken, R., Dixon, R., & Lamb, C. (1994). H_2O_2 from the oxidative burst orchestrates the plant hypersensitive disease resistance response. *Cell, 79*, 583–593.

Lewandowska, M., Wawrzynska, A., Moniuszko, G., Lukomska, J., Zientara, K., Piecho, M., ... Sirko, A. (2010). A contribution to identification of novel regulators of plant response to sulfur deficiency: Characteristics of a tobacco gene *UP9C*, its protein product and the effects of *UP9C* silencing. *Molecular Plant, 3*, 347–360.

Li, Z., Zhang, Y., Zhang, X. Q., Peng, Y., Merewitz, E., Ma, X., ... Yan, Y. (2016). The alterations of endogenous polyamines and phytohormones induced by exogenous application of spermidine regulate antioxidant metabolism, metallothionein and relevant genes conferring drought tolerance in white clover. *Environmental and Experimental Botany, 124*, 22–38.

Li, Z. S., Alfenito, M., Rea, P. A., Walbot, V., & Dixon, R. A. (1997). Vacuolar uptake of the phytoalexin medicarpin by the glutathione conjugate pump. *Phytochemistry, 45*, 689–693.

Liang, J., Shohag, M. J. I., Yang, X. E., Tian, S. K., Zhang, Y. B., Feng, Y., & He, Z. (2014). Role of sulfur assimilation pathway in cadmium hyperaccumulation by *Sedum alfredii* Hance. *Ecotoxicology and Environmental Safety, 100*, 159–165.

Liang, T. S., Ding, H., Wang, G. D., Kang, J. Q., Pang, H. X., & Lv, J. Y. (2016). Sulfur decreases cadmium translocation and enhances cadmium tolerance by promoting sulfur assimilation and glutathione metabolism in *Brassica chinensis* L. *Ecotoxicology and Environmental Safety, 124*, 129–137.

Lim, B., Meyer, A. J., & Cobbett, C. S. (2011). Development of glutathione-deficient embryos in *Arabidopsis* is influenced by the maternal level of glutathione. *Plant Biology, 13*, 693–697.

Liu, J., Shi, X. T., Qian, M., Zheng, L. Q., Lian, C. L., Xia, Y., & Shen, Z. (2015). Copper-induced hydrogen peroxide upregulation of a metallothionein gene, *OsMT2c*, from *Oryza sativa* L. confers copper tolerance in *Arabidopsis thaliana*. *Journal of Hazardous Materials, 294*, 99–108.

Liu, J.-H., Wang, W., Wu, H., Gong, X., & Moriguchi, T. (2015). Polyamines function in stress tolerance: From synthesis to regulation. *Frontiers in Plant Science, 6*, 1–10.

Lo Cicero, L., Madesis, P., Tsaftaris, A., & Lo Piero, A. R. (2015). Tobacco plants over-expressing the sweet orange tau glutathione transferases (CsGSTUs) acquire tolerance to the diphenyl ether herbicide fluorodifen and to salt and drought stresses. *Phytochemistry, 116*, 69–77.

Loyall, L., Uchida, K., Braun, S., Furuya, M., & Frohnmeyer, H. (2000). Glutathione and a UV light-induced glutathione S-transferase are involved in signaling to chalcone synthase in cell cultures. *The Plant Cell, 12*, 1939–1950.

Lyubenova, L., & Schroder, P. (2011). Plants for waste water treatment – Effects of heavy metals on the detoxification system of *Typha latifolia*. *Bioresource Technology, 102*, 996–1004.

Ma, Y., Oliveira, R. S., Freitas, H., & Zhang, C. (2016). Biochemical and molecular mechanisms of plant-microbe-metal interactions: Relevance for phytoremediation. *Frontiers in Plant Science, 7*, 1–19.
Maksymiec, W. (2011). Effects of jasmonate and some other signalling factors on bean and onion growth during the initial phase of cadmium action. *Biologia Plantarum, 55*, 112–118.
Mandavian, K., Ghaderian, S. M., & Schat, H. (2016). Pb accumulation, Pb tolerance, antioxidants, thiols, and organic acids in metallicolous and non-metallicolous *Peganum harmala* L. under Pb exposure. *Environmental and Experimental Botany, 126*, 21–31.
Mannervik, B., & Danielson, U. H. (1988). Glutathione transferases – Structure and catalytic activity. *CRC Critical Reviews in Biochemistry, 23*, 283–337.
Marrs, K. A. (1996). The functions and regulation of glutathione S-transferases in plants. *Annual Review of Plant Physiology and Plant Molecular Biology, 47*, 127–158.
Marrs, K. A., Alfenito, M. R., Lloyd, A. M., & Walbot, V. (1995). A glutathione S-transferase involved in vacuolar transfer encoded by the maize gene *Bronze-2*. *Nature, 375*, 397–400.
Martínez, M., Bernal, P., Almela, C., Vélez, D., García-Agustín, P., Serrano, R., & Navarro-Aviñó, J. (2006). An engineered plant that accumulates higher levels of heavy metals than *Thlaspi caerulescens*, with yields of 100 times more biomass in mine soils. *Chemosphere, 64*, 478–485.
Martinoia, E., Grill, E., Tommasini, R., Kreuz, K., & Amrhein, N. (1993). ATP-dependent glutathione S-conjugate export pump in the vacuolar membrane of plants. *Nature, 364*, 247–249.
Maruyama-Nakashita, A., Nakamura, Y., Watanabe-Takahashi, A., Inoue, E., Yamaya, T., & Takahashi, H. (2005). Identification of a novel cis-acting element conferring sulfur deficiency response in *Arabidopsis* roots. *Plant Journal, 42*, 305–314.
Masood, A., Iqbal, N., & Khan, N. A. (2012). Role of ethylene in alleviation of cadmium-induced photosynthetic capacity inhibition by sulphur in mustard. *Plant Cell and Environment, 35*, 524–533.
Mauch, F., & Dudler, R. (1993). Differential induction of distinct glutathione-S-transferases of wheat by xenobiotics and by pathogen attack. *Plant Physiology, 102*, 1193–1201.
Maughan, S. C., Pasternak, M., Cairns, N., Kiddle, G., Brach, T., Jarvis, R., … Cobbett, C. S. (2010). Plant homologs of the *Plasmodium falciparum* chloroquine-resistance transporter, PfCRT, are required for glutathione homeostasis and stress responses. *Proceedings of the National Academy of Sciences of the United States of America, 107*, 2331–2336.
Mazen, A. M. A. (2004). Accumulation of four metals in tissues of *Corchorus olitorius* and possible mechanisms of their tolerance. *Biologia Plantarum, 48*, 267–272.
Mengoni, A., Gonnelli, C., Hakvoort, H. W. J., Galardi, F., Bazzicalupo, M., Gabbrielli, R., & Schat, H. (2003). Evolution of copper-tolerance and increased expression of a 2b-type metallothionein gene in *Silene paradoxa* L. populations. *Plant and Soil, 257*, 451–457.
Meyer, R. C., Goldsbrough, P. B., & Woodson, W. R. (1991). An ethylene-responsive flower senescence-related gene from carnation encodes a protein homologous to glutathione S-transferases. *Plant Molecular Biology, 17*, 277–281.
Minocha, R., Majumdar, R., & Minocha, S. C. (2014). Polyamines and abiotic stress in plants: A complex relationship. *Frontiers in Plant Science, 5*, 1–17.
Mir, G., Domenech, J., Huguet, G., Guo, W. J., Goldsbrough, P., Atrian, S., & Molinas, M. (2004). A plant type 2 metallothionein (MT) from cork tissue responds to oxidative stress. *Journal of Experimental Botany, 55*, 2483–2493.
Miteva, L. P. E., Ivanov, S. V., & Alexieva, V. S. (2010). Alterations in glutathione pool and some related enzymes in leaves and roots of pea plants treated with the herbicide glyphosate. *Russian Journal of Plant Physiology, 57*, 131–136.

Mohanasundaram, K. A., Haworth, N. L., Grover, M. P., Crowley, T. M., Goscinski, A., & Wouters, M. A. (2015). Potential role of glutathione in evolution of thiol-based redox signaling sites in proteins. *Frontiers in Pharmacology, 6*, 1–15.

Moniuszko, G., Skoneczny, M., Zientara-Rytter, K., Wawrzynska, A., Glow, D., Cristescu, S. M., … Sirko, A. (2013). Tobacco LSU-like protein couples sulphur-deficiency response with ethylene signalling pathway. *Journal of Experimental Botany, 64*, 5173–5182.

Montero-Palmero, M. B., Martin-Barranco, A., Escobar, C., & Hernandez, L. E. (2014). Early transcriptional responses to mercury: A role for ethylene in mercury-induced stress. *New Phytologist, 201*, 116–130.

Morgenstern, R., Depierre, J. W., & Jornvall, H. (1985). Microsomal glutathione transferase – Primary structure. *Journal of Biological Chemistry, 260*, 3976–3983.

Moyle, R., Fairbairn, D. J., Ripi, J., Crowe, M., & Botella, J. R. (2005). Developing pineapple fruit has a small transcriptome dominated by metallothionein. *Journal of Experimental Botany, 56*, 101–112.

Murr, D. P., & Yang, S. F. (1975). Conversion of 5′-methylthioadenosine to methionine by apple tissue. *Phytochemistry, 14*, 1291–1292.

Na, G., & Salt, D. E. (2011). The role of sulfur assimilation and sulfur-containing compounds in trace element homeostasis in plants. *Environmental and Experimental Botany, 72*, 18–25.

Nath, S., Panda, P., Mishra, S., Dey, M., Choudhury, S., Sahoo, L., & Panda, S. K. (2014). Arsenic stress in rice: Redox consequences and regulation by iron. *Plant Physiology and Biochemistry, 80*, 203–210.

Nazar, R., Khan, M. I. R., Iqbal, N., Masood, A., & Khan, N. A. (2014). Involvement of ethylene in reversal of salt-inhibited photosynthesis by sulfur in mustard. *Physiologia Plantarum, 152*, 331–344.

Ning, H. X., Zhang, C. H., Yao, Y., & Yu, D. Y. (2010). Overexpression of a soybean O-acetylserine (thiol) lyase-encoding gene *GmOASTL4* in tobacco increases cysteine levels and enhances tolerance to cadmium stress. *Biotechnology Letters, 32*, 557–564.

Nocito, F. F., Lancilli, C., Crema, B., Fourcroy, P., Davidian, J. C., & Sacchi, G. A. (2006). Heavy metal stress and sulfate uptake in maize roots. *Plant Physiology, 141*, 1138–1148.

Noctor, G., Gomez, L., Vanacker, H., & Foyer, C. H. (2002). Interactions between biosynthesis, compartmentation and transport in the control of glutathione homeostasis and signalling. *Journal of Experimental Botany, 53*, 1283–1304.

Noctor, G., Mhamdi, A., Chaouch, S., Han, Y., Neukermans, J., Marquez-Garcia, B., … Foyer, C. H. (2012). Glutathione in plants: An integrated overview. *Plant, Cell & Environment, 35*, 454–484.

Noguera-Oviedo, K., & Aga, D. S. (2016). Lessons learned from more than two decades of research on emerging contaminants in the environment. *Journal of Hazardous Materials, 316*, 242–251.

Nutricati, E., Miceli, A., Blando, F., & De Bellis, L. (2006). Characterization of two *Arabidopsis thaliana* glutathione S-transferases. *Plant Cell Reports, 25*, 997–1005.

Onouchi, H., Nagami, Y., Haraguchi, Y., Nakamoto, M., Nishimura, Y., Sakurai, R., … Naito, S. (2005). Nascent peptide-mediated translation elongation arrest coupled with mRNA degradation in the *CGS1* gene of *Arabidopsis*. *Genes & Development, 19*, 1799–1810.

Oztetik, E. (2008). A tale of plant Glutathione S-transferases: Since 1970. *Botanical Review, 74*, 419–437.

Pandey, S., Ranade, S. A., Nagar, P. K., & Kumar, N. (2000). Role of polyamines and ethylene as modulators of plant senescence. *Journal of Biosciences, 25*, 291–299.

Parisy, V., Poinssot, B., Owsianowski, L., Buchala, A., Glazebrook, J., & Mauch, F. (2007). Identification of PAD2 as a gamma-glutamylcysteine synthetase highlights the importance of glutathione in disease resistance of *Arabidopsis*. *Plant Journal, 49*, 159–172.

Pavlíková, D., Zemanová, V., Procházková, D., Pavlík, M., Száková, J., & Wilhelmová, N. (2014). The long-term effect of zinc soil contamination on selected free amino acids playing an important role in plant adaptation to stress and senescence. *Ecotoxicology and Environmental Safety, 100*, 166–170.

Pflugmacher, S., Schroder, P., & Sandermann, H. (2000). Taxonomic distribution of plant glutathione S-transferases acting on xenobiotics. *Phytochemistry, 54*, 267–273.

Prade, L., Huber, R., & Bieseler, B. (1998). Structures of herbicides in complex with their detoxifying enzyme glutathione S-transferase – Explanations for the selectivity of the enzyme in plants. *Structure, 6*, 1445–1452.

Queval, G., Thominet, D., Vanacker, H., Miginiac-Maslow, M., Gakiere, B., & Noctor, G. (2009). H_2O_2-activated up-regulation of glutathione in *Arabidopsis* involves induction of genes encoding enzymes involved in cysteine synthesis in the chloroplast. *Molecular Plant, 2*, 344–356.

Radhakrishnan, R., & Lee, I. J. (2013). Spermine promotes acclimation to osmotic stress by modifying antioxidant, abscisic acid, and jasmonic acid signals in soybean. *Journal of Plant Growth Regulation, 32*, 22–30.

Rady, M. M., & Hemida, K. A. (2015). Modulation of cadmium toxicity and enhancing cadmium-tolerance in wheat seedlings by exogenous application of polyamines. *Ecotoxicology and Environmental Safety, 119*, 178–185.

Ramassamy, S., Olmos, E., Bouzayen, M., Pech, J. C., & Latche, A. (1998). 1-Aminocyclopropane-1-carboxylate oxidase of apple fruit is periplasmic. *Journal of Experimental Botany, 49*, 1909–1915.

Ravanel, S., Gakiere, B., Job, D., & Douce, R. (1998). The specific features of methionine biosynthesis and metabolism in plants. *Proceedings of the National Academy of Sciences of the United States of America, 95*, 7805–7812.

Ravilious, G. E., & Jez, J. M. (2012). Structural biology of plant sulfur metabolism: From assimilation to biosynthesis. *Natural Product Reports, 29*, 1138–1152.

Razem, F. A., & Bernards, M. A. (2002). Hydrogen peroxide is required for poly(phenolic) domain formation during wound-induced suberization. *Journal of Agricultural and Food Chemistry, 50*, 1009–1015.

Reichenauer, T. G., & Germida, J. J. (2008). Phytoremediation of organic contaminants in soil and groundwater. *ChemSusChem, 1*, 708–717.

Remans, T., Opdenakker, K., Guisez, Y., Carleer, R., Schat, H., Vangronsveld, J., & Cuypers, A. (2012). Exposure of *Arabidopsis thaliana* to excess Zn reveals a Zn-specific oxidative stress signature. *Environmental and Experimental Botany, 84*, 61–71.

Ren, Y. J., Liu, Y., Chen, H. Y., Li, G., Zhang, X. L., & Zhao, J. (2012). Type 4 metallothionein genes are involved in regulating Zn ion accumulation in late embryo and in controlling early seedling growth in *Arabidopsis*. *Plant Cell and Environment, 35*, 770–789.

Roje, S. (2006). S-Adenosyl-L-methionine: Beyond the universal methyl group donor. *Phytochemistry, 67*, 1686–1698.

Rombaldi, C., Lelievre, J. M., Latche, A., Petitprez, M., Bouzayen, M., & Pech, J. C. (1994). Immunocytolocalization of 1-aminocyclopropane-1-carboxylic acid oxidase in tomato and apple fruit. *Planta, 192*, 453–460.

Romero, L. C., Aroca, M. A., Laureano-Marin, A. M., Moreno, I., Garcia, I., & Gotor, C. (2014). Cysteine and cysteine-related signaling pathways in *Arabidopsis thaliana*. *Molecular Plant, 7*, 264–276.

Roosens, N. H., Bernard, C., Leplae, R., & Verbruggen, N. (2004). Evidence for copper homeostasis function of metallothionein (MT3) in the hyperaccumulator *Thlaspi caerulescens*. *FEBS Letters, 577*, 9–16.

Rouached, H., Wirtz, M., Alary, R., Hell, R., Arpat, A. B., Davidian, J. C., … Berthomieu, P. (2008). Differential regulation of the expression of two high-affinity sulfate transporters, SULTR1.1 and SULTR1.2, in *Arabidopsis*. *Plant Physiology, 147*, 897–911.

Rouhier, N. (2010). Plant glutaredoxins: Pivotal players in redox biology and iron-sulphur centre assembly. *New Phytologist, 186*, 365–372.

Ruiz, O. N., Alvarez, D., Torres, C., Roman, L., & Daniell, H. (2011). Metallothionein expression in chloroplasts enhances mercury accumulation and phytoremediation capability. *Plant Biotechnology Journal, 9*, 609–617.

Saito, K. (2000). Regulation of sulfate transport and synthesis of sulfur-containing amino acids. *Current Opinion in Plant Biology, 3*, 188–195.

Sauter, M., Moffatt, B., Saechao, M. C., Hell, R., & Wirtz, M. (2013). Methionine salvage and S-adenosylmethionine: Essential links between sulfur, ethylene and polyamine biosynthesis. *Biochemical Journal, 451*, 145–154.

Savvides, A., Ali, S., Tester, M., & Fotopoulos, V. (2016). Chemical priming of plants against multiple abiotic stresses: Mission possible? *Trends in Plant Science, 21*, 329–340.

Scalla, R., & Roulet, A. (2002). Cloning and characterization of a glutathione S-transferase induced by a herbicide safener in barley (*Hordeum vulgare*). *Physiologia Plantarum, 116*, 336–344.

Schellingen, K., Van Der Straeten, D., Remans, T., Loix, C., Vangronsveld, J., & Cuypers, A. (2015). Ethylene biosynthesis is involved in the early oxidative challenge induced by moderate Cd exposure in *Arabidopsis thaliana*. *Environmental and Experimental Botany, 117*, 1–11.

Schiavon, M., Galla, G., Wirtz, M., Pilon-Smits, E. A. H., Telatin, V., Quaggiotti, S., … Malagoli, M. (2012). Transcriptome profiling of genes differentially modulated by sulfur and chromium identifies potential targets for phytoremediation and reveals a complex S-Cr interplay on sulfate transport regulation in *B. juncea*. *Journal of Hazardous Materials, 239*, 192–205.

Schiavon, M., Pilon-Smits, E. A. H., Wirtz, M., Hell, R., & Malagoli, M. (2008). Interactions between chromium and sulfur metabolism in *Brassica juncea*. *Journal of Environmental Quality, 37*, 1536–1545.

Schiavon, M., Wirtz, M., Borsa, P., Quaggicitti, S., Hell, R., & Malagoli, M. (2007). Chromate differentially affects the expression of a high-affinity sulfate transporter and isoforms of components of the sulfate assimilatory pathway in *Zea mays* (L.). *Plant Biology, 9*, 662–671.

Schiller, M., Hegelund, J. N., Pedas, P., Kichey, T., Laursen, K. H., Husted, S., & Schjoerring, J. K. (2014). Barley metallothioneins differ in ontogenetic pattern and response to metals. *Plant Cell and Environment, 37*, 353–367.

Schnaubelt, D., Queval, G., Dong, Y. P., Diaz-Vivancos, P., Makgopa, M. E., Howell, G., … Foyer, C. H. (2015). Low glutathione regulates gene expression and the redox potentials of the nucleus and cytosol in *Arabidopsis thaliana*. *Plant Cell and Environment, 38*, 266–279.

Schröder, P. (2006). Enzymes transferring biomolecules to organic foreign compounds: A role for glucosyltransferase and glutathione S-transferase in phytoremediation. In M. Mackova, D. Dowling, & T. Macek (Eds.), *Phytoremediation and rhizoremediation* (pp. 133–142). Netherlands: Dordrecht: Springer.

Schroder, P., & Collins, C. (2002). Conjugating enzymes involved in xenobiotic metabolism of organic xenobiotics in plants. *International Journal of Phytoremediation, 4*, 247–265.

Schroder, P., & Pflugmacher, S. (1996). Induction of glutathione S-transferase activity in Norway spruce by xenobiotics and herbicide safeners. *Journal of Applied Botany-Angewandte Botanik, 70*, 97–100.

Sekhar, K., Priyanka, B., Reddy, V. D., & Rao, K. V. (2011). Metallothionein 1 (CcMT1) of pigeonpea (*Cajanus cajan*, L.) confers enhanced tolerance to copper and cadmium in *Escherichia coli* and *Arabidopsis thaliana*. *Environmental and Experimental Botany, 72*, 131–139.

Seth, C. S., Remans, T., Keunen, E., Jozefczak, M., Gielen, H., Opdenakker, K., ... Cuypers, A. (2012). Phytoextraction of toxic metals: A central role for glutathione. *Plant, Cell & Environment, 35*, 334–346.

Shu, S., Yuan, L. Y., Guo, S. R., Sun, J., & Yuan, Y. H. (2013). Effects of exogenous spermine on chlorophyll fluorescence, antioxidant system and ultrastructure of chloroplasts in *Cucumis sativus* L. under salt stress. *Plant Physiology and Biochemistry, 63*, 209–216.

Singh, B. R., & Shaw, R. W. (1988). Selective-inhibition of oat glutathione-S-transferase activity by tetrapyrroles. *FEBS Letters, 234*, 379–382.

Sinha, R., Kumar, D., Datta, R., Hazra, S., Bhattacharyya, D., Mazumdar, A. B., ... Chattopadhyay, S. (2015). Integrated transcriptomic and proteomic analysis of *Arabidopsis thaliana* exposed to glutathione unravels its role in plant defense. *Plant Cell Tissue and Organ Culture, 120*, 975–988.

Smith, I. K., Kendall, A. C., Keys, A. J., Turner, J. C., & Lea, P. J. (1985). The regulation of the biosynthesis of glutathione in leaves of barley (*Hordeum vulgare* L.). *Plant Science, 41*, 11–17.

Sommer, A., & Boger, P. (1999). Characterization of recombinant corn glutathione S-transferase isoforms I, II, III, and IV. *Pesticide Biochemistry and Physiology, 63*, 127–138.

Song, Y. F., Cui, J., Zhang, H. X., Wang, G. P., Zhao, F. J., & Shen, Z. G. (2013). Proteomic analysis of copper stress responses in the roots of two rice (*Oryza sativa* L.) varieties differing in Cu tolerance. *Plant and Soil, 366*, 647–658.

Stuiver, C. E. E., Posthumus, F. S., Parmar, S., Shahbaz, M., Hawkesford, M. J., & De Kok, L. J. (2014). Zinc exposure has differential effects on uptake and metabolism of sulfur and nitrogen in Chinese cabbage. *Journal of Plant Nutrition and Soil Science, 177*, 748–757.

Sun, X., & Guo, L. Z. (2013). Relationship between cadmium-induced root subapical hair development and ethylene biosynthesis in oilseed rape seedlings. *Acta Biologica Cracoviensia Series Botanica, 55*, 68–75.

Takahashi, H., Kopriva, S., Giordano, M., Saito, K., & Hell, R. (2011). Sulfur assimilation in photosynthetic organisms: Molecular functions and regulations of transporters and assimilatory enzymes. *Annual Review of Plant Biology, 62*, 157–184.

Takahashi, H., Watanabe-Takahashi, A., Smith, F. W., Blake-Kalff, M., Hawkesford, M. J., & Saito, K. (2000). The roles of three functional sulphate transporters involved in uptake and translocation of sulphate in *Arabidopsis thaliana*. *Plant Journal, 23*, 171–182.

Takahashi, Y., & Nagata, T. (1992). parB: An auxin-regulated gene encoding glutathione S-transferase. *Proceedings of the National Academy of Sciences of the United States of America, 89*, 56–59.

Talukdar, D., & Talukdar, T. (2014). Coordinated response of sulfate transport, cysteine biosynthesis, and glutathione-mediated antioxidant defense in lentil (*Lens culinaris* Medik.) genotypes exposed to arsenic. *Protoplasma, 251*, 839–855.

Tennstedt, P., Peisker, D., Bottcher, C., Trampczynska, A., & Clemens, S. (2009). Phytochelatin synthesis is essential for the detoxification of excess zinc and contributes significantly to the accumulation of zinc. *Plant Physiology, 149*, 938–948.

Thao, N. P., Khan, M. I. R., Thu, N. B. A., Hoang, X. L. T., Asgher, M., Khan, N. A., & Tran, L. S. (2015). Role of ethylene and its cross talk with other signaling molecules in plant responses to heavy metal stress. *Plant Physiology, 169*, 73–84.

Theodoulou, F. L. (2000). Plant ABC transporters. *Biochimica et Biophysica Acta–Biomembranes, 1465*, 79–103.

Thijs, S., Langill, T., & Vangronsveld, J. (2017). The bacterial and fungal microbiota of hyperaccumulator plants: Small organisms, large influence. In A. Cuypers, & J. Vangronsveld (Eds.), *Phytoremediation* (Vol. 83, pp. 43–86).

Tsuchisaka, A., & Theologis, A. (2004). Unique and overlapping expression patterns among the *Arabidopsis* 1-amino-cyclopropane-1-carboxylate synthase gene family members. *Plant Physiology, 136*, 2982–3000.

Tsuchisaka, A., Yu, G. X., Jin, H. L., Alonso, J. M., Ecker, J. R., Zhang, X. M., ... Theologis, A. (2009). A combinatorial interplay among the 1-aminocyclopropane-1-carboxylate isoforms regulates ethylene biosynthesis in *Arabidopsis thaliana*. *Genetics, 183*, 979–1003.

Usha, B., Keeran, N. S., Harikrishnan, M., Kavitha, K., & Parida, A. (2011). Characterization of a type 3 metallothionein isolated from *Porteresia coarctata*. *Biologia Plantarum, 55*, 119–124.

Van de Poel, B., & Van Der Straeten, D. (2014). 1-Aminocyclopropane-1-carboxylic acid (ACC) in plants: More than just the precursor of ethylene! *Frontiers in Plant Science, 5*, 11.

Vangronsveld, J., Herzig, R., Weyens, N., Boulet, J., Adriaensen, K., Ruttens, A., ... Mench, M. (2009). Phytoremediation of contaminated soils and groundwater: Lessons from the field. *Environmental Science and Pollution Research, 16*, 765–794.

Venkatachalam, P., Srivastava, A. K., Raghothama, K. G., & Sahi, S. V. (2009). Genes induced in response to mercury-ion-exposure in heavy metal hyperaccumulator *Sesbania drummondii*. *Environmental Science & Technology, 43*, 843–850.

Vercampt, H., Koleva, L., Vassilev, A., Vangronsveld, J., & Cuypers, A. (2016). Short-term phytotoxicity in *Brassica napus* (L.) in response to pre-emergently applied metazachlor: A microcosm study. *Environmental Toxicology and Chemistry*.

Verkleij, J. A. C., Golan-Goldhirsh, A., Antosiewisz, D. M., Schwitzguébel, J.-P., & Schröder, P. (2009). Dualities in plant tolerance to pollutants and their uptake and translocation to the upper plant parts. *Environmental and Experimental Botany, 67*, 10–22.

Vernoux, T., Wilson, R. C., Seeley, K. A., Reichheld, J. P., Muroy, S., Brown, S., ... Sung, Z. R. (2000). The *ROOT MERISTEMLESS1/CADMIUM SENSITIVE2* gene defines a glutathione-dependent pathway involved in initiation and maintenance of cell division during postembryonic root development. *The Plant Cell, 12*, 97–110.

Vestergaard, M., Matsumoto, S., Nishikori, S., Shiraki, K., Hirata, K., & Takagi, M. (2008). Chelation of cadmium ions by phytochelatin synthase: Role of the cysteine-rich C-terminal. *Analytical Sciences, 24*, 277–281.

Vrbova, M., Kotrba, P., Horacek, J., Smykal, P., Svabova, L., Vetrovcova, M., ... Griga, M. (2013). Enhanced accumulation of cadmium in *Linum usitatissimum* L. plants due to overproduction of metallothionein alpha-domain as a fusion to beta-glucuronidase protein. *Plant Cell Tissue and Organ Culture, 112*, 321–330.

Wachter, A., Wolf, S., Steininger, H., Bogs, J., & Rausch, T. (2005). Differential targeting of GSH1 and GSH2 is achieved by multiple transcription initiation: Implications for the compartmentation of glutathione biosynthesis in the *Brassicaceae*. *Plant Journal, 41*, 15–30.

Walters, D. (2003). Resistance to plant pathogens: Possible roles for free polyamines and polyamine catabolism. *New Phytologist, 159*, 109–115.

Wawrzynska, A., Moniuszko, G., & Sirko, A. (2015). Links between ethylene and sulfur nutrition — A regulatory interplay or just metabolite association? *Frontiers in Plant Science, 6*, 1–7.

Weber, M., Trampczynska, A., & Clemens, S. (2006). Comparative transcriptome analysis of toxic metal responses in *Arabidopsis thaliana* and the Cd^{2+}-hypertolerant facultative metallophyte *Arabidopsis halleri*. *Plant Cell and Environment, 29*, 950–963.

Wirtz, M., Birke, H., Heeg, C., Muller, C., Hosp, F., Throm, C., ... Hell, R. (2010). Structure and function of the hetero-oligomeric cysteine synthase complex in plants. *Journal of Biological Chemistry, 285*, 32810–32817.

Wirtz, M., & Droux, M. (2005). Synthesis of the sulfur amino acids: Cysteine and methionine. *Photosynthesis Research, 86*, 345–362.

Wirtz, M., & Hell, R. (2006). Functional analysis of the cysteine synthase protein complex from plants: Structural, biochemical and regulatory properties. *Journal of Plant Physiology, 163*, 273–286.

Wójcik, M., Gonnelli, C., Selvi, F., Dresler, S., Rostański, A., Vangronsveld, J., et al. (2017). Metallophytes of serpentine and calamine soils – Their unique ecophysiology and potential for phytoremediation. In A. Cuypers, & J. Vangronsveld (Eds.), *Phytoremediation* (Vol. 83, pp. 1–42).

Xue, T. T., Li, X. Z., Zhu, W., Wu, C. G., Yang, G. G., & Zheng, C. C. (2009). Cotton metallothionein GhMT3a, a reactive oxygen species scavenger, increased tolerance against abiotic stress in transgenic tobacco and yeast. *Journal of Experimental Botany, 60*, 339–349.

Yoshimoto, N., Inoue, E., Saito, K., Yamaya, T., & Takahashi, H. (2003). Phloem-localizing sulfate transporter, Sultr1;3, mediates re-distribution of sulfur from source to sink organs in *Arabidopsis*. *Plant Physiology, 131*, 1511–1517.

Zagorchev, L., Seal, C. E., Kranner, I., & Odjakova, M. (2013). A central role for thiols in plant tolerance to abiotic stress. *International Journal of Molecular Sciences, 14*, 7405–7432.

Zechmann, B. (2014). Compartment specific importance of glutathione during abiotic and biotic stress. *Frontiers in Plant Science, 5*, 1–11.

Zettl, R., Schell, J., & Palme, K. (1994). Photoaffinity-labeling of *Arabidopsis thaliana* plasma membrane vesicles by 5-azido-[7-^{3}H]indole-3-acetic acid: Identification of a glutathione S-transferase. *Proceedings of the National Academy of Sciences of the United States of America, 91*, 689–693.

Zhang, J., Zhang, M., Tian, S. K., Lu, L. L., Shohag, M. J. I., & Yang, X. E. (2014). Metallothionein 2 (SaMT2) from *Sedum alfredii* Hance confers increased Cd tolerance and accumulation in yeast and tobacco. *PLoS One, 9*, 1–10.

Zhang, L. W., Song, J. B., Shu, X. X., Zhang, Y., & Yang, Z. M. (2013). miR395 is involved in detoxification of cadmium in *Brassica napus*. *Journal of Hazardous Materials, 250*, 204–211.

Zhang, Q., Xu, F., Lambert, K. N., & Riechers, D. E. (2007). Safeners coordinately induce the expression of multiple proteins and MRP transcripts involved in herbicide metabolism and detoxification in *Triticum tauschii* seedling tissues. *Proteomics, 7*, 1261–1278.

Zhang, Y. Y., He, Q. Q., Zhao, S. Y., Huang, L. L., & Hao, L. (2014). *Arabidopsis ein2-1* and *npr1-1* response to Al stress. *Bulletin of Environmental Contamination and Toxicology, 93*, 78–83.

Zhang, Z. H., Ren, J. S., Clifton, I. J., & Schofield, C. J. (2004). Crystal structure and mechanistic implications of 1-aminocyclopropane-1-carboxylic acid oxidase – The ethylene-forming enzyme. *Chemistry & Biology, 11*, 1383–1394.

Zhao, H. J., Wu, L. Q., Chai, T. Y., Zhang, Y. X., Tan, J. J., & Ma, S. W. (2012). The effects of copper, manganese and zinc on plant growth and elemental accumulation in the manganese-hyperaccumulator *Phytolacca americana*. *Journal of Plant Physiology, 169*, 1243–1252.

Zhu, W., Zhao, D. X., Miao, Q., Xue, T. T., Li, X. Z., & Zheng, C. C. (2009). *Arabidopsis thaliana* metallothionein, AtMT2a, mediates ROS balance during oxidative stress. *Journal of Plant Biology, 52*, 585–592.

CHAPTER SEVEN

Bio- and Phytoremediation of Pesticide-Contaminated Environments: A Review

Nele Eevers*, Jason C. White§, Jaco Vangronsveld*,[1] and Nele Weyens*

*Hasselt University, Diepenbeek, Belgium
§Connecticut Agricultural Experiment Station, New Haven, CT, United States
[1]Corresponding author: E-mail: jaco.vangronsveld@uhasselt.be

Contents

Abstract

Pesticide-contaminated fields can be found worldwide due to excessive use of insecticides, herbicides and fungicides. Many of the pesticides that were once used intensively are now forbidden and have been shown to have deleterious health effects.

Plants, bacteria and fungi have been shown to possess pesticide-degrading capacities, which can be applied in the successful remediation of contaminated fields and water.

This article will first provide an overview of the different **types of pesticides**, their application and their key characteristics, followed by an analysis of their behaviour in the environment. Pesticides that are introduced into the environment seldom stay where they were applied. A complex system of **transport, transfer and transformation of pesticides** throughout different environmental compartments often takes place. These processes all influence the possible remediation of the

Advances in Botanical Research, Volume 83
ISSN 0065-2296
http://dx.doi.org/10.1016/bs.abr.2017.01.001

pesticide-contaminated media. We will then review several possible remediation strategies that are currently available. **Bioremediation** is the first technology that is reviewed. With bioremediation, the focus is on the remediation of pesticides by microorganisms in bulk soil, without the aid or presence of plants. Second, **plant-associated remediation** is discussed. When focussing on plant-associated remediation, a distinction has to be made between **rhizoremediation** in the rhizosphere and **phytoremediation** within the plant tissues. While rhizoremediation and phytoremediation processes are possible solely with the use of plants, many of these processes are optimized by associations between plants and microorganisms. Plants and bacteria or fungi often live in a symbiotic relationship that aids them in surviving contaminated environments, as well as with the degradation of the contaminants they encounter.

In the last part of the review, we discuss the advantages and disadvantages of "natural" remediation strategies as compared to more classical industrial approaches.

1. INTRODUCTION

Pesticides have been extensively used worldwide for crop protection in agriculture and gardening as well as in the management of insect-borne diseases such as malaria and typhus (McKone & Ryan, 1989). This widespread use has led to the contamination of many agricultural soils, natural water reservoirs and rural areas by persistent organic pesticides (Chaudry, Blom-Zandstra, Gupta, & Joner, 2005; Pascal-Lorber & Laurent, 2011). For a long period, the primary goal in farming was to protect crops against pests and thus gain the highest crop yields possible. In the meantime, the toxicity of the compounds used to both the farmers and consumers of the crops, as well as the environment, were likely underestimated and not always the primary concern when applying pesticides (Mackay & Fraser, 2000). Many of the pollutants were applied globally for years before it was discovered that they possessed unacceptable toxicity and hazard with regard to human health (Berdowski, Baas, Bloos, Visschedijk, & Zandveld, 1997). Often, these chemicals were also persistent in natural environments. Long after their original use, the analytes remain in soils and sediments, from which they can subsequently enter the food chain and surface and ground water (Gavrilescu, 2004; Li, Scholdz, & Van Heyst, 2000). One primary concern for these persistent organic pollutants (POPs) is that their hydrophobicity can lead to accumulation in adipose tissues of animals, which can cause biomagnification in higher trophic levels. These increasing levels of toxic compounds in the body may cause health problems over time (Kaufman, 1983; Moerner, Bos, & Fredrix, 2002).

Classical remediation technologies for areas contaminated with POPs include physicochemical methods such as incineration, burning, land filling, composting and chemical amendments (Kempa, 1997; Wehtje, Walker, & Shaw, 2000). As these methods are mainly ex situ, there is a high cost associated with excavation and transportation. Furthermore, since a significant part of the soil is removed, these methods are invasive and destructive to the overall ecosystem. Consequently, over the last decades there has been increasing interest for in situ remediation technologies, since they are noninvasive, low-cost, low-maintenance and often solar-driven (Chaudry et al., 2005). Bioremediation and phytoremediation, either with or without the assistance of plant-associated bacteria, are two "natural" remediation technologies that have been proven successful in many instances. Both strategies rely on the natural capacities of soil microorganisms and plants to take up contaminants as they do with nutrients and to metabolize, store, or even co-metabolize them (McGuinness & Dowling, 2009). The efficient application of these technologies is complicated by the wide variety of contaminants that are present in soils. Since every contaminated soil has its own specific physicochemical properties and contaminant profile, each remediation process has to be optimized or even tailored accordingly (Glick, 2003). Depending on the situation, a choice has to be made whether to utilize bioremediation, which relies solely on soil microorganisms such as bacteria and fungi, or for phytoremediation, which relies on plants for the remediation of contaminated soils. A third option is to exploit the symbiotic relationships between plants and their associated microorganisms for an enhanced phytoremediation efficiency of POPs and other contaminants (Weyens, Van Der Lelie, Taghavi, Newman, & Vangronsveld, 2009).

In this review, an overview will be given on the common remedial options for different pesticides. Bioremediation and phytoremediation, with or without plant-associated microorganisms, are both discussed in detail and a summary of plant and microbial species that have proven effective in remediation is provided. Lastly, the future challenges and perspectives are described.

2. PESTICIDES: DIFFERENT TYPES AND CHARACTERISTICS

"Pesticides" is the collective term for all chemicals that are used to counteract a certain group of organisms. The primary classes of pesticides are insecticides (against insects), herbicides (against plants) and fungicides (against fungi). However, inside these different classes, there still is a wide variety between the compounds and their chemical properties. They are defined by their ionic or nonionic properties, hydrophobic properties, mechanism of action and their molecular structure (Gavrilescu, 2005).

The four main groups of insecticides are: organochlorines, organophosphates, carbamate esters and pyrethroids. When considering insecticide contamination, organochlorines are often the chemicals of greatest concern. Commonly known members of this group are the DDTs (dichlorodiphenyltrichloroethane), DDE (dichlorodiphenyldichloroethylene), DDD (dichlorodiphenyldichloroethane), the HCHs (hexacyclochlorohexanes: α, β, γ, δ, t) and chlordane (Pascal-Lorber & Laurent, 2011). No less than eight compounds of the "Dirty Dozen" that were defined by the Stockholm Convention on persistent organic pollutants are organochlorine insecticides (Annex A, Stockholm Convention, 2001). The Stockholm Convention states that these compounds are banned and that remediation for their presence in the environment is needed. Many soils are contaminated with low to moderate levels of DDT and its breakdown products, although the use of this compound has been forbidden for decades (Thomas, Ou, & Al-Agely, 2008). These molecules are highly hydrophobic, with log K_{OW} values between 5.5 and 6.9. When these compounds reside in soils for decades, a significant weathering effect can be observed (Lunney, Zeeb, & Reimer, 2004). Weathered DDTs strongly adsorb to soil particles, further enhanced by alternate drying and wetting. Due to their hydrophobic and lipophilic nature, DDTs naturally accumulate in adipose tissue and often get magnified in the food chain. Several higher trophic level animals have been shown to experience deleterious effects of DDT exposure, e.g., egg shell thinning in birds and endocrine disruptors in mammals (Sharpe, 1995).

The most widespread herbicide contamination occurs with atrazine (Pascal-Lorber & Laurent, 2011). Atrazine (2-chloro-4-(aminoethyl)-6-(aminoisopropyl)-*s*-1,3,5-triazine) is a photosynthesis-inhibiting herbicide and is used in agriculture for the control of annual grasses and broad-leaved weeds, as well as in industrial sites and along railroads (Garmouna, Teil, Blanchard, & Chevreuil, 1998). This caused widespread contamination of surface and groundwater reservoirs. Atrazine is known to be an endocrine disruptor, and significant toxicity has been noted in amphibians (Forson & Storfer, 2006; Hayes et al., 2002).

Less is known about fungicide contamination, but the most important contaminants are hexachlorobenzene (HCB) and pentachlorophenol. HCB is a hydrophobic organic compound that is known for its bioaccumulation and the analyte has been detected in air, soil, fish, birds and even human milk (Qiu et al., 2004). Although HCB has not been produced since the 1970s, it is still being released into the environment as a by-product of simazine and thus, overall contamination has been increasing (Bailey, 2001).

3. BEHAVIOUR OF PESTICIDES IN THE ENVIRONMENT

The behaviour of pesticides in the environment is influenced by a combination of natural processes. When in the environment, pesticides interact with soils, water and organisms and the interactions are controlled by a complex collection of biological, physical and chemical reactions (Gavrilescu, 2005). Generally speaking, the processes that influence pesticide contamination can be classified into three types: (1) transport processes, which move the pesticide from the original point of introduction, (2) transfer processes that control the pesticides movement through environmental compartments such as water, sediments, the atmosphere and biota and (3) transformation processes that change the structure/nature of a pesticide or even completely degrade it to its constituent elements. All these processes are in turn influenced by different soil and climatic factors, as well as the characteristics of the pesticides themselves. The most important soil factors are the following:

- Soil texture and the distribution of soil particles. Soils that have a fine texture have a larger surface to volume ratio and a lower permeability. Therefore, the water and pesticides tend to diffuse more slowly, giving the pesticide a longer time for sorption to soil particles. Soils containing larger particles will retain considerably less pesticide residues.
- Soil depth influences the amount of time the pesticides spend in contact with soil particles before being washed out to deeper layers.
- Soil pH has an effect on the adsorption potential of soils and the rate of the biological processes that can remove pesticide residues.
- Soil organic matter content can change the sorption potential towards pesticide molecules as well as the number of biologically active microorganisms (Fukushima & Tatsumi, 2001). Organic matter provides an energy source for the microorganisms during the possible degradation of pesticide residues and is a highly complex medium that can interact significantly with organic and inorganic analytes.

- Soil slope is an external geological factor that can change the time that water and pesticides have to infiltrate the soil. Steep areas have considerably more runoff, giving pesticides less time to seep through.

The climate can also have an effect on the remediation of pesticides in the environment. A well-studied example is weathering of pesticides. Weathering is the process where pesticides molecules become more firmly attached to soil particles through the process of becoming wet by rain or dew and drying again (White et al., 2005). Weathered pesticides are known to be accumulated to a lesser extent easily than their recently applied equivalent.

In addition to soil and climatic factors, the chemical characteristics of the pesticide itself will greatly influence the fate of these compounds in the environment. Important characteristics include water solubility, tendency to adsorb to the soil particles and the half-life period or persistency in the environment.

The contaminant's availability is determined by combining the above mentioned soil and pesticide characteristics. Contaminant availability refers to the rate and the extent that the pesticide molecules will be released to and remain in the environment and greatly influences the possible remediation potential of a certain technology.

3.1 Transport of Pesticides

Transport of pesticides is the movement of the compounds from their point of introduction to the environment (Gavrilescu, 2005). Most pesticides are applied by spraying, causing a partial evaporation of the analytes into the air. Second, pesticides can evaporate from the soil and from plant surfaces. When the pesticides interact with soil, the fraction of the analyte that does not adsorb to soil particles can leach through the matrix into surface waters and cause contamination there.

3.2 Transfer of Pesticides

In addition to transport through the environment, pesticides can also be transferred between the different individual compartments: soil, water, atmosphere and biota (Gavrilescu, 2005).

Originally, most pesticides are applied in solid or liquid state to soils and plant surfaces, but *volatilization* can take place. Volatilization is a process where solid and liquid pesticides are converted into the gas phase, with subsequent transfer to the air (Frazar, 2000; Whitford et al., 1995). Once in the air, the pesticides can be transported over long distances and contamination can be spread throughout larger areas.

Precipitation events will partially convert the solid and liquid pesticides into dissolved from. Part of this water will run off the soil, which can cause pesticide contamination in the bodies of water where the *runoff* accumulates. However, a significant part of the rain water will also enter the soil system at distance from the site of application; in this case, *leaching* occurs. The amount of leaching that takes place is highly dependent on both the pesticide properties and the geological conditions.

From soil, air or water, pesticides can also be transferred into plants, microorganisms and animals. This capacity of plants and microorganisms to accumulate pesticides is the foundation of the natural remediation technologies that will be discussed in detail below.

3.3 Transformation

A third process to which pesticides are susceptible in the environment is transformation (Gavrilescu, 2005). Pesticide transformation or degradation is the oxidation of pesticide molecules. When a pesticide is introduced into the environment, it is prone to different transformation pathways. First, there is a *chemical degradation*, where the analytes react with organisms or enzymes in the environment and degradation occurs. Secondly, the molecules can be degraded by exposure to light; *photodegradation*. Thirdly, *microbial degradation* can occur in both bulk and rhizosphere soil. The microbial degradation process is the basic process for bioremediation of pesticide-contaminated soils and will be further explored below.

4. REMEDIATION OF PESTICIDES

Remediation strategies for soils and water contaminated with pesticides can be physical, chemical, biological or a combination of some or all of these approaches (Pascal-Lorber & Laurent, 2011). Traditionally, pesticide contamination has been remediated using physico-chemical technologies where soils are excavated and subsequently transported to specialized landfills; the material may also be incinerated or stabilized on site (McGuinness & Dowling, 2009). Although efficient, these technologies have significant limitations. The excavation and transport of contaminated soils is both labour-intensive and costly. Furthermore, when soils are treated in this manner, ecosystem disruption is significant and recovery may take years (Dijkgraaf & Vollebergh, 2004). Another limitation to these technologies is the area on which they can be applied. Since they are so costly and intensive, they are only cost-effective for small areas of high contamination; use on large-scale areas with moderate contamination is not feasible (Jin, Wang, & Ran, 2006).

Viable alternatives to these traditional remediation technologies are bio- and phytoremediation. These approaches are innovative technologies that show promise for alleviating pesticide contamination in both soils and water (Hussain, Siddique, Arshad, & Saleem, 2009). An overview of the main concepts underlying bio- and phytoremediation follows below.

4.1 Bioremediation

Bioremediation is the partial or complete conversion of the contaminant of interest to its elemental constituents by soil microorganisms (Megharaj, Ramakrishnan, Venkateswarlu, Sethunathan, & Naidu, 2011). It is estimated that 1 g of bulk soil contains more than 1 million bacterial cells of 5000–7000 different species and more than 10,000 fungal colonies (Dindal, 1990; Melling, 1993). These metabolic potential of the indigenous microbial community can be used for the detoxification of pesticide residues in soil (Karpouzas, Fotopoulou, Menkissoglu-Spiroudi, & Singh, 2005; Kumar & Philip, 2006; Siddique, Okeke, Arshad, & Frankenberger, 2003).

The efficiency of bioremediation depends on the bioavailability of the contaminant (related to analyte adsorption to solid materials and to surface complexation) and on the degradation potential of the microorganisms (Chaudry et al., 2005). Here, we review the relevant bioremediation and in these articles, many microorganisms showing pesticide-degrading capacities are described as shown in Table 1 (bacteria) and Table 2 (fungi).

When considering the bacteria, a dominant presence of the Proteobacteria is clearly evident. Of the 35 bacterial species that were recently noted to have the ability to remediate pesticide contamination, 21 belong to the Proteobacteria (6 Alphaproteobacteria, 4 Betaproteobacteria and 11 Gammaproteobacteria). When focussing on the species level, *Pseudomonas* sp. is the most abundant group present in literature. This species was mentioned in 16 publications and was shown to facilitate the remediation of 25 different pesticide residues and metabolites. The *Pseudomonas* species that were described were isolated from bulk soils, as well as the rhizosphere and from plants themselves, proving that this species is both omnipresent and adaptable.

Table 1 Overview of the Bacterial Genera Proven to Show Pesticide Degrading Capacities and Their Source of Isolation

Bacterial Genus	Pesticide	Isolated From	References
Aeromonas sp.	Chlorpyrifos, fenpropathrin	*Phragmites communis*	Chen et al. (2012)
	Carbaryl	Bulk soil	Hamada, Matar, and Bashir (2015)
Agrobacterium sp.	Atrazine, metalachlor	Rhizosphere *Andropogon gerardii*	Zhao, Arthur, Moorman, and Coats (2005)
Anabaena sp.	Butachlor	Bulk soil	Agrawal, Sen, Yadav, Rai, and Rai (2015)
Arthrobacter sp.	Cyhalothrin, cypermethrin, DDT, HCH	Rhizosphere *Nicotiana tabacum*	Wang, Wang, Li, and Gao (2015)
Azospirillum sp.	Ethion	Sludge	Foster, Kwan, and Vancov (2004)
	Chlorpyrifos, cyanophos	Bulk soil	Romeh (2014)
Bacillus sp.	Acibenzolar-*S*-methyl, metribuzin, napropamide, propamocarb hydrochloride, thiamethoxam	Bulk soil	Myresiotis et al. (2012)
	Bifenthrin, chlorpyrifos	*Phragmites communis*	Chen et al. (2012)
	Carbaryl	Bulk soil	Hamada et al. (2015)
	Chlorpyrifos	Bulk soil	Pailan, Gupta, Apte, Krishnamurthi, and Saha (2015)
		Lolium multiflorum	Ahmad et al. (2012)
	Cypermethrin	Bulk soil	Pankaj et al. (2016)
	Methyl parathion	Bulk soil	Sreenivasulu and Aparna (2001)
	Monocrotophos	Bulk soil	Acharya, Shilpkar, Shah, and Chellapandi (2015)
	Trifluralin	Bulk soil	Bellinaso, Greer, Peralba, Henriques, and Gaylarde (2003)

(*Continued*)

Table 1 Overview of the Bacterial Genera Proven to Show Pesticide Degrading Capacities and Their Source of Isolation—cont'd

Bacterial Genus	Pesticide	Isolated From	References
Bordetella sp.	HCHs	Bulk soil	Kengara et al. (2010)
Brevibacterium sp.	Phorate	Bulk soil	Jariyal, Gupta, Mandal, and Jindal (2015)
Chryseobacterium sp.	DDT	Bulk soil	Qu, Xu, Ai, Liu, and Liu (2015)
Corynebacterium sp.	Carbaryl	Bulk soil	Hamada et al. (2015)
Cupriavidus sp.	Azoxystrobin	Bulk soil	Howell, Semple, and Bending (2014)
Enterobacter sp.	Bifenthrin	*Phragmites communis*	Chen et al. (2012)
	Chlorpyrifos	Bulk soil	Singh et al. (2006) and Awas, Sabit, Abo-Aba, and Bayoumi (2011)
	DDE	*Cucurbita pepo*	Eevers, Van Hamme, Bottos, Weyens, and Vangronsveld (2015a, 2015b, 2015c)
	Fenamiphos	Bulk soil	Singh et al. (2006)
Flavobacterium sp.	Bifenthrin	*Phragmites communis*	Chen et al. (2012)
Herbaspirillum sp.	Trifluralin	Bulk soil	Bellinaso et al. (2003)
Klebsiella sp.	Bifenthrin, fenpropathrin, Naphtalene	*Nymphaea tetragona*	Chen et al. (2012)
	Chlorpyrifos	Sludge	Ghanem, Orfi, and Shamma (2007)
	Fenpropathrin	*Spirodela polyrhiza*	Xu et al. (2015)
	Trifluralin	Bulk soil	Bellinaso et al. (2003)
Lactobacillus sp.	Chlorpyrifos	Rice straw	Wang et al. (2016)
	Chlorpyrifos, coumaphos, diazonin, methyl parathion, parathion	Kimchi (food dish)	Islam et al. (2010)
Lactococcus sp.	Fenpropathrin	*Nymphaea tetragona*	Chen et al. (2012)
Mesorhizobium sp.	Chlorpyrifos	*Lollium multiflorum*	Jabeen, Iqbal, Ahmad, Afzal, and Firdous (2015)

Methylobacterium sp.	DDE	*Cucurbita pepo*	Eevers et al. (2015a, 2015b, 2015c)
Microbacterium sp.	Chlorpyrifos, fenpropathrin	*Phragmites communis*	Chen et al. (2012)
Morganella sp.	Carbaryl	Bulk soil	Hamada et al. (2015)
Nocardia sp.	Sulfentrazone	Bulk soil	Martinez et al. (2010)
Novosphingobium sp.	2,4-Dichlorophenoxyacetic acid	Bulk soil	Dai, Li, Zhao, and Xie (2015)
Paenibacillus sp.	Chlorpyrifos	*Nymphaea tetragona*	Chen et al. (2012)
	Chlorpyrifos, cyanophos	Bulk soil	Romeh (2014)
	Fenpropathrin	*Najas marina*	Chen et al. (2012)
Paracoccus sp.	Alachlor, Acetochlor, butachlor, metolachlor, propisochlor	Bulk soil	Zhang et al. (2011)
	Chlorpyrifos	Sludge	Xu, Li, Wang, Zhang, and Yan (2008)
Pseudomonas sp.	Aldrin, dieldrin, heptachlor, heptachlor epoxide	Bulk soil	Bandala, Andres-Octaviano, Pastrana, and Torres (2006)
	Atrazine, metolachlor	Rhizosphere *Sorghastrum nutans*	Zhao et al. (2005)
	Bifenthrin	*Phragmites communis*	Chen et al. (2012)
	Carbaryl	Bulk soil	Hamada et al. (2015)
	Chlorpyrifos	Bulk soil	Gilani et al. (2016) and Awas et al. (2011)
	Chlorpyrifos, cypermethrin, fenitrothion, fenpropathrin, methyl parathion, permethrin	Bulk soil	Zuo et al. (2015)
	Cyhalothrin, cypermethrin, fenpropathrin, fenvalerate	Bulk soil	Tang et al. (2015)
	2,4-Dichlorophenoxyacetic acid	*Populus trichocarpa x deltoides*	Germaine et al. (2006)
	Dimethoate, malathion	Bulk soil	Al-Qurainy and Abdel-Megeed (2009)
	Endosulfan	Bulk soil	Narkhede et al. (2015)

(Continued)

Table 1 Overview of the Bacterial Genera Proven to Show Pesticide Degrading Capacities and Their Source of Isolation—cont'd

Bacterial Genus	Pesticide	Isolated From	References
	Ethion	Sludge	Foster et al. (2004)
	Fenpropathrin	*Spirodela polyrhiza*	Xu et al. (2015)
	Glyphosate	Bulk soil	Al-Arfaj, Abdel-Megeed, Ali, and Al-Shahrani (2013)
	Phorate	Bulk soil	Jariyal, Gupta, Jindal, and Mandal (2015)
	Quinalphos	Bulk soil	Nair, Rebello, Rishad, Asok, and Jisha (2015)
Pseudoxanthomonas sp.	Profenos	Bulk soil	Talwar and Ninnekar (2015)
Psychrobacter sp.	Chlorpyrifos	Bulk soil	Khalid and Hashmi (2016)
Rhodanobacter sp.	Azoxystrobin	Bulk soil	Howell et al. (2014)
Rhodococcus sp.	HCHs	*Cytisus striatus*	Becerra-Castro et al. (2012)
	Simazine	Bulk soil	Eker and Uyar (2016)
Serratia sp.	Monocrotophos	Bulk soil	Abraham and Silambarasan (2015)
	Quinalphos	Bulk soil	Nair et al. (2015)
Sphingomonas sp.	Acetachlor, alachlor, butachlor	Sludge	Chen et al. (2015)
	DDE	*Cucurbita pepo*	Eevers et al. (2015a, 2015b, 2015c)
	Diclofop-methyl	Bulk soil	Adkins (1999)
	HCHs	Bulk soil	Becerra-Castro et al. (2012)

Stenotrophomonas sp.	Chlorpyrifos, diazonin, methyl parathion, parathion, phoxim, profenofos, triazophos	Sludge	Deng et al. (2015)
	DDT	Bulk soil	Zhu et al. (2012) and Pan et al. (2016)
Streptomyces sp.	Atrazine	*Saccharum officinarum*	Mesquini, Sawaya, Lopez, Oliveira, and Miyasaki (2015)
	Chlordane, HCHs, methoxychlor	Bulk soil	Fuentes, Benimeli, Cuozzo, and Amoroso (2010)
	HCHs	Bulk soil	Alvarez, Benimeli, Saez, Giuliano, and Amoroso (2015)
Variovorax sp.	Linuron	Bulk soil	Horemans, Vandermaesen, Vanhaecke, Smolders, and Springael (2013)

Table 2 Overview of the Fungal Genera Known to Degrade Pesticides and Their Source of Isolation

Fungal Genus	Pesticide	Isolated From	References
Aspergillus sp.	Pentachlorophenol	*Didemnun ligulum*	Vacondio et al. (2015)
Bjerkandera sp.	Azinphos methyl, phosmet, terbufos, tribufos	Wood	Jauregui et al. (2003)
Cladosporium sp.	Pentachlorophenol	*Didemnun ligulum*	Vacondio et al. (2015)
Cordyceps sp.	Dieldrin	Bulk soil	Xiao and Kondo (2013)
Coriolus sp.	Aldicarb, alachlor, atrazine	Bulk soil	Hai et al. (2012)
Fusarium sp.	Pentachlorophenol	*Didemnun ligulum*	Vacondio et al. (2015)
Glomus sp.	Phoxim	*Medicago sativa*	Wang et al. (2011)
Lentinula sp.	Difenoconazole, pendimethalin, terbuthylazine	Bulk soil	Pinto et al. (2012)
Penicillium sp.	Sulfentrazone	Bulk soil	Martinez et al. (2010)
Phanerochaete sp.	Azinphos methyl, DDT, phosmet, terbufos, tribufos	Wood	Jauregui et al. (2003) and Thomas and Gohil (2011)
Pleurotus sp.	Azinphos methyl, phosmet, terbufos, tribufos	Wood	Jauregui et al. (2003)
Trichoderma sp.	Pentachlorophenol	*Didemnun ligulum*	Vacondio et al. (2015)
Verticillium sp.	Chlorpyrifos	*Brassica chinensis*	Fang et al. (2008)

Significantly less has been published on the possible remediation of pesticides using fungal species (Table 2). A total of 13 fungal species were mentioned, of which seven belonged to the Ascomycota, five to the Basidiomycota and one to the Glomeromycota. Many of these species were isolated from decaying wood (white rot fungi) and bulk soils. The large difference in the number of studies focussing on bacteria and fungi could be the result of (1) fungi showing less pesticide-degrading potential or (2) fungi that are simply less studied in the context of natural remediation technologies.

The pesticides described in this literature are quite broad and we review the most common analytes here. Chlorpyrifos is an organophosphorus pesticide that has been used in agriculture and can be very persistent in soils. However, many bacterial and fungal strains have shown potential for degrading the compound (Gilani et al., 2016). *Serratia* sp. was shown to completely degrade 100 mg/L of chlorpyrifos in as little as 18 h (Xu, Li, Zheng, & Peng, 2007); *Stenotrophomonas* sp. displayed similar results in 28 h (Yang, Liu, Guo, & Qiao, 2006). Gilani et al. (2016) identified 14 different *Pseudomonas* strains isolated from soil that degraded chlorpyrifos. Many soils, such as those from former pesticide production facilities, are contaminated with mixtures of pollutants (Moklyachuk, Gorodiska, Slobodenyuk, & Petryshyna, 2010; Nurzhanova et al., 2010). To remediate these contaminations scenarios, researchers may have to use consortia of different bacterial strains such as that employed by Fan and Song (2014) for the degradation of atrazine and deisopropylatrazine. Alternatively, one may apply bacterial strains that can degrade several pesticides, such as the *Bacillus* strain used by Myresiotis, Vryzas, and Papadopoulou-Mourkidou (2012) to remediate soil contaminated with acidobenzolar-*S*-methyl, metribuzin, napropamide, propamocarb hydrochloride and thiamethoxam.

The efficiency of bioremediation processes depends largely on the local environmental conditions, such as soil moisture, redox status, temperature, pH and organic matter content (Singh, Walker, & Wright, 2006). The *soil moisture* is determined by the soil water content and influences not only the availability of water to soil microorganisms, but also the redox conditions that can impact possible biochemical degradation reactions (Schroll et al., 2006). Furthermore, high soil water content can create anoxic conditions, which can alter the microbial activity (Phillips, Lee, Trevors, & Seech, 2006). For example, anoxic conditions can enhance the degradation of pesticides, as demonstrated by Phillips et al. (2006) showed for HCH and Wu, Chen, Li, and Li (2014) for DDT .

Temperature and *pH* are also two major parameters that influence biodegradation processes in soils (Arshad, Hussain, & Saleem, 2007). Like most other enzymes, those molecules shown to be active in the degradation of pesticides are known to be temperature dependent (Alberty, 2006, pp. 43–70). Temperatures ranging between 15°C and 40°C were shown to be optimal conditions for the degradation of pesticides such as fenitrothion (Hong, Zhang, Hong, & Li, 2007) and fenamiphos (Singh et al., 2006). Enzymatic activity is also dependent on pH. Most bacteria function optimally in a pH range between 6.5 and 7.5; conditions that approximate intracellular pH (Hussain et al., 2009). In addition to affecting enzyme activity, soil pH can also influence abiotic adsorption and desorption processes of pesticides in soils. Lowering soil pH can increase pesticide desorption from soil particles (Ferrell, Witt, & Vencill, 2003), which enhances the bioavailability and bioremediation efficiency.

Another factor influencing the adsorption and desorption of pesticide molecules to soil particles is the *soil organic matter content*. When soils contain higher levels of organic matter, two competing effects are possible. On the one hand, pesticides may bind more strongly to organic soil particles and thus become less available for biodegradation. Alternatively, in high organic matter soils more nutrients may be available for the soil microorganisms, which can then stimulate microbial growth and an increase in pesticide degradation (Briceno, Palma, & Duran, 2007). Zhang, Sheng, Feng, and Miller (2005) made a comparison between the inhibition by adsorption and the stimulation by nutrient presence in benzonitrile-contaminated soils and concluded that the contaminant degradation increased with the addition of wheat-derived char. However, the more hydrophobic compounds with a higher log K_{OW} were likely to become more strongly associated with the solid phase of the organic matter (Aronstein, Calvillo, & Alexander, 1991). Many soil microorganisms produce organic acids, which may cause the desorption of pesticide molecules from organic matter particles and thus increase biodegradation potential (Chen & Zhu, 2005; Mata-Sandoval, Karns, & Torrents, 2000).

Since microorganisms are capable of excreting a large number of surfactants and enzymes, it is possible that the pesticides are also degraded extracellularly by these released enzymes (Hussain et al., 2009). This strategy could be exploited and serve as a rapid method for the remediation of pesticide-contaminated soils since no energy-demanding processes to transport the analytes into biota are needed. In fact, some have reported the successful bioremediation experiments using free enzymes (Seffernick et al., 2002;

Sutherland et al., 2004; Wackett, 2004; Wackett, Sadowsky, Martinez, & Shapir, 2002). However, their application under field circumstances and optimization for site-specific environmental factors has yet to be deployed on a large scale.

4.2 Plant-Associated Remediation

Phytoremediation is based on the same principles as bioremediation, although in this case pesticide degradation takes place in the plant or its rhizosphere. Similar to bioremediation, phytoremediation is also to be considered to be an innovative, cost-effective and ecologically beneficial technology (Hussain et al., 2009). Phytoremediation is a collection of processes, including phytotransformation, phytodegradation, phytovolatilization and rhizoremediation (Fig. 1). For the first three processes, uptake of the contaminant into the plant tissues is necessary; rhizoremediation takes place at the soil-root interface or rhizosphere.

All of the above-mentioned phytoremediation processes can be influenced by plant-associated microorganisms. Both rhizospheric and endophytic microorganisms can play a role in the remediation through pesticide-degrading and plant growth—promoting capacities (Glick, 2003, 2010; Weyens, Monchy, Vangronsveld, Taghavi, & Van Der Lelie, 2010).

4.2.1 Rhizoremediation

The degradation of pesticides is often greater in rhizosphere soil than it is in bulk soil. This can be explained by a phenomenon known as the rhizosphere effect (Hussain et al., 2009). The rhizosphere is the soil volume directly around the roots and is heavily influenced by the activities of the plant. This activity makes the rhizosphere a more complex environment than bulk soil, supporting large numbers of metabolically active microbial communities. These numbers can be 10—100 times larger than the number of microorganisms in unvegetated or bulk soil (Lynch, 1990; Pascal-Lorber & Laurent, 2011; Weyens et al., 2009), reaching up to 10^{12} cells per gram of soil (Whipps, 1990). Notably, the presence of plants with a large rhizospheric community can even increase the number of microbial cells in surrounding bulk soils (Leigh et al., 2006; Siciliano, Germida, Banks, & Greer, 2003). The presence of microbial communities in the soil can be beneficial to the plant by producing protective or beneficial compounds such as 1-aminocyclopropane-1-carboxylate (ACC) deaminase (Glick, 2003). ACC is the precursor molecule for ethylene, which is a stress hormone that plants often produce in contaminated soils. When

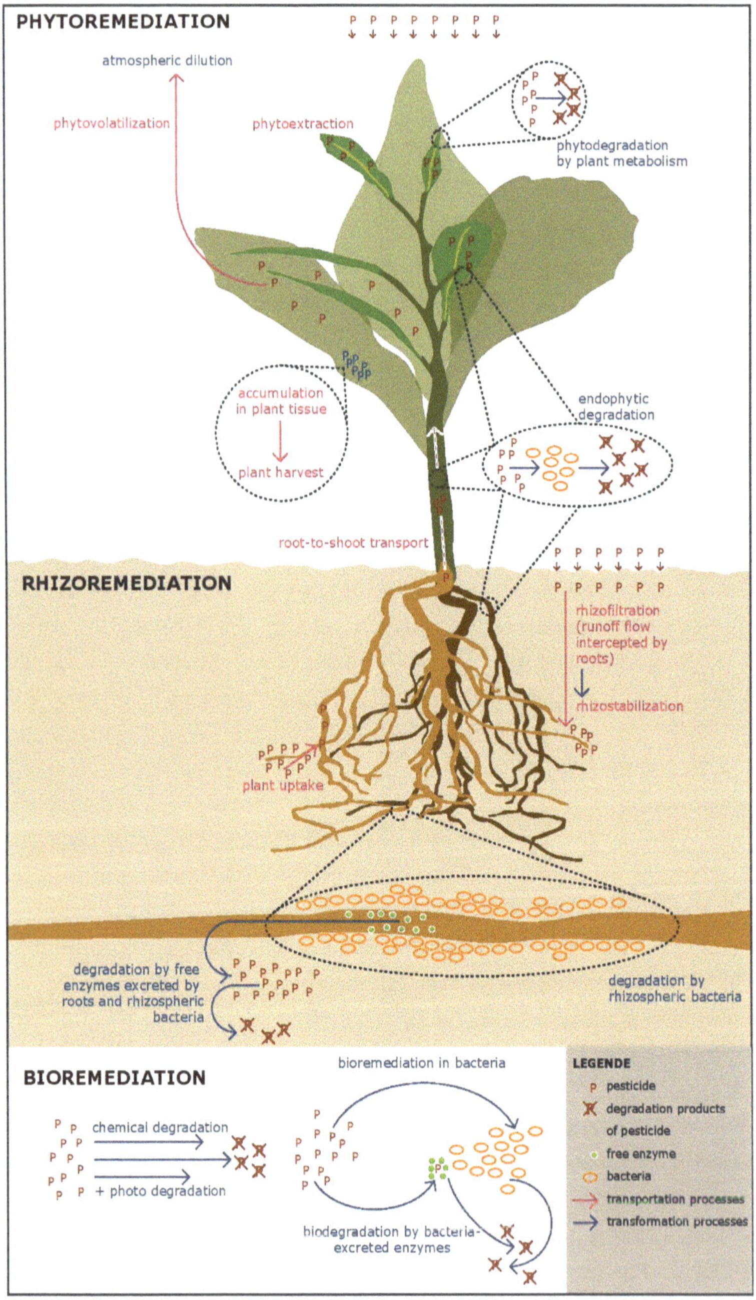

Figure 1 Overview of processes included in bio- and phytoremediation.

ACC-deaminase degrades ACC, ethylene levels are lowered and the plants experience less stress from contaminant exposure. All of these mechanisms can decrease plant phytotoxicity, increase plant growth and increase phytoremediation potential. Furthermore, microorganisms are capable facilitating the uptake of essential nutrients by plants through the production of organic acids (Lunney, Rutter, & Zeeb, 2010); the microbiota may also protect plants against pathogens by competing for a position in the plant microbiome and degrade contaminants before they negatively affect the plant (Gerhardt, Huang, Glick, & Greenberg, 2009).

Rhizodegradation is a process that occurs naturally, but that can be enhanced by planting the most appropriate plant species, or by adding pesticide-degrading bacteria through inoculation. Plant root systems can excrete enzymes that degrade pesticides in the rhizosphere (Gerhardt et al., 2009), but they also release photosynthetic products that can serve as nutrients for rhizospheric bacteria. Several researchers have isolated rhizospheric bacteria that show pesticide-degrading capacities (Table 1). If these pesticide-degrading bacteria are enriched by means of inoculation, the process is considered to be bacteria-enhanced rhizodegradation. Kidd, Prieto-Fernández, Monterroso, and Acea (2008) showed a higher dissipation of HCHs in the rhizosphere when *Cytisus striatus* and *Holcus lanatus* were inoculated with HCH-degrading bacteria. Ahmad et al. (2012) reported a 50% increase in chlorpyrifos degradation in the *Lolium multiflorum* rhizosphere when the plant roots were inoculated with *Bacillus pumilus C2A1*.

Wang, Tong, Shi, Xu, and He (2011) described the successful degradation of phoxim when carrot (*Daucus carota*) and green onion (*Allium fistulosum*) were inoculated with the arbuscular mycorrhizal fungi *Glomus intraradices* and *Glomus mosseae*; notably, contaminant degradation was negligible in noninoculated plants. The fungus *Trichoderma harzianum* was isolated from a marine plant *Didemnum ligulum* and had the capacity to degrade 50 mg/L phoxim in liquid medium in 7 days (Vacondio et al., 2015). Jauregui, Valderrama, Albores, and Vazquez-Duhalt (2003) conducted a large experiment on pesticide degradation by fungi. These researchers tested 17 white rot fungi; 16 were able to degrade the pesticides parathion, terbufos, azinphos-methyl, phosmet and tribufos after a four-day growth period.

Similar to bioremediation, rhizoremediation processes are also heavily influenced by plant, soil and pesticide characteristics. Factors such as temperature, pH and soil organic matter content will influence pesticide bioavailability, as well as the bacterial and enzymatic degradation potential in the rhizosphere. However, the presence of the root system can moderate some of these effects and in general, the degradation potential is higher in comparison to bulk soil (Hussain et al., 2009).

The rhizosphere can also play another role in phytoremediation processes: rhizostabilization. As explained earlier, pesticides can move through the soil as runoff from agricultural fields and subsequently contaminate surface waters. However, root systems and their associated microbial communities may intercept the pesticide molecules and thus stabilize them in the soil (Pascal-Lorber & Laurent, 2011).

4.2.2 Phytoremediation

In contrast to rhizoremediation, the accumulation of the pesticide is a prerequisite for phytotransformation, phytodegradation and phytovolitalization. Many plants have been reported to efficiently accumulate pesticides; an overview is given in Table 3. *Cucurbita pepo* and *Zea mays* are the plant species most frequently used in research paper addressing the phytoremediation of pesticides (11 and 7, respectively). These plants are often considered for phytoremediation because of their high number of cultivars as a result of their important role in agriculture and gardening, as well as their good accumulation potential of a wide range of organic contaminants (White, 2010; White et al., 2005). *Ricinus communis* is the plant shown to accumulate the greatest number of different contaminants. Rissato et al. (2015) and Huang et al. (2011) described the uptake of 11 different pesticides with a wide variety of characteristics.

The plant's uptake efficiency of these compounds is determined by many soil and plant characteristics (Hussain et al., 2009). The soil factors that can influence the pesticide availability to microorganisms and plants were discussed earlier in this review. In addition to soil moisture, temperature, pH and organic matter content, the time that the pesticide resides in soils can also influence the pesticide uptake; time-dependent decreases in availability are often described during weathering or ageing of the residues (Lunney et al., 2010, 2004).

The potential of plants to take up pesticide residues varies greatly between plant species (Dosnon-Olette, Trotel-Aziz, Couderchet, & Eullaffroy, 2010; Gent et al., 2007; Lunney et al., 2004; Mitton, Miglioranza, Gonzalez, Shimabukuro, & Monserrat, 2014; Moklyachuk et al., 2010; Moore & Locke, 2012; Mukherjee & Kumar, 2012; Nurzhanova et al., 2010; Romeh, 2015; Sun, Xu, Yang, Liu, & Dai, 2004; Wang, Zhang, Li, & Xiao, 2012; White et al., 2005) and even

Table 3 Overview of the Plant Species That Have Shown Pesticide Phytoremediation Potential

Plant Species	Pesticide
Achillea millefolium	DDT (Moklyachuk, Petryshyna, Slobodenyuk, & Zatsarinna, 2012)
Acorus calamus	Atrazine (Wang et al., 2012)
Allium fistulosum	Phoxim (Wang et al., 2011)
Amaranthus caudate	Dimethoate, malathion (Al-Qurainy & Abdel-Megeed, 2009)
	Glyphosate (Al-Arfaj et al., 2013)
Andropogon gerardii	Atrazine, metolachlor, pendimethalin (Zhao et al., 2005 and Henderson, Belden, Zhao, & Coats, 2006)
Arachis hypogaea	DDE (White et al., 2005)
Avena sativa	HCHs (Calvelo Pereira, Camps-Arbestain, Rodríguez-Garrido, Macias, & Monterroso, 2006)
Brassica campestris	Endosulfan (Mukherjee & Kumar, 2012)
Brassica juncea	DDE (White et al., 2005)
Brassica napus	DDE (White et al., 2005)
Cabomba aquatica	Copper sulphate, dimethomorph, flazasulfuron (Olette, Couderchet, Biagianti, & Eullaffroy, 2008)
Cajanus cajan	DDE (White et al., 2005)
Canna x hybrida	Simazine (Knuteson, Whitwell, & Klaine, 2002)
Chenopodium spp.	HCHs (Calvelo Pereira et al., 2006)
Cirsium arvense	Chlordane (Mattina et al., 2003)
Cucumis sativus	Chlordane (Gent et al., 2007)
	DDE (Gent et al., 2007)
	Endosulfan sulphate (Somtrakoon, Kruatrachue, & Lee, 2014)
Cucurbita pepo	Chlordane (Mattina et al., 2003)
	DDD, DDE, DDT (Bogdevich & Cadocinicov, 2010; Chhikara et al., 2010; Gent et al., 2007; Lunney et al., 2010; Lunney et al., 2004; Matsuo, Yamazaki, Gion, Eun, & Inui, 2011; Moklyachuk et al., 2010; Whitfield Aslund & Zeeb, 2010)
	Dieldrin, endrin (Matsumoto, Kawanaka, Yun, & Oyaizu, 2009; Otani et al., 2007)
	HCHs (Bogdevich & Cadocinicov, 2010; Moklyachuk et al., 2010)
Cytisus striatus	HCHs (Becerra-Castro et al., 2012; Calvelo Pereira et al., 2006; Kidd et al., 2008)
Daucus carota	DDD, DDE, DDT (Bogdevich & Cadocinicov, 2010; Moklyachuk et al., 2012)
	HCHs (Bogdevich & Cadocinicov, 2010)
	Phoxim (Wang et al., 2011)

(Continued)

Table 3 Overview of the Plant Species That Have Shown Pesticide Phytoremediation Potential—cont'd

Plant Species	Pesticide
Digitaria sp.	Atrazine, metolachlor, trifluralin (Anderson, Kruger, & Coats, 1994)
Eichhornia crassipes	Aldrin, chlorpyrifos, DDT, dieldrin, endosulfan, malathion, methyl parathion (Mercado-Borrayo, Heydrich, Perez, Quiroz, & Hill, 2015)
	Ethion (Xia & Ma, 2006)
Elodea canadensis	DDT (Gao, Garrison, Hoehamer, Mazur, & Wolfe, 2000)
	Copper sulphate, dimethomorph, flazasulfuron (Olette et al., 2008)
Erigeron canadensis	DDD, DDE, DDT (Moklyachuk et al., 2012)
Festuca arundinacea	DDT (Lunney et al., 2004)
Glycine max	Azoxystrobin (Romeh, 2015)
	DDD, DDE, DDT (Moklyachuk et al., 2010; Mitton et al., 2014)
	HCHs (Moklyachuk et al., 2010)
Halimione portulacoides	DDD, DDE, DDT (Carvalho, Rodrigues, Evangelista, Basto, & Vasconcelos, 2011)
Helianthus annus	Azoxystrobin (Romeh, 2015)
	DDD, DDE, DDT (Mitton et al., 2014)
Holcus lanatus	HCHs (Kidd et al., 2008)
Hordeum vulgare	DDD, DDE, DDT, HCHs (Moklyachuk et al., 2010)
	Dodemorph, tridemorph (Chamberlain, Patel, & Bromilow, 1998)
Iris Pseudacorus	Atrazine (Wang et al., 2012)
	Chlorpyrifos (Wang, Yang, Cui, Xiao, & Xiaoe, 2013)
Jathropa curcas	HCHs (Abhilash, Singh, Srivastava, Schaeffer, & Singh, 2013)
Juncus effusus	Chlorpyrifos (Lytle & Lytle, 2002)
	Gamma-cyhalothrin (Bouldin et al., 2006)
	Atrazine (Bouldin et al., 2006; Lytle & Lytle, 2002)
Juncus maritimus	DDD, DDE, DDT (Carvalho et al., 2011)
Kanthium strumarium	DDD, DDE, DDT (Moklyachuk et al., 2012)
Kochia sp.	Atrazine (Anderson et al., 1994; Kruger et al., 1997)
	Metolachlor, trifluralin (Anderson et al., 1994)
Lactuca sativa	Chlordane (Mattina et al., 2003)
	Dimethoate, malathion (Al-Qurainy & Abdel-Megeed, 2009)
Leersia oryzoides	Atrazine, diazonin, permethrin (Moore & Locke, 2012)
Lemna minor	Copper sulphate, dimethomorph, flazasulfuron (Olette et al., 2008)

Table 3 Overview of the Plant Species That Have Shown Pesticide Phytoremediation Potential—cont'd

Plant Species	Pesticide
	Atrazine, clofibric acid, 2,4-dichlorophenoxyacetic acid, picloram (Reinhold, Vishwanathan, Park, Oh, & Saunders, 2010)
	Glyphosate, isoproturon (Dosnon-Olette, Couderchet, Oturan, Oturan, & Eullaffroy, 2011)
Lemna punctate	Atrazine, clofibric acid, 2,4-dichlorophenoxyacetic acid, picloram (Reinhold et al., 2010)
Lolium multiflorum	Atrazine (Merini, Bobillo, Cuadrado, Corach, & Giulietti, 2009)
	Chlorpyrifos (Ahmad et al., 2012; Jabeen et al., 2015)
	DDE (White et al., 2005)
	DDT (Lunney et al., 2004)
	Terbuthylazine (Mimmo, Bartucca, Del Buono, & Cesco, 2015)
Lolium perenne	2,4-Dichlorophenoxyacetic acid (Shaw & Burns, 2004)
	Pentachlorophenol (He, Xu, Tang, & Wu, 2005)
Ludwigia peploides	Atrazine, gamma-cyhalothrin (Bouldin et al., 2006)
Lupinus albus	Chlordane (Mattina et al., 2003)
	DDE (White et al., 2005)
Lupinus angustifolius	Atrazine, fenamiphos, isoproturon, simazine (Garcinuño, Fernandez-Hernando, & Camara, 2003)
	Carbaryl, linuron, permethrin (Garcinuño et al., 2003; Garcinuño, Fernandez-Hernando, & Camara, 2006)
Lycopersicon esculentum	Chlordane (Mattina et al., 2003)
Lythrum salicaria	Atrazine (Wang et al., 2012)
Medicago sativa	DDD, DDE, DDT (Lunney et al., 2004; Mitton et al., 2014)
	Napropamide (Cui & Yang, 2011)
Myriophyllum aquaticum	Atrazine, cycloxidim, tertbutryn, trifularin (Turgut, 2005)
	DDT (Gao et al., 2000)
	Simazine (Knuteson et al., 2002)
Najas marina	Bifenthrin, chlorpyrifos, fenpropathrin, naphtalene (Chen et al., 2012)
Nasturtium officinale	Dimethoate, malathion (Al-Qurainy & Abdel-Megeed, 2009)
Nicotiana tabacum	HCHs (Singh, Sherkhane, Kale, & Eapen, 2011)
	Sulfentrazone (Ferrell et al., 2003)
Nymphaea tetragona	Bifenthrin, chlorpyrifos, fenpropathrin, naphtalene (Chen et al., 2012)
Oenotheria biennis	DDD, DDE, DDT (Moklyachuk et al., 2012)

(*Continued*)

Table 3 Overview of the Plant Species That Have Shown Pesticide Phytoremediation Potential—cont'd

Plant Species	Pesticide
Orychophragmus violaceus	DDD, DDE, DDT, HCHs (Sun et al., 2015)
Oryza sativa	DDT (Chu, Wong, & Zhang, 2006)
Panicum virgatum	Atrazine, metolachlor, pendimethalin (Zhao et al., 2005; Henderson et al., 2006)
Phaseolus vulgaris	DDD, DDE, DDT, HCHs (Moklyachuk et al., 2010)
	Dimethoate, malathion (Al-Qurainy & Abdel-Megeed, 2009)
Phragmites australis	DDT (Chu et al., 2006)
	Hexachlorbenzene (Zhou et al., 2013)
Phragmites communis	Bifenthrin, chlorpyrifos, fenpropathrin, naphtalene (Chen et al., 2012)
Pisum sativum	2,4-Dichlorophenoxyacetic acid (Germaine et al., 2006)
Plantago lagopus	DDD, DDE, DDT (Moklyachuk et al., 2012)
Plantago major	Azoxystrobin (Romeh, 2015)
	Chlorpyrifos (Romeh & Hendawi, 2013)
	Cyanophos (Romeh, 2014)
Polygonum sp.	Atrazine, metolachlor, trifluralin (Anderson et al., 1994)
Populus deltoides x nigra	Atrazine (Burken & Schnoor, 1997)
Potamogeton crispus	Bifenthrin, chlorpyrifos, fenpropathrin, naphtalene (Chen et al., 2012)
Potentilla argentae	DDD, DDE, DDT (Moklyachuk et al., 2012)
Raphanus sativus	DDT, HCHs (Mikes, Cupr, Trapp, & Klanova, 2009)
Ricinus communis	Aldrin, chlordane, chlorpyrifos, DDE, diclofop methyl, dieldrin, endrin, HCHs, heptachlor, methoxychlor (Rissato et al., 2015)
	DDT (Huang et al., 2011; Rissato et al., 2015)
Salix alba	Metalaxyl, trifluralin (Warsaw et al., 2012)
Salix humboldtiana	DDD, DDE, DDT (Mitton et al., 2012)
Sambucus nigra	Metalaxyl, trifluralin (Warsaw et al., 2012)
Scenedesmus obliquus	Isoproturon, dimethomorph, pyrimethanil (Dosnon-Olette et al., 2010)
Scenedesmus quadricauda	Isoproturon, dimethomorph, pyrimethanil (Dosnon-Olette et al., 2010)
Schoenoplectus californicus	Aldrin, chlordane, DDD, DDE, DDT, dieldrin, endosulfan, HCHs, heptachlor, heptachlor epoxide (Miglioranza, de Moreno, & Moreno, 2004)
Scirpus maritimus	DDD, DDE, DDT (Carvalho et al., 2011)
Sedum alfredii	DDD, DDE, DDT (Zhu et al., 2012)
Sesamum indicum	HCHs (Abhilash & Singh, 2010)
Solanum lycopersicum	DDD, DDE, DDT (Mitton et al., 2014)

Table 3 Overview of the Plant Species That Have Shown Pesticide Phytoremediation Potential—cont'd

Plant Species	Pesticide
Solanum nigrum	HCHs (Calvelo Pereira et al., 2006)
Sorghastrum nutans	Atrazine, metolachlor, pendimethalin (Henderson et al., 2006; Zhao et al., 2005)
Sorghum bicolor	DDD, DDE, DDT, HCHs (Bogdevich & Cadocinicov, 2010)
Sparganium americanum	Atrazine, diazonin, permethrin (Moore & Locke, 2012)
Spinacia oleracea	Chlordane (Mattina et al., 2003)
	HCHs (Dubey, Tripathi, Singh, & Abhilash, 2014)
Spirodela polyrhiza	Fenpropathrin (Xu et al., 2015)
Taraxacum officinalis	DDD, DDE, DDT (Moklyachuk et al., 2012)
Trifolium incarnatum	DDE (White et al., 2005)
Trifolium pretense	2,4-Dichlorophenoxyacetic acid (Shaw & Burns, 2004)
Triticum vulgare	Butachlor (Yu et al., 2003)
	DDD, DDE, DDT, HCHs (Moklyachuk et al., 2010)
Typha latifolia	Atrazine, diazonin, permethrin (Moore & Locke, 2012)
	Hexachlorobenzene (Zhou et al., 2013)
	Methyl parathion (Amaya-Chavez, Martinez-Tabche, Lopez-Lopez, & Galar-Martinez, 2006)
Vicia villosa	DDE (White et al., 2005)
Vigna radiate	Aldicarb (Sun et al., 2004)
Vigna sinensis	Endosulfan sulphate (Somtrakoon et al., 2014)
Vigna unguiculata	Aldicarb (Sun et al., 2004)
Withania somnifera	HCHs (Abhilash & Nandita, 2010)
Zea mays	Aldicarb (Sun et al., 2004)
	DDD, DDE, DDT (Bogdevich & Cadocinicov, 2010)
	Endosulfan (Mukherjee & Kumar, 2012)
	Endosulfan sulphate (Somtrakoon et al., 2014)
	HCHs (Alvarez et al., 2015; Bogdevich & Cadocinicov, 2010)
	Napthalene (Sun et al., 2004)

between different subspecies (Otani, Seike, & Sakata, 2007; White, 2010). To maximize the phytoremediation potential in a contaminated field, the optimum combination of soil, plant and possibly endophytes has to be established. Bouldin, Farris, Moore, Smith, and Cooper (2006) tested two different plants (*Juncus effusus* and *Ludwigia peploides*) for their uptake potential of two different pesticides (atrazine and lambda-cyhalothrin) and noticed a higher uptake of atrazine in *J. effuses*, while *L. peploides* accumulated more lambda-cyhalothrin. Atrazine was efficiently translocated to the shoots of *J. effuses*, but *L. peploides* showed a 98.2% retention of the analyte in the roots. The differences in uptake and translocation efficiency of certain plants towards pesticides depends on pesticide and plant characteristics. The log K_{OW} or octanol—water partitioning coefficient greatly influences the bioavailability and translocation of pesticides in plants (Bromilow & Chamberlain, 1995). Turgut (2005) investigated the uptake of trifluralin, atrazine, terbutryn and cycloxidim in *Myriophyllum aquaticum* and observed an increasing root concentration factor and submerged shoot concentration factor with an increasing log K_{OW}. The more polar (hydrophilic) a compound is, the more difficulties the analyte has in crossing biological membranes, which causes a lower uptake in comparison to lipophilic compounds that easily cross biomembranes (Burken & Schnoor, 1997; Trapp, 2000, 2004).

In environments that are contaminated with several pollutants, significant interactions between the contaminants may occur. Su, Zhu, Lin, and Zhang (2005) documented that the interaction between Cd^{2+} and atrazine reduced the individual toxicities of the contaminants to *Oryza sativa* seedlings and increased the uptake and translocation of atrazine into the plant tissues.

Plants mainly accumulate pesticides through a soil-to-plant pathway, although deposition from the air is also possible in the form of straight deposition from the gas phase or contaminants that are sorbed to particles that are subsequently deposited. Lee, Iannuci-Berger, Eitzer, White, and Mattina (2003) studied the uptake of chlordane and reported different profiles of chlordane in *Cucurbita pepo* when taken up through air or soil in parallel studies.

In case of the soil-to-plant uptake, several plant characteristics such as water uptake potential and root depth/structure can influence the accumulation potential. Once pesticides are retained by plant root tissues, they can be immobilized in the roots or translocated to the aerial plant parts where the analytes can be stored, metabolized or volatilized (Fig. 1). Generally, pesticide accumulation in roots is inefficient for remediation purposes; although the soil contaminant concentration decreases, the root tissues are not cost-effectively harvested. Aquatic plant-based remediation systems are an obvious exception to this, where contaminant removal by plant roots can be significant. *Eichhornia crassipes* (water hyacinth) can accumulate the insecticide ethion more efficiently in its root system than in its shoot system. Given that the roots make up over 50% of the plant mass and that the total plant, including roots, can easily be harvested, this system can be efficiently used for the phytoremediation of ethion-contaminated waters (Xia & Ma, 2006).

After being accumulated by plant roots, pesticide molecules can be transported to the xylem vessels and translocated with the transpiration stream of the plants. Many studies have been devoted to the transport of pesticides within crop species, largely because of the plants' high growth rates and easy cultivability (Vila, Lorber-Pascal, & Laurent, 2007). Several crops have been reported to have significant accumulation potential for a range of pesticides (Table 3). White et al. (White, 2010; White et al., 2005; Chhikara, Paulose, White, & Dhankher, 2010; Mattina, Lannucci-Berger, Musante, & White, 2003) studied the uptake of DDE by different *Cucurbita pepo* cultivars and reported shoot bioconcentration factors up to 23.7 for the Raven cultivar. A fraction of the pesticide molecules that are translocated to shoots can be adsorbed within vessel macromolecules, such as lignin or cellulose. The use of trees, mainly poplar and willow, takes this mechanism into account for phytoremediation as well as for phytopumping (Burken & Schnoor, 1997; Mitton, Gonzalez, Peña, & Miglioranza, 2012; Warsaw et al., 2012). Phytoremediation using solely phytoaccumulation requires harvest of the shoots after the uptake period. The shoot tissues can subsequently be burned or composted or otherwise disposed of by other means (Pascal-Lorber & Laurent, 2011).

Pesticides that are volatile such as triflutrin can be transported to the shoot system and subsequently be volatilized to the atmosphere (Fig. 1). This process is often an unwanted side effect of the transport processes in the plant, since the end result is merely a relocation of the contamination from soil to air; the exception would be if the plant evaporation is captured (Pascal-Lorber & Laurent, 2011).

Generally, the aim of efficient phytoremediation is not solely phytoaccumulation, but also the metabolic breakdown of the contaminant within the plant tissues; this can be done with or without the aid of endophytic bacteria. Plants often metabolize pesticides into more polar molecular structures that can be stored in vacuoles or bound as residues in the cell walls (Sandermann, 1994). In rape (*Brassica rapa*), atrazine residues were shown to be incorporated into plant cell walls as hydroxyatrazine (HO-A) (Dupont, Khan, & Bound, 1993). In roots, this process can lead to stabilization of the pesticides and therefore limit redispersion of the contaminants after plant death; in this scenario, the degrading capacity of white-rot fungi for lignin becomes very important (Jauregui et al., 2003; Thomas & Gohil, 2011).

Few compounds can be completely degraded by only plant metabolic pathways since most pesticides contain one or more aromatic cycles that are inherently difficult to break. In this case, bacteria- or fungi-enhanced phytoremediation may become important. As discussed earlier, many bacteria and fungi show pesticide-degrading capacities in soils; the same is true for endophytic microorganisms (Tables 1 and 2). Xu, Sun, Nie, and Wu (2015) showed that *Spirodela polyrhiza* stimulated the growth of their endophytes, which in their turn led to the degradation of fenpropathrin inside the plant tissues. Chen, Tang, Mori, and Wu (2012) also concluded that the efficient remediation of waters polluted by a mixture of chlorpyrifos, fenpropathrin, naphthalene and bifenthrin was mainly due to the endophytic bacteria residing in the aquatic plants, *Phragmites communis, Potamogeton crispus, Nymphaea tetragona* and *Najas marina*, as opposed to the plants themselves. Endophytic bacteria that show pesticide-degrading capacities have an advantage over their nondegrading competitors in contaminated environments and tend to dominate the community in those scenarios (Megharaj et al., 2011).

Using endophytes in the phytoremediation process not only has an advantage when it comes to pesticide degradation, but they also often possess plant growth–promoting properties. This plant growth promotion is shown through enhanced cycling of nutrients such as nitrogen and phosphate (Ryan, Germaine, Franks, Ryan, & Dowling, 2008). Endophytes have also been shown to have phosphate solubilisation abilities (Verma, Ladha, & Tripathi, 2001; Wakelin, Warren, Harvey, & Ryder, 2004), indole-3-acetic acid production (Lee et al., 2004), iron-binding siderophores production (Costa & Loper, 1994) and ACC-deaminase production (Glick, 2003); all will enhance plant growth.

Furthermore, the bacteria can also indirectly protect the plants. Symbiotic plant endophytes can prevent or lessen the deleterious effects of certain pathogens, often by outcompeting these organisms within the microbial community (Compant, Clement, & Sessitsch, 2010).

5. ADVANTAGES AND DISADVANTAGES OF NATURAL REMEDIATION TECHNOLOGIES

Bio- and phytoremediation show many advantages, which make them preferred over the classical physico-chemical remediation strategies. These alternative remediation technologies are based on the natural abilities of plants and microorganisms to utilize a wide range of organic compounds and to metabolize these analytes into harmless products such as carbon dioxide and water. These technologies exploit the natural ability of an environment to restore itself. The greatest advantage is that these technologies remediate the soils in situ without major disruptions in the environment. No excavation and transportation is needed, which makes the technologies inherently less costly and labour intensive, as well as more readily acceptable by the public.

Phytoremediation has some additional advantages over bioremediation. In the case of microbe-enhanced phytoremediation, the plants and microorganisms provide protection and nutrients for each other. Therefore, it is easier to stabilize the remediation system when compared to bulk soil, where microorganisms have to compete to a greater extent to become established in the community. In addition, the plants that are grown during phytoremediation provide stabilization of the soil and could potentially be used for green energy purposes.

Although the advantages of these technologies are obvious, some disadvantages do also exist. First, not all contaminants are susceptible to biodegradation. The contaminants can also be toxic to plants and microorganisms that may not possess the required degradation pathways. Secondly, if the parent compound is only partially degraded, products might appear that are more toxic and persistent than the original contaminant. And even though bio- and phytoremediation require less resources in the field than the classical remediation strategies, greater effort is needed to address the site-specific requirements of each contamination, plant and/or microorganisms scenario. Also, if degradation does not occur in the plants, the contaminant might be released into the environment again through evapotranspiration or through decaying tissues associated with natural senescence.

Lastly, the period that is needed for efficient decontamination of the soil has to be taken in consideration. As is the case when considering the remediation strategy, the treatment is also heavily dependent on the specific circumstances. The traditional thinking here is that bio and phytoremediation approaches are slower (Gerhardt et al., 2009), although some studies have shown otherwise. Compernolle et al. (2012) investigated a case study where a BTEX plume was remediated using poplar trees. In this study, the cost-effectiveness of phytoremediation is compared to that of classical remediation strategies such as pump-and-treat. These researchers concluded that both phytoremediation and pump-and-treat reached the remediation goal within 1 year of treatment, with phytoremediation being of lower cost.

Therefore, though depending on the situation, natural remediation technologies can be the cost-effective approach.

6. CONCLUSION, FUTURE PERSPECTIVES AND CHALLENGES

The contamination of soil with pesticides is a major concern to the environment and public health. Although bioremediation and phytoremediation are efficient technologies that have significant potential, physicochemical remediation strategies are still being applied more frequently. Soils and plants host a wide range of microbial communities with many metabolic pathways that can be applied for the efficient degradation of pesticide residues in soil and water. Recent progress in both plant biotechnology and microbiology, such as next generation sequencing to identify and utilize total microbial communities, have made these technologies more and more efficient. The findings that are discussed in this review show that bio- and phytoremediation have been successfully applied in several field trials and that these technologies should see greater use in the remediation of pesticide-contaminated field sites.

To expand the scale and efficiency of these technologies, greater focus is needed on unravelling the elucidating mechanisms of bio-, rhizo- and phytoremediation in the relevant biota. New technologies such as next generation sequencing might be useful in this regard; by investigating the composition of the total microbial community that is present in soils, rhizospheres and plants, an efficient method to introduce new microorganisms to the community, may be developed. However, for every specific contaminated site, a specific plan has to be constructed to appropriately address the location-specific characteristics. Additional research is needed to expand the knowledge base on how to efficiently translate successful lab trials into robust field applications.

REFERENCES

Abhilash, P. C., & Nandita, C. (2010). *Withania somnifera* Dunal-mediated dissipation of lindane from simulated soil: Implications for rhizoremediation of contaminated soil. *Journal of Soils and Sediments, 10*, 272–282.

Abhilash, P. C., Singh, B., Srivastava, P., Schaeffer, A., & Singh, N. (2013). Remediation of lindane by *Jatropha curcas* L.: Utilization of multipurpose species for rhizoremediation. *Biomass and Bioenergy, 51*, 189–193.

Abhilash, P. C., & Singh, N. (2010). Effect of growing *Sesamum indicum* L. on enhanced dissipation of lindane (1,2,3,4,5,6-hexachlorocyclohexane) from soil. *International Journal of Phytoremediation, 12*, 440–453.

Abraham, J., & Silambarasan, S. (2015). Bacterial degradation of monocrotophos and phyto- and cyto-toxicological evaluation of metabolites. *Toxicological and Environmental Chemistry, 97*(9), 1202–1216.

Acharya, K. P., Shilpkar, P., Shah, M. C., & Chellapandi, P. (2015). Bioremediation using *Novosphingobium* strain DY4 for 2,4-dichlorophenoxyacetic acid-contaminated soil and impact on microbial community structure. *Applied Biochemistry and Biotechnology, 175*(4), 1789–1804.

Adkins, A. (1999). Degradation of the phenoxy acid herbicide diclofop-methyl by *Sphingomonas paucimobilis* isolated from a Canadian prairie soil. *Journal of Industrial Microbiology and Biotechnology, 23*(4–5), 332–335.

Agrawal, C., Sen, S., Yadav, S., Rai, S., & Rai, L. C. (2015). A novel aldo-keto reductase (AKR17A1) of *Anabaena* sp PCC 7120 degrades the rice field herbicide butachlor and confers tolerance to abiotic stresses in *E. coli*. *PLoS One, 10*(9).

Ahmad, F., Iqbal, S., Anwar, S., Afzal, M., Islam, E., Mustafa, T., & Khan, Q. M. (2012). Enhanced remediation of chlorpyrifos from soil using ryegrass (*Lollium multiflorum*) and chlorpyrifos-degrading bacterium *Bacillus pumilus* C2A1. *Journal of Hazardous Materials, 237–238*, 110–115.

Al-Arfaj, A., Abdel-Megeed, A., Ali, H. M., & Al-Shahrani, O. (2013). Phyto-microbial degradation of glyphosate in Riyadh area. *Journal of Pure and Applied Microbiology*, 7(2), 1351–1365.

Alberty, R. A. (2006). *Biochemical reactions at specified temperature and various pHs, in biochemical thermodynamics*. Cambridge, MA, USA: John Wiley & Sons Inc.

Al-Qurainy, F., & Abdel-Megeed, A. (2009). Phytoremediation and detoxification of two organophosphorous pesticides residues in Riyadh area. *World Applied Sciences Journal, 6*(7), 987–998.

Alvarez, A., Benimeli, C. S., Saez, J. M., Giuliano, A., & Amoroso, M. J. (2015). Lindane removal using *Streptomyces* strains and maize plants: A biological system for reducing pesticides in soils. *Plant and Soil, 395*, 401–413.

Amaya-Chavez, A., Martinez-Tabche, L., Lopez-Lopez, E., & Galar-Martinez, M. (2006). Methyl parathion toxicity to and removal efficiency by *Typha latifolia* in water and artificial sediments. *Chemosphere, 63*, 1124–1129.

Anderson, T. A., Kruger, E. L., & Coats, J. R. (1994). Enhanced degradation of a mixture of three herbicides in the rhizosphere of a herbicide-tolerant plant. *Chemosphere, 28*(8), 1551–1557.

Aronstein, B. N., Calvillo, Y. M., & Alexander, M. (1991). Effect of surfactants at low concentrations on the desorption and biodegradation of sorbed aromatic compounds in soil. *Environmental Science and Technology, 25*, 1728–1731.

Arshad, M., Hussain, S., & Saleem, M. (2007). Optimization of environmental parameters for biodegradation of alpha and beta endosulfan in soil slurry by *Pseudomonas aeruginosa*. *Journal of Applied Microbiology, 104*, 364–370.

Awas, N. S., Sabit, H. H., Abo-Aba, S. E. M., & Bayoumi, R. A. (2011). Isolation, characterization and fingerprinting of some chlorpyrifos – degrading bacterial strains isolated from Egyptian pesticides-polluted soils. *African Journal of Microbiology Research, 5*(18), 2855–2862.

Bailey, R. E. (2001). Global hexachlorobenzene emissions. *Chemosphere, 43*, 167–182.

Bandala, E. R., Andres-Octaviano, J., Pastrana, P., & Torres, L. G. (2006). Removal of aldrin, dieldrin, heptachlor, and heptachlor epoxide using activated carbon and/or *Pseudomonas fluorescens* free cell cultures. *Journal of Environmental Science and Health, 41*(5), 553–569.

Becerra-Castro, C., Prieto-Fernández, A., Kidd, P. S., Weyens, N., Rodríguez-Garrido, B., Touceda González, M., & Acea, M. J. (2012). Improving performance of *Cytisus striatus* on substrates contaminated with hexachlorocyclohexane (HCH) isomers using bacterial inoculants: Developing a phytoremediation strategy. *Plant and Soil.* http://dx.doi.org/10.1007/s11104-012-1276-6.

Bellinaso, M. D., Greer, C. W., Peralba, M. D., Henriques, J. A. P., & Gaylarde, C. C. (2003). Biodegradation of the herbicide trifluralin by bacteria isolated from soil. *FEMS Microbiology Ecology, 43*(2), 191–194.

Berdowski, J. J. M., Baas, J., Bloos, J. P. J., Visschedijk, A. J. H., & Zandveld, P. Y. J. (1997). *The European emission inventory of heavy metals and persistent organic pollutants for 1990.* UBA-FB. Apeldoorn, The Netherlands: Institute of Environmental Science.

Bogdevich, O., & Cadocinicov, O. (2010). Elimination of acute risks from obsolete pesticides in Moldova: Phytoremediation experiment at a former pesticide storehouse. In P. A. Kulakow, & V. V. Pidlisnyuk (Eds.), *Application of phytotechnologies for cleanup of industrial, agricultural and wastewater contamination.* Springer Science + Business Media B.V.

Bouldin, J. L., Farris, J. L., Moore, M. T., Smith, S. J., & Cooper, C. M. (2006). Hydroponic uptake of atrazine and lambda-cyhalothrin in *Juncus effusus* and *Ludwigia peploides. Chemosphere, 65*, 1049–1057.

Briceno, G., Palma, G., & Duran, N. (2007). Influence of organic amendment on the biodegradation and movement of pesticides. *Critical Reviews of Environmental Science and Technology, 37*, 233–271.

Bromilow, R. H., & Chamberlain, K. (1995). Principles governing uptake and transport of chemicals. In S. Trapp, & J. C. McFarlane (Eds.), *Plant contamination: Modeling and simulation of organic chemical processes.* Boca Raton, Fla: Lewis Publishers.

Burken, J. G., & Schnoor, J. L. (1997). Uptake and metabolism of atrazine by poplar trees. *Environmental Science and Technology, 31*, 1399–1406.

Calvelo Pereira, R., Camps-Arbestain, M., Rodríguez-Garrido, B., Macias, F., & Monterroso, C. (2006). Behaviour of alpha-, beta-, gamma-, and delta-hexachlorocyclohexane in the soil-plant system of a contaminated site. *Environmental Pollution, 144*, 210–217.

Carvalho, P. N., Rodrigues, P. N. R., Evangelista, R., Basto, M. C. P., & Vasconcelos, M. T. S. D. (2011). Can salt marsh plants influence levels and distribution of DDTs in estuarine areas? *Estuarine, Coastal and Shelf Science, 93*, 415–419.

Chamberlain, K., Patel, S., & Bromilow, R. H. (1998). Uptake by roots and translocation to shoots of two morpholine fungicides in barley. *Pesticide Science, 54*, 1–7.

Chaudry, Q., Blom-Zandstra, M., Gupta, S., & Joner, E. J. (2005). Utilising the synergy between plants and rhizosphere microoganisms to enhance breakdown of organic pollutants in the environment. *Environmental Science and Pollution Research, 12*(1), 34–48.

Chen, G., & Zhu, H. L. (2005). Lindane affinity to silica sand as related to surface properties. *Separation Science and Technology, 40*, 1277–1291.

Chen, K., Chen, Q., Wang, G. X., Ni, H. Y., He, J., Yan, X., … Li, S. P. (2015). *Sphingomonas chloroacetimidivorans* sp nov., a chloroacetamide herbicide-degrading bacterium isolated from activated sludge. *Antonie van Leeuwenhoek International Journal of General and Molecular Microbiology, 108*(3), 703–710.

Chen, W. M., Tang, Y. Q., Mori, K., & Wu, X. L. (2012). Distribution of culturable endophytic bacteria in aquatic plants and their potential for bioremediation in polluted waters. *Aquatic Biology, 15*, 99–110.

Chhikara, S., Paulose, B., White, J. C., & Dhankher, O. P. (2010). Understanding the physiological and molecular mechanism of persistent organic pollutant uptake and detoxification in *Cucurbit* species (zucchini and squash). *Environmental Science and Technology, 44*, 7295–7301.

Chu, W. K., Wong, M. H., & Zhang, J. (2006). Accumulation, distribution and transformation of DDT and PCBs by *Phragmites australis* and *Oryza sativa* L.: I. Whole plant study. *Environmental Geochemistry and Health, 28*, 159–168.

Compant, S., Clement, C., & Sessitsch, A. (2010). Plant growth-promoting bacteria in the rhizo- and endosphere of plants: Their role, colonization, mechanisms involved and prospects for utilization. *Soil Biology and Biochemistry, 42*, 669–678.
Compernolle, T., Van Passel, S., Weyens, N., Vangronsveld, J., Lebbe, L., & Thewys, T. (2012). Groundwater remediation and the cost effectiveness of phytoremediation. *International Journal of Phytoremediation, 14*(9), 861–877.
Costa, J. M., & Loper, J. E. (1994). Characterization of siderophore production by the biological-control agent *Enterobacter cloacae*. *Molecular Plant-Microbe Interactions, 7*, 440–448.
Cui, L. E., & Yang, H. (2011). Accumulation and residue of napropamide in alfalfa (*Medicago sativa*) and soil involved in toxic response. *Journal of Hazardous Materials, 190*, 81–86.
Dai, Y., Li, N. N., Zhao, Q., & Xie, S. G. (2015). Bioremediation using *Novosphingobium* strain DY4 for 2,4-dichlorophenoxyacetic acid-contaminated soil and impact on microbial community structure. *Biodegradation, 26*(2), 161–170.
Deng, S. Y., Chen, Y., Wang, D. S., Shi, T. Z., Wu, X. W., Ma, X., … Li, Q. X. (2015). Rapid biodegradation of organophosphorus pesticides by *Stenotrophomonas* sp. G1. *Journal of Hazardous Materials, 297*, 17–24.
Dijkgraaf, E., & Vollebergh, H. R. J. (2004). Burn or bury? A social cost comparison of final waste disposal methods. *Ecological Economy, 50*, 233–247.
Dindal, D. L. (1990). *Soil biology guide*. New York, USA: John Wiley & Sons Inc.
Dosnon-Olette, R., Couderchet, M., Oturan, M. A., Oturan, N., & Eullaffroy, P. (2011). Potential use of *Lemna minor* for the phytoremediation of isoproturon and glyphosate. *International Journal of Phytoremediation, 13*, 601–612.
Dosnon-Olette, R., Trotel-Aziz, P., Couderchet, M., & Eullaffroy, P. (2010). Fungicides and herbicide removal in *Scenedesmus* cell suspensions. *Chemosphere, 79*, 117–123.
Dubey, R. K., Tripathi, V., Singh, N., & Abhilash, P. C. (2014). Phytoextraction and dissipation of lindane by *Spinacia oleracea* L. *Ecotoxicology and Environmental Safety, 109*, 22–26.
Dupont, S., & Khan, S. U. (1993). Bound and extractable ^{14}C residues in canola (*Brassica napus*) plants treated with radiolabelled atrazine. *Weed Research, 33*, 9–16.
Eevers, N., Van Hamme, J. D., Bottos, E. M., Weyens, N., & Vangronsveld, J. (2015a). *Enterobacter aerogenes*, a DDE degrading and plant growth promoting strain isolated from *Cucurbita pepo*, a draft genome. *Genome Announcements, 3*. http://dx.doi.org/10.1128/genomeA.00317-15. e00317–00315.
Eevers, N., Van Hamme, J. D., Bottos, E. M., Weyens, N., & Vangronsveld, J. (2015b). Draft genome sequence of *Methylobacterium radiotolerans*, a DDE-degrading and plant growth-promoting strain isolated from *Cucurbita pepo*. *Genome Announcements, 3*(3). e00488–00415.
Eevers, N., Van Hamme, J. D., Bottos, E. M., Weyens, N., & Vangronsveld, J. (2015c). *Sphingomonas taxi*, isolated from *Cucurbita pepo*, proves to be a DDE-degrading and plant growth-promoting strain. *Genome Announcements, 3*(3). e00489–00415.
Eker, S., & Uyar, C. C. (2016). Biodegradation of simazine by *Rhodococcus rhodocrous*: Kinetic modeling. *Fresenius Environmental Bulletin, 25*(1), 113–120.
Fan, X., & Song, F. (2014). Bioremediation of atrazine: Recent advances and promises. *Journal of Soils and Sediments, 14*, 1727–1737.
Fang, H., Xiang, Y. Q., Hao, Y. J., Chu, X. Q., Pan, X. D., Yu, J. Q., & Yu, Y. L. (2008). Fungal degradation of chlorpyrifos by *Verticillium* sp. DSP in pure cultures and its use in bioremediation of contaminated soil and pakchoi. *International Biodeterioration and Biodegradation, 61*(4), 294–303.
Ferrell, J. A., Witt, W. W., & Vencill, W. K. (2003). Sulfentrazone absorption by plant roots increases as soil or solution pH decreases. *Weed Science, 51*, 826–830.
Forson, D. D., & Storfer, A. (2006). Atrazine increases ranavirus susceptibility in the tiger salamander, *Ambystoma tigrinum*. *Ecological Applications, 16*, 2325–2332.

Foster, L. J. R., Kwan, B. H., & Vancov, T. (2004). Microbial degradation of the organophosphate pesticide, Ethion. *FEMS Microbiology Letters, 240*(1), 49–53.
Frazar, C. (2000). *The bioremediation and phytoremediation of pesticide contaminated sites*. Washington: National Network of Environmental Studies (NNEMS).
Fuentes, M. S., Benimeli, C. S., Cuozzo, S. A., & Amoroso, M. J. (2010). Isolation of pesticide-degrading actinomycetes from a contaminated site: Bacterial growth, removal and dechlorination of organochlorine pesticides. *International Biodeterioration and Biodegradation, 64*, 434–441.
Fukushima, M., & Tatsumi, K. (2001). Functionalities of humic acid for the remedial processes of organic pollutants. *Analytical Science, 17*, 821–823.
Gao, J., Garrison, A. W., Hoehamer, C., Mazur, C. S., & Wolfe, N. L. (2000). Uptake and phytotransformation of o,p′-DDT and p,p′-DDT by axenically cultivated aquatic plants. *Journal for Agriculture and Food Chemistry, 48*, 6121–6127.
Garcinuño, R. M., Fernandez-Hernando, P., & Camara, C. (2003). Evaluation of pesticide uptake by *Lupinus* seeds. *Water Research, 37*, 3481–3489.
Garcinuño, R. M., Fernandez-Hernando, P., & Camara, C. (2006). Removal of carbaryl, linuron, and permethrin by *Lupinus angustifolius* under hydroponic conditions. *Journal of Agricultural and Food Chemistry, 54*, 5034–5039.
Garmouna, M., Teil, M. J., Blanchard, M., & Chevreuil, M. (1998). Spatial and temporal variations of herbicide (triazines and phenylureas) concentrations in the catchment basin of the Marne river (France). *Science of the Total Environment, 224*, 93–107.
Gavrilescu, M. (2004). Removal of heavy metals from the environment by biosorption. *English Journal of Life Sciences, 4*, 219–232.
Gavrilescu, M. (2005). Fate of pesticides in the environment and its bioremediation. *Engineering in Life Sciences, 5*(6), 497–526.
Gent, M. P. N., White, J. C., Parrish, Z., Isleyen, M., Eitzer, B., & Incorvia Mattina, M. J. (2007). Uptake and translocation of p,p′-dichlorodiphenyldichloroethylene supplied in hydroponics solution to *Cucurbita*. *Environmental Toxicology and Chemistry, 26*(12), 2467–2475.
Gerhardt, K. E., Huang, X. D., Glick, B., & Greenberg, B. M. (2009). Phytoremediation and rhizoremediation of organic soil contaminants: Potential and challenges. *Plant Science, 176*, 20–30.
Germaine, K., Liu, X., Cabellos, G. G., Hogan, J. P., Ryan, D., & Dowling, D. (2006). Bacterial endophyte-enhanced phytoremediation of the organochlorine herbicide 2,4-dichlorophenoxyacetic acid. *FEMS Microbiology Ecology, 57*, 302–310.
Ghanem, I., Orfi, M., & Shamma, M. (2007). Biodegradation of chlorpyrifos by *Klebsiella* sp isolated from an activated sludge sample of waste water treatment plant in Damascus. *Folia Microbiologica, 52*(4), 423–427.
Gilani, R. A., Rafique, M., Rehman, A., Munis, M. F. H., Rehman, S. U., & Chaudhary, H. J. (2016). Biodegradation of chlorpyrifos by bacterial genus *Pseudomonas*. *Journal of Basic Microbiology, 56*(2), 105–119.
Glick, B. (2003). Phytoremediation: Synergistic use of plants and bacteria to clean up the environment. *Biotechnology Advances, 21*, 383–393.
Glick, B. (2010). Using soil bacteria to facilitate phytoremediation. *Biotechnology Advances, 28*, 367–374.
Hai, F. I., Modin, O., Yamamoto, K., Fukushi, K., Nakajima, F., & Nghiem, L. D. (2012). Pesticide removal by a mixed culture of bacteria and white-rot fungi. *Journal of the Taiwan Institute of Chemical Engineers, 43*(3), 459–462.
Hamada, M., Matar, A., & Bashir, A. (2015). Carbaryl degradation by bacterial isolates from a soil ecosystem of the Gaza Strip. *Brazilian Journal of Microbiology, 46*(4), 1087–1091.

Hayes, T., Haston, K., Tsui, M., Hoang, A., Haeffele, C., & Honk, A. (2002). Feminization of male frogs in the wild. Water-born herbicide threatens amphibian populations in part of the United States. *Nature, 419*, 895–896.

He, Y., Xu, J., Tang, C., & Wu, Y. (2005). Facilitation of pentachlorophenol degradation in the rhizosphere of ryegrass (*Lolium perenne* L.). *Soil Biology and Biochemistry, 37*, 2017–2024.

Henderson, K. L. D., Belden, J. B., Zhao, S., & Coats, J. R. (2006). Phytoremediation of pesticide wastes in soil. *Verslag der Zeitschrift für Naturforschung, 61*(c), 213–221.

Hong, Q., Zhang, Z. H., Hong, Y. F., & Li, S. P. (2007). A microcosm study on bioremediation of fenitrothion-contaminated soil using *Burkholderia* sp. FDS-1. *International Journal on Biodeteriation and Biodegradation, 59*, 55–61.

Horemans, B., Vandermaesen, J., Vanhaecke, L., Smolders, E., & Springael, D. (2013). *Variovorax* sp.-mediated biodegradation of the phenyl urea herbicide linuron at micropollutant concentrations and effects of natural dissolved organic matter as supplementary carbon source. *Applied Microbiology and Biotechnology, 97*(22), 9837–9846.

Howell, C. C., Semple, K. T., & Bending, G. D. (2014). Isolation and characterization of azoxystrobin degrading bacteria from soil. *Chemosphere, 95*, 370–378.

Huang, H., Yu, N., Wang, L., Gupta, D. K., He, Z., Wang, K., ... Yang, X. (2011). The phytoremediation potential of bioenergy crop *Ricinus communis* for DDTs and cadmium co-contaminated soil. *Bioresource Technology, 102*, 11034–11038.

Hussain, S., Siddique, T., Arshad, M., & Saleem, M. (2009). Bioremediation and phytoremediation of pesticides: Recent advances. *Critical Reviews in Environmental Science and Technology, 39*(10), 843–907.

Islam, S. M. A., Math, R. K., Cho, K. M., Lim, W. J., HOng, S. Y., Kim, J. M., ... Yun, H. D. (2010). Organophosphorus hydrolase (OpdB) of *Lactobacillus brevis* WCP902 from Kimchi is able to degrade organophosphorus pesticides. *Journal of Agricultural and Food Chemistry, 58*(9), 5380–5386.

Jabeen, H., Iqbal, S., Ahmad, F., Afzal, M., & Firdous, S. (2015). Enhanced remediation of chlorypyrifos by ryegrass (*Lolium multiflorum*) and a chlorpyrifos degrading bacterial endophyte *Mezorhizobium* sp. HN3. *International Journal of Phytoremediation*. http://dx.doi.org/10.1080/15226514.2015.1073666.

Jariyal, M., Gupta, V. K., Jindal, V., & Mandal, K. (2015). Isolation and evaluation of potent *Pseudomonas* species for bioremediation of phorate in amended soil. *Ecotoxicology and Environmental Safety, 122*, 24–30.

Jariyal, M., Gupta, V. K., Mandal, K., & Jindal, V. (2015). *Brevibacterium frigoritolerans* as a novel organism for the bioremediation of phorate. *Bulletin of Environmental Contamination and Toxicology, 95*(5), 680–686.

Jauregui, J., Valderrama, B., Albores, A., & Vazquez-Duhalt, R. (2003). Microsomal transformation of organophosphorus pesticides by white rot fungi. *Biodegradation, 14*(6), 397–406.

Jin, J. J., Wang, Z. S., & Ran, S. H. (2006). Solid waste management in Macao practices and challenges. *Waste Management, 26*, 1045–1051.

Karpouzas, D. G., Fotopoulou, A., Menkissoglu-Spiroudi, U., & Singh, B. K. (2005). Non-specific biodegradation of the organophosphorus pesticides, cadusafos and ethoprophos, by two bacterial isolates. *FEMS Microbiology Ecology, 53*, 369–378.

Kaufman, D. D. (1983). Fate of toxic organic compounds in land-applied wastes. In J. F. Parr, P. B. Marsh, & J. M. Kla (Eds.), *Land treatment of hazardous wastes*. New York, USA: Noyes Data Corp.

Kempa, E. S. (1997). Hazardous wastes and economic risk reduction: Case study, Poland. *International Journal of Environmental Pollution, 7*, 221–248.

Kengara, F. O., Schramm, K. W., Doerfler, U., Munch, J. C., Henkelmann, B., Welzl, G., ... Schroll, R. (2010). Degradation capacity of a 1,2,4-trichlorobenzene mineralizing microbial community for traces of organochlorine pesticides. *Science of the Total Environment, 408*, 3359–3366.

Khalid, S., & Hashmi, I. (2016). Biotreatment of chlorpyrifos in a bench scale bioreactor using *Psychrobacter alimentarius* T14. *Environmental Technology, 37*(3), 316–325.

Kidd, P. S., Prieto-Fernández, A., Monterroso, C., & Acea, M. J. (2008). Rhizosphere microbial community and hexachlorocyclohexane degradative potential in contrasting plant species. *Plant and Soil, 302*, 233–247.

Knuteson, S. L., Whitwell, T., & Klaine, S. J. (2002). Wetlands and aquatic processes: Influence of plant age & size on simazine toxicity and uptake. *Journal of Environmental Quality, 31*(6), 2096–2103.

Kruger, E. L., Anhalt, J. C., Sorenson, D., Nelson, B., Chouhy, A. L., Anderson, T. A., & Coats, J. R. (1997). Atrazine degradation in pesticide-contaminated soils: Phytoremediation potential. In E. L. Kruger (Ed.), *Phytoremediation of soil and water contaminants* (pp. 55–64). Washington, DC: American Chemical Society.

Kumar, M., & Philip, L. (2006). Enrichment and isolation of a mixed bacterial culture for complete mineralization of endosulfan. *Jounal of Environmental Science and Health, 41*, 81–96.

Lee, S., Flores-Encarnacion, M., Contreras-Zentella, M., Garcia-Flores, L., Escamilla, J. E., & Kennedy, C. (2004). Indole-3-acetic acid biosynthesis is deficient in *Gluconacetobacter diazotrophicus* strains with mutations in cytrochrome C biogenesis genes. *Journal of Bacteriology, 186*, 5384–5391.

Lee, W. Y., Iannuci-Berger, W. A., Eitzer, B., White, J. C., & Mattina, M. J. I. (2003). Plant uptake and translocation of air-borne chlordane and comparison with the soil-to-plant route. *Chemosphere, 53*, 111–121.

Leigh, M. B., Prouzova, P., Mackova, M., Macek, T., Nagle, D. P., & Fletcher, J. S. (2006). Polychlorinated biphenyl (PCB)-degrading bacteria associated with trees in a PCB-contaminated site. *Applied and Environmental Microbiology, 72*, 2331–2342.

Li, Y. F., Scholdz, M. T., & Van Heyst, B. J. (2000). Globar gridded emission inventory of alfa-hexachlorocyclohexane. *Journal of Geophysical Research, 105*, 6621–6632.

Lunney, A. I., Rutter, A., & Zeeb, B. A. (2010). Effect of organic matter additions on uptake of weathered DDT by *Cucurbita pepo* ssp. *pepo* cv. Howden. *International Journal of Phytoremediation, 12*, 404–417.

Lunney, A. I., Zeeb, B. A., & Reimer, K. J. (2004). Uptake of weathered DDT in vascular plants: Potential for phytoremediation. *Environmental Science and Technology, 38*(22), 6147–6154.

Lynch, J. M. (1990). *The rhizosphere*. New York, USA: Wiley.

Lytle, J. S., & Lytle, T. F. (2002). Uptake and loss of chlorpyrifos and atrazine by *Juncus effusus* L. in a mecocosm study with a mixture of pesticides. *Environmental Toxicology and Chemistry, 21*(9), 1817–1825.

Mackay, D., & Fraser, A. (2000). Bioaccumulation of persistent organic chemicals: Mechanisms & models. *Environmental Pollution, 2000*(110), 375–391.

Martinez, C. O., Silva, C. M. M. S., Fay, E. F., Abakerli, R. B., Maia, A. H. N., & Durrant, L. R. (2010). Microbial degradation of sulfentrazone in a Brazilian rhodic hapludox soil. *Brazilian Journal of Microbiology, 41*(1), 209–217.

Mata-Sandoval, J. C., Karns, J., & Torrents, A. (2000). Effect of rhamuolipids produced by *Pseudomonas aeruginosa* UG2 on the solubilization of pesticides. *Environmental Science and Technology, 34*, 4923–4930.

Matsumoto, E., Kawanaka, Y., Yun, S. J., & Oyaizu, H. (2009). Bioremediation of the organochlorine pesticides, dieldrin and endrin, and their occurence in the environment. *Applied Microbiology and Biotechnology, 84*, 205–216.

Matsuo, S., Yamazaki, K., Gion, K., Eun, H., & Inui, H. (2011). Structure-selective accumulation of polychlorinated biphenyls in *Cucurbita pepo*. *Journal of Pesticide Science, 36*(3), 363–369.

Mattina, M. J. I., Lannucci-Berger, W., Musante, C., & White, J. C. (2003). Concurrent plant uptake of heavy metals and persistent organic pollutants from soil. *Environmental Pollution, 124*, 375–378.

McGuinness, M., & Dowling, D. (2009). Plant-associated bacterial degradation of toxic organic compounds in soil. *International Journal of Environmental Research and Public Health, 6*, 2226–2247.

McKone, T. E., & Ryan, R. B. (1989). Human exposure of chemicals through food chains: An uncentainty analysis. *Environmental Science and Technology, 23*, 1154–1163.

Megharaj, M., Ramakrishnan, B., Venkateswarlu, K., Sethunathan, N., & Naidu, R. (2011). Bioremediation approaches for organic pollutants: A critical perspective. *Environmental International, 37*, 1362–1375.

Melling, F. B. (1993). *Soil microbial ecology: Applications in agricultural and environmental management*. New York, USA: Marcel Dekker.

Mercado-Borrayo, B. M., Heydrich, S. M., Perez, I. R., Quiroz, M. H., & Hill, C. P. L. (2015). Organophosphorus and organochlorine pesticides bioaccumulation by *Eichhornia crassipes* in irrigation canals in an urban agricultural system. *International Journal of Phytoremediation, 17*(7), 701–708.

Merini, L. J., Bobillo, C., Cuadrado, V., Corach, D., & Giulietti, A. M. (2009). Phytoremediation potential of the novel atrazine tolerant *Lolium multiflorum* and studies on the mechanisms involved. *Environmental Pollution, 157*, 3059–3063.

Mesquini, J. A., Sawaya, A. C. H. F., Lopez, B. G. C., Oliveira, V. M., & Miyasaki, N. R. S. (2015). Detoxification of atrazine by endophytic *Streptomyces* sp. isolated from sugarcane and detection of nontoxic metabolite. *Bulletin of Environmental Contamination and Toxicology, 95*(6), 803–809.

Miglioranza, K. S. B., de Moreno, J. E. A., & Moreno, V. J. (2004). Organochlorine pesticides sequestered in the aquatic macrophyte *Schoenoplectus californicus* (C.A. Meyer) Sojak from a shallow lake in Argentina. *Water Research, 38*, 1765–1772.

Mikes, O., Cupr, P., Trapp, S., & Klanova, J. (2009). Uptake of polychlorinated biphenyls and organochlorine pesticides from soil and air into radishes (*Raphanus sativus*). *Environmental Pollution, 157*, 488–496.

Mimmo, T., Bartucca, M. L., Del Buono, D., & Cesco, S. (2015). Italian ryegrass for the phytoremediation of solutions polluted with tertbuthylazine. *Chemosphere, 119*, 31–36.

Mitton, F. M., Gonzalez, M., Peña, A., & Miglioranza, K. S. B. (2012). Effect of amendments on soil availability and phytoremediation potential of aged p,p′-DDT, p,p′-DDE and p,p′-DDD residues by willow plants (*Salix* sp.). *Journal of Hazardous Materials, 203–204*, 62–68.

Mitton, F. M., Miglioranza, K. S. B., Gonzalez, M., Shimabukuro, V. M., & Monserrat, J. M. (2014). Assessment of tolerance and efficiency of crop species in the phytoremediation of DDT polluted soils. *Ecological Engineering, 71*, 501–508.

Moerner, J., Bos, R., & Fredrix, M. (2002). *Reducing and eliminating the use of persistent organic pollutants, in guidance on alternative strategies for sustainable pest and vector management*. Geneva: United Nations Environment Programme.

Moklyachuk, L., Gorodiska, I., Slobodenyuk, O., & Petryshyna, V. (2010). Phytoremediation of soil polluted with obsolete pesticides in Ukraine. In P. A. Kulakow, & V. V. Pidlisnyuk (Eds.), *Application of phytotechnologies for cleanup of industrial, agricultural and wastewater contamination*. Springer Science + Business Media B.V.

Moklyachuk, L., Petryshyna, V., Slobodenyuk, O., & Zatsarinna, Y. (2012). Sustainable strategies of phytoremediation of the sites polluted with obsolete pesticides. In K. Vitale (Ed.), *Environmental and food Safety and security for South-East Europe*. Springer Science + Business Media B.V.

Moore, M. T., & Locke, M. A. (2012). Phytotoxicity of atrazine, *S*-metolachlor, and permethrin to *Typha latifolia* (Linnaeus) germination and seedling growth. *Bulletin of Environmental Contamination and Toxicology, 89*, 292–295.

Mukherjee, I., & Kumar, A. (2012). Phytoextraction of endosulfan: A remediation technique. *Bulletin of Environmental Contamination and Toxicology, 88*, 250–254.

Myresiotis, C. K., Vryzas, Z., & Papadopoulou-Mourkidou, E. (2012). Biodegradation of soil-applied pesticides by selected strains of plant growth-promoting rhizobacteria (PGPR) and their effects on bacterial growth. *Biodegradation, 23*(2), 297–310.

Nair, A. M., Rebello, S., Rishad, K. S., Asok, A. K., & Jisha, M. S. (2015). Biosurfactant facilitated biodegradation of quinalphos at high concentrations by *Pseudomons aeruginosa* Q10. *Soil and Sediment Contamination, 24*(5), 542–553.

Narkhede, C. P., Patil, A. R., Koli, S., Suryawanshi, R., Wagh, N. D., Salunke, B. K., & Patil, S. V. (2015). Studies on endosulfan degradation by local isolate *Pseudomonas aeruginosa*. *Biocatalysis and Agricultural Biotechnology, 4*(2), 259–265.

Nurzhanova, A., Kulakow, P. A., Rubin, E., Rakhimbayev, I., Sedlovsky, A., Zhambakin, K., … Erickson, L. (2010). Obsolete pesticides pollution and phytoremediation of contaminated soil in Kazakhstan. In P. A. Kulakow, & V. V. Pidlisnyuk (Eds.), *Application of phytotechnologies for cleanup of industrial, agricultural and wastewater contamination*. Springer Science + Business Media B.V.

Olette, R., Couderchet, M., Biagianti, S., & Eullaffroy, P. (2008). Toxicity and removal of pesticides by selected aquatic plants. *Chemosphere, 70*, 1414–1421.

Otani, T., Seike, N., & Sakata, Y. (2007). Differential uptake of dieldrin and endrin from soil by several plant families and *Cucurbita genera*. *Soil Science and Plant Nutrition, 53*, 86–94.

Pailan, S., Gupta, D., Apte, S., Krishnamurthi, S., & Saha, P. (2015). Degradation of organophosphate insecticide by a novel *Bacillus aryabhattai* strain SanPS1, isolated from soil of agricultural field in Burdwan, West Bengal, India. *International Biodeteriation and Biodegradation, 103*, 191–195.

Pan, X., Lin, D. L., Zheng, Y., Zhang, Q., Yin, Y. M. C. L., Fang, H., & Yu, Y. L. (2016). Biodegradation of DDT by *Stenotrophomonas* sp. DDT-1: Characterization and genome functional analysis. *Scientific Reports, 6*, 21332.

Pankaj, P., Sharma, A., Gangola, S., Khati, P., Kumar, G., & Srivastava, A. (2016). Novel pathway of cypermethrin biodegradation in a *Bacillus* sp. strain SG2 isolated from cypermethrin-contaminated agriculture field. *3 Biotech, 6*, 45.

Pascal-Lorber, S., & Laurent, F. (2011). Phytoremediation techniques for pesticide contaminations. In E. Lichtfouse (Ed.), *Alternative farming systems, biotechnology, drought stress and ecological fertilisation*. Springer.

Phillips, T. M., Lee, H., Trevors, J. T., & Seech, A. G. (2006). Full-scale in situ bioremediation of hexachlorocyclohexane contaminated soil. *Journal of Chemical Technology and Biotechnology, 81*, 289–298.

Pinto, A. P., Serrano, C., Pires, T., Mestrinho, E., Dias, L., Teixeira, D. M., & Caldeira, A. T. (2012). Degradation of terbuthylazine, difenoconazole, and pendimethalin pesticides by selected fungi cultures. *Science of the Total Environment, 435*, 402–410.

Qiu, X. H., Zhu, T., Li, J., Pan, H. S., Li, Q. L., Miao, G. F., & Gong, J. C. (2004). Organochlorine pesticides in the air around the Taihu Lake, China. *Environmental Science and Technology, 38*, 1368–1374.

Qu, J., Xu, Y., Ai, G. M., Liu, Y., & Liu, Z. P. (2015). Novel *Chryseobacterium* sp. PYR degrades various organochlorine pesticides (OCPs) and achieves enhancing removal and complete degradation of DDT in highly contaminated soil. *Journal of Environmental Management, 161*, 350–357.

Reinhold, D., Vishwanathan, S., Park, J. J., Oh, D., & Saunders, F. M. (2010). Assessment of plant-driven removal of emerging organic pollutants by duckweed. *Chemosphere, 80*(7), 687–692.

Rissato, S. R., Galhiane, M. S., Fernandes, J. R., Gerenutti, M., Gomes, H. M., Ribeiro, R., & Vinicius de Almeida, M. (2015). Evaluation of *Ricinus communis* L. for the phytoremediation of polluted soil with organochlorine pesticides. *Biomed Research International, 2015*, 1–8.
Romeh, A. A. (2014). Phytoremediation of cyanophos insecticide by *Plantago major* L. in water. *Journal of Environmental Health Science and Engineering, 12*(38), 8.
Romeh, A. A. (2015). Evaluation of the phytoremediation potential of three plant species for azoxystrobin-contaminated soil. *International Journal for Environmental Science and Technology, 12*, 3509–3518.
Romeh, A. A., & Hendawi, M. Y. (2013). Chlorpyrifos insecticide uptake by plantain from polluted water and soil. *Environmental Chemistry Letters, 11*, 163–170.
Ryan, R. P., Germaine, K., Franks, A., Ryan, D. J., & Dowling, D. (2008). Bacterial endophytes: Recent developments and applications. *FEMS Microbiology Letters, 278*(1), 1–9.
Sandermann, H. (1994). Higher plant metabolism of xenobiotics: The "green liver" concept. *Pharmacogenetics, 4*, 225–241.
Schroll, R., Becher, H. H., Dorfler, U., Gayler, S., Grundmann, S., Hartmann, H. P., & Ruoss, J. (2006). Quantifying the effect of soil moisture on the aerobic microbial mineralization of selected pesticides in different soils. *Environmental Science and Technology, 40*, 3305–3312.
Seffernick, J. L., Shapir, N., Schoeb, M., Johnson, G., Sadowsky, M. J., & Wackett, L. P. (2002). Enzymatic degradation of chlorodiamino-*s*-triazine. *Applied and Environmental Microbiology, 68*, 4672–4675.
Sharpe, R. (1995). Reproductive biology: Another DDT connection. *Nature, 375*, 538–539.
Shaw, L. J., & Burns, R. G. (2004). Enhanced mineralization of [U-^{14}C]2,4-dichlorophenoxyacetic acid in soil from the rhizosphere of *Trifolium pratense*. *Applied and Environmental Microbiology, 70*(8), 4766–4774.
Siciliano, S. D., Germida, J. J., Banks, K., & Greer, C. W. (2003). Changes in microbial community composition and function during a polyaromatic hydrocarbon phytoremediation field trial. *Apllied Environmental Microbiology, 69*, 483–489.
Siddique, T., Okeke, B. C., Arshad, M., & Frankenberger, W. T. (2003). Enrichment and isolation of endosulfan-degrading microorganisms. *Journal of Environmental Quality, 32*, 47–54.
Singh, B. K., Walker, A., & Wright, D. J. (2006). Bioremedial potential of fenamiphos and chlorpyrifos degrading isolates: Influence of different environmental conditions. *Soil Biology and Biochemistry, 38*, 2682–2693.
Singh, S., Sherkhane, P. D., Kale, S. P., & Eapen, S. (2011). Expression of a human cytochrome P4502E1 in *Nicotiana tabacum* enhances tolerance and remediation of gamma-hexachlorocyclohexane. *New Biotechnology, 28*(4), 423–429.
Somtrakoon, K., Kruatrachue, M., & Lee, H. (2014). Phytoremediation of endosulfan sulfate-contaminated soil by single and mixed plant cultivations. *Water, Air and Soil Pollution, 225*.
Sreenivasulu, C., & Aparna, Y. (2001). Bioremediation of methylparathion by free and immobilized cells of *Bacillus* sp. isolated from soil. *Bulletin of Environmental Contamination and Toxicology, 67*(1), 98–105.
Su, Y. H., Zhu, Y. G., Lin, A. J., & Zhang, X. H. (2005). Interaction between cadmium and atrazine during uptake by rice seedlings (*Oryza sativa* L.). *Chemosphere, 60*, 802–809.
Sun, G., Zhang, X., Hu, Q., Zhang, H., Zhang, D., & Li, G. (2015). Biodegradation of dichlorophenyltrichloroethanes (DDTs) and hexachlorocyclohexanes (HCHs) with plant and nutrients and their effects on the microbial ecological kinetics. *Environmental Microbiology, 69*, 281–292.
Sun, H., Xu, J., Yang, S., Liu, G., & Dai, S. (2004). Plant uptake of aldicarb from contaminated soil and its enhanced degradation in the rhizosphere. *Chemosphere, 54*, 569–574.

Sutherland, T. D., Horne, I., Weir, K. M., Coppin, C. W., Williams, M. R., Selleck, M., ... Oakeshott, J. G. (2004). Enzymatic bioremediation: From enzyme discovery to applications. *Clinical and Experimental Pharmacology and Physiology, 31*(11), 817–821.
Talwar, M. P., & Ninnekar, H. Z. (2015). Biodegradation of pesticide profenos by the free and immobilized cells of *Pseudoxanthomonas suwonensis* strain HNM. *Journal of Basis Microbiology, 55*(9), 1094–1103.
Tang, A. X., Wang, B. W., Liu, Y. Y., Li, Q. Y., Tong, Z. F., & Wei, Y. J. (2015). Biodegradation and extracellular enzymatic activities of *Pseudomonas aeruginosa* strain GF31 on beta-cypermethrin. *Environmental Science and Pollution Research, 22*(17), 13049–13057.
Thomas, J. E., & Gohil, H. (2011). Microcosm studies on the degradation of o,p′-and p,p′-DDT, DDE, and DDD in a muck soil. *World Journal of Microbiology and Biotechnology, 27*, 619–625.
Thomas, J. E., Ou, L., & Al-Agely, A. (2008). DDE remediation and degradation. *Reviews of Environmental Contamination and Toxicology*, 55–70.
Trapp, S. (2000). Modeling uptake into roots and subsequent translocation of neutral and ionisable organic compounds. *Pest Management Science, 56*, 767–778.
Trapp, S. (2004). Plant uptake and transport models for neutral and ionic chemicals. *Environmental Science and Pollution Research, 11*, 33–39.
Turgut, C. (2005). Uptake and modeling of pesticides by roots and shoots of parrotfeather (*Myriophyllum aquaticum*). *Environmental Science and Pollution Research, 12*(6), 342–346.
Vacondio, B., Birolli, W. G., Ferreira, I. M., Seleghim, M. H. R., Goncalves, S., Vasconcellos, S. P., & Porto, A. L. M. (2015). Biodegradation of pentachlorophenol by marine-derived fungus *Trichoderma harziam* CBMAI 1677 isolated from ascidian *Didemnun ligulum*. *Biocatalysis and Agricultural Biotechnology, 4*(2), 266–275.
Verma, S. C., Ladha, J. K., & Tripathi, A. K. (2001). Evaluation of plant growth promoting and colonization ability of endophytic diazotrophs from deep water rice. *Journal of Biotechnology, 91*, 127–141.
Vila, M., Lorber-Pascal, S., & Laurent, F. (2007). Fate of RDX and TNT in agronomic plants. *Environmental Pollution, 148*, 148–154.
Wackett, L. P. (2004). Evolution of enzymes for the metabolism of new chemical inputs into the environment. *Journal of Biology and Chemistry, 279*, 41259–41262.
Wackett, L. P., Sadowsky, M. J., Martinez, B., & Shapir, N. (2002). Biodegradation of atrazine and related triazine compounds: From enzymes to field studies. *Applied Microbiology and Biotechnology, 58*, 39–45.
Wakelin, S., Warren, R., Harvey, P., & Ryder, M. (2004). Phosphate solubilization by *Penicillium* spp. closely associated with wheat roots. *Biology and Fertility of Soils, 40*, 36–43.
Wang, F. Y., Tong, R. J., Shi, Z. Y., Xu, X. F., & He, X. H. (2011). Inoculations with arbuscular mycorrhizal fungi increase vegetable yields and decrease phoxim concentrations in carrot and green onion and their soils. *Plos One, 6*(2), e16949.
Wang, Q., Yang, J., Cui, L., Xiao, B., & Xiaoe, Q. (2013). Influence of initial pesticide concentrations in water on chlorpyrifos toxicity and removal by *Iris pseudacorus*. *Water Science and Technology, 67*(9), 1908–1915.
Wang, Q., Zhang, W. X., Li, C., & Xiao, B. (2012). Phytoremediation of atrazine by three emergent hydrophytes in a hydroponic system. *Water Science and Technology, 66*(6), 1282–1288.
Wang, Y., Wang, C., Li, A., & Gao, J. (2015). Biodegradation of pentachloronitrobenzene by *Arthrobacter nicotianae* DH19. *Letters in Applied Microbiology, 61*(4), 403–410.
Wang, Y. S., Wu, T. H., Yang, Y., Zhu, C. L., Ding, C. L., & Dai, C. C. (2016). Binding and detoxification of chlorpyrifos by lactic acid bacteria on rice straw silage fermentation. *Journal of Environmental Science and Health, 51*(5), 316–325.

Warsaw, A. L., Fernandez, R. T., Kort, D. R., Cregg, B. M., Rowe, B., & Vandervoort, C. (2012). Remediation of metalaxyl, trifluralin, and nitrate from nursery runoff using container-grown woody ornamentals and phytoremediation areas. *Ecological Engineering, 47*, 254–263.

Wehtje, G., Walker, R. H., & Shaw, J. N. (2000). Pesticide retention by inorganic soil amendments. *Weed Science, 48*, 248–254.

Weyens, N., Monchy, S., Vangronsveld, J., Taghavi, S., & Van Der Lelie, D. (2010). Plant-microbe partnerships. In K. N. Timmis (Ed.), *Handbook of hydrocarbon and lipid microbiology* (pp. 2545–2574). Berlin, Heidelberg: Springer-Verslag.

Weyens, N., Van Der Lelie, D., Taghavi, S., Newman, L., & Vangronsveld, J. (2009). Exploiting plant-microbe partnerships to improve biomass production and remediation. *Trends in Biotechnology, 10*, 591–598.

Whipps, J. M. (1990). Carbon economy. In J. M. Lynch (Ed.), *The rhizosphere* (pp. 59–97). New York, USA: Wiley.

White, J. C. (2010). Inheritance of p,p′-DDE phytoextraction ability in hybridized *Cucurbita pepo* cultivars. *Environmental Science and Technology, 44*, 5165–5169.

White, J. C., Parrish, Z., Isleyen, M., Gent, M., Iannucci-Berger, W., Eitzer, B., & Mattina, M. (2005). Uptake of weathered p,p′-DDE by plant species effective at accumulating soil elements. *Microchemical Journal, 81*, 148–155.

Whitfield Aslund, M., & Zeeb, B. A. (2010). A review of recent research developments into the potential for phytoextraction of persistent organic pollutants (POPs) from weathered, contaminated soil. In P. A. Kulakow, & V. V. Pidlisnyuk (Eds.), *Application of phytotechnologies for cleanup of industrial, agricultural, and wastewater contamination*. Springer Science + Business Media B.V.

Whitford, F., Wolt, J., Nelson, H., Barret, M., Brichford, S., & Turco, R. (1995). *Pesticides and water quality principles, policies and programs*. West Lafayette: Purdue University Cooperative Extension Service.

Wu, C. Y., Chen, N., Li, H., & Li, Q. F. (2014). *Kocuria rosea* HN01, a newly alkaliphilic humus-reducing bacterium isolated from cassava dreg compost. *Journal of Soils and Sediments, 14*(2), 423–431.

Xia, H., & Ma, X. (2006). Phytoremediation of ethion by water hyachinth (*Eichhornia crassipes*) from water. *Bioresource Technology, 97*, 1050–1054.

Xiao, P. F., & Kondo, R. (2013). Biodegradation of dieldrin by *Cordyceps* fungi and detection of metabolites. Pts 1–4. In X. Y. Tang, W. Zhong, D. Zhuang, C. Li, & Y. Liu (Eds.), *Progress in environmental Protection and Processing of resource* (pp. 30–34). Guangzhou, China: Guangdong University.

Xu, X. J., Sun, J. Q., Nie, Y., & Wu, X. L. (2015). *Spirodela polyrhiza* stimulates the growth of its endophytes but differentially increases their fenpropathrin-degradation capabilities. *Chemosphere, 125*, 33–40.

Xu, G., Li, Y., Zheng, W., & Peng, X. (2007). Mineralization of chlorpyrifos by co-culture of *Serratia* and *Trichosporon* spp. *Journal of Biotechnology Letters, 29*, 1469–1473.

Xu, G., Li, Z. W., Wang, Y., Zhang, S., & Yan, Y. C. (2008). Biodegradation of chlorpyrifos and 3,5,6-trichloro-2-pyridinol by a newly isolated *Paracoccus* sp strain TRP. *International Biodeterioration and Biodegradation, 62*, 51–56.

Yang, C., Liu, N., Guo, X., & Qiao, C. (2006). Cloning of mpd gene from a chlorpyrifos-degrading bacterium and use of this strain in bioremediation of contaminated soil. *FEMS Microbiology Letters, 265*, 67–73.

Yu, Y. L., Chen, Y. X., Luo, Y. M., Pan, X. D., He, Y. F., & Wong, M. H. (2003). Rapid degradation of butachlor in wheat rhizosphere soil. *Chemosphere, 50*, 771–774.

Zhang, J., Zheng, J. W., Liang, B., Wang, C. H., CAi, S., Ni, Y. Y., … Li, S. P. (2011). Biodegradation of chloroacetamide herbicides by *Paracoccus* sp. FLY-8 in vitro. *Journal of Agricultural and Food Chemistry, 59*(9), 4614–4621.

Zhang, P., Sheng, G. Y., Feng, Y. C., & Miller, D. M. (2005). Role of wheat-residue-derived char in the biodegradation of benzonitrile in soil: Nutritional stimulation versus adsorptive inhibition. *Environmental Science and Technology, 39*, 5442–5448.

Zhao, S., Arthur, E. L., Moorman, T. B., & Coats, J. R. (2005). Evaluation of microbial inoculation and vegetation to enhance the dissipation of atrazine and metolachlor in soil. *Environmental Toxicology and Chemistry, 24*(10), 2428–2434.

Zhou, Y., Tigane, T., Li, X., Truu, M., Truu, J., & Mander, U. (2013). Hexachlorobenzene dechlorination in constructed wetland mecocosms. *Water Research, 47*, 102–110.

Zhu, Z., Yang, X., Wang, K., Huang, H., Zhang, X., Fang, H., ... He, Z. (2012). Bioremediation of Cd-DDT co-contaminated soil using the Cd-hyperaccumulatior *Sedum alfredii* and DDT-degrading microbes. *Journal of Hazardous Materials, 235–236*, 144–151.

Zuo, Z. Q., Gong, T., Che, Y., Liu, R. H., Xu, P., Jiang, H., ... Yang, C. (2015). Engineering *Pseudomonas putida* KT2440 for simultaneous degradation of organophosphates and pyrethroids and its application in bioremediation of soil. *Biodegradation, 26*(3), 223–233.

CHAPTER EIGHT

Plants in Air Phytoremediation

Stanislaw W. Gawronski*,[1], Helena Gawronska*, Slawo Lomnicki[§], Arne Sæbo[¶] and Jaco Vangronsveld[||]

*Warsaw University of Life Sciences, Warsaw, Poland
[§]Louisiana State University, Baton Rouge, LA, United States
[¶]Norwegian Institute of Bioeconomy Research (NIBIO), Ås, Norway
[||]Hasselt University, Diepenbeek, Belgium
[1]Corresponding author: E-mail: stanislaw_gawronski@sggw.pl

Contents

Abstract

Air pollution has become a global problem and affects nearly all of us. Most of the pollution is of anthropogenic origin and therefore we are obliged to improve this situation. In solving this problem basically our only partners are plants with their enormous biologically active surface area. Plants themselves are also victims of air pollution but because they are sedentary they developed very efficient defence mechanisms, which can also be exploited to improve the humanosphere. For their life processes plants require intensive gas exchange, during which air contaminants are accumulated on leaf surfaces or absorbed into the tissues. Some of the pollutants are included by plants in their own metabolism while others are sequestered. In some plant species, the processes of removing pollutants from the air is conducted in a very efficient way and therefore they are used in the environmental friendly biotechnology called phytoremediation. For urban areas, outdoor phytoremediation is recommended while indoor phytoremediation can be applied in our homes and workplaces. Because in near future purifying outdoor air to protect human health and well-being does not look the most promising, an important and increasing role will be played by indoor phytoremediation.

Advances in Botanical Research, Volume 83
ISSN 0065-2296
http://dx.doi.org/10.1016/bs.abr.2016.12.008

1. INTRODUCTION

Impacts of air pollution on human health and well-being are well documented worldwide by the World Health Organization. Every year 4.3 million premature deaths occur from exposure to indoor air pollution, and 3.7 million deaths are attributed to outdoor air pollution (WHO, 2015), with the major health effects being respiratory and cardiovascular diseases. A striking finding of Feigin et al. (2016) showed that between 10.2% (high-income country) and 33.7% (low- and middle-income countries) of stroke burden are due to exposure to air pollution. Others indicate the relationship between air pollution and neurodegenerative diseases (Calderon-Garciduenas et al., 2008). Air pollution is associated with the presence of excess amounts of chemicals and aerosols, compared to the pristine conditions, such as for example remote ocean locations. The origin of the pollutants is typically of the anthropogenic origin: particulate matter (PM), carbon monoxide (CO), nitrogen oxides (NO_x), volatile organic compounds (VOC) with most often listed in this group the polycyclic aromatic hydrocarbons (PAHs), heavy metals (HMs) and sulphur dioxide (SO_2), or the result of their secondary atmospheric reactions such as peroxyacetyl nitrates (PAN) or ozone (O_3). A special group is the persistent organic pollutants (POPs) which are not released in large amounts, however, due to their environmental stability they persist in the environment for extensive periods of time. These include synthetic organic compounds, used on a large scale in human daily life such as: polychlorinated biphenyls (PCBs) that were commonly used as heat transfer liquids, polybrominated diphenyl ethers (PBDEs), used as fire retardants. The Stockholm Convention has banned the production of many of these POPs production, but nevertheless their stockpiles and abundance in already manufactured products still results in significant emissions to the environment (U.N. Stockholm Convention C.N.524.2009.TREATIES-4). Phthalates used as plasticizers in PVC plastics are not chemically bound to PVC and are slowly released from the plastic matrix by leaching into fluids or evaporation into both indoor and outdoor air. Their high analogy to human hormones makes them suspected as endocrine disruptors. These chemicals, due to their vapour pressure and gas–solid partitioning coefficient can easily adhere to the ambient air PM and undergo gas phase transport with these particulates. As a result, the main mechanism of their removal from air is particle dry deposition or alternatively, incorporation within rain droplets and precipitation (Fang & Huang, 2013; Sun et al., 2014). Deposited PM particles on surfaces, such as for example plant leaves are subject to further migration with precipitation, as a substantial fraction of PM deposited on leaves is continuously washed off to the soil or on sealed urban surfaces with every rainfall (Nowak, Crane, & Stevens, 2006; Sun et al., 2014). Rain removes PM from both the air and from the surfaces of leaves contributing to air purification; however, between rainfalls plants accumulation of PM on their surface and in wax (sPM and wPM, respectively) (Popek, Gawronska, Wrochna, Gawronski, & Sæbo, 2013) by far exceeds other dry deposition rates and ensures a much more efficient removal of pollutants from the air (McDonald et al., 2007). It is thus possible to exploit higher plants with their microbiome in an environmental friendly biotechnology for the removal of airborne pollutants associated with PM in outdoor and often also indoor locations — an example of the general process of phytoremediation.

The most important physical parameter of plants in the context of PM phytoremediation is their enormous biologically active surface area. The calculated surface area of above ground plants' organs on earth is estimated to be at least 200 million km^2 and exceeds the earth land area of 149 million km^2 (Varma, Abbott, Werner, & Hampp, 2008). Plants' leaves, needles and young twigs contribute the most to this surface area and all of them capture air pollutants. Plants' leaf areas exceed by a few factors the projected by the canopies ground surface area, and this characteristic is referred to as a Leaf Area Index (LAI), which depending on the group of plants equals 2—3 for grasses, 4—5 for shrubs and up to 10 or more for trees. Special leaf features and structures such as leaf folding, hairs (pubescence), trichomes and wax layers have specific roles in pollutant removal and all favour the accumulation of PM.

Plants control the impact of the environment on their organs through the outermost layer of cells called the epidermis. The outer wall of this epidermis is covered with a lipid layer (wax) that forms a protective barrier for the cells from the surroundings. Plant waxes are very diverse and Barthlott et al. (1998) distinguish 23 types of waxes. Waxes are typical and at a fairly uniform structure for certain taxa, with different wax structures varying in shape, roughness, thickness of the layer, and chemical composition. Trichomes, hairs, veins and rough wax are conducive for the accumulation of PM on the leaf surface. During raining and snowing, PMs are removed from air as well as from the needles and leaves of plants which to a large extent is due to the surface electrostatic charge conducive to the accumulation. PMs are localized on the surface itself, but also embedded in the wax coating leaves or needles. Deep surface penetration into the interior of the leaf tissue is very limited even for fine and ultrafine particles; however, due to their size they could only penetrate into the plant through the stomata. The stomata, however, comprise only 0.5%–1.5% of the leaf surface and usually are situated on the lower side of leaves.

Plants are higher, autotrophic organisms and for their life processes require intensive gas exchange, during which gaseous air contaminants can also be accumulated/absorbed internally. The intensity of gas exchange depends on a number of factors such as temperature, humidity, carbon dioxide concentration, light intensity, specific photosynthetic system and air pollution. During evolution, to adapt to the changing environmental conditions, plants developed different photosynthetic systems (PSs) such as C_3, C_4, CAM and facultative CAM and plants with specific PS occupy different ecosystems, mainly depending on the climate. C_3 plants are dominant in the low temperate zone and cold areas, and their growth and biomass accumulation is usually lower. Such plants keep their stomata constantly open during the day to maximize CO_2 uptake (together with gaseous pollutants). C_4 plants perform the process of photosynthesis in spatially separate cells, with higher intensity of gas exchange and accumulate biomass much faster and in greater amounts. On the contrary, CAM plants conduct the process of photosynthesis in temporally separated steps. Uptake of gasses occurs during the night (open stomata) while during the day photosynthesis proceeds with the stomata closed. Such temporal separation of the CAM system steps allows plants to survive in very dry conditions. Epiphytic plants species also sometimes are exposed to dry periods and thanks to their ability to run photosynthesis via the CAM system they survive in difficult conditions, though at low efficiency of photosynthesis and growth. Interesting is the existence of facultative CAM plants that under optimal or near optimal conditions function as C_3 plants while under stress conditions they can run photosynthesis via the CAM pathway. The intensity of this process differs from typical CAM plants with closed stomata during the day to C_3 pathway (Winter & Smith, 1996, p. 449). Facultative CAM plant species are very interesting for air phytoremediation, as they have stomata open at night and take up CO_2 along with pollutants, and during the day they carry out the other steps of photosynthesis. They can be useful for indoor phytoremediation in bedrooms; during the night CAM plants would improve the air quality absorbing CO_2 released both by these plants and humans during their sleep. This group of plants should be recommended also for cultivation in the extensive green roofs, with hot conditions during the day and the plants carrying out gas exchange at night.

The waxy surface of leaves is colonized by microorganisms, creating a phyllomicrobiome consisting on average out of 10^6—10^7 microorganisms cm^2 (Vorholt, 2012). De Kempeneer, Sercu, Vanbrabant, Van Lanenhove, and Verstraete (2004) reported that these bacteria are actively involved in the degradation of available organic pollutants such as for example toluene. This phenomenon further contributes to the air remediation by the surface of plants extending it to plant microbiome. Gaseous pollutants penetrate the leaves through stomata while surface deposited lipophilic compounds can penetrate the cuticular wax and become degraded by the endobacteria. It is difficult to precisely determine the role of epibacteria (surface) and endobacteria (internalized). However, washing the leaves resulting in the removal of epiphytic bacteria decreased the degradation process of organic pollutants as for example benzene confirming their substantial role in the degradation of organic air pollutants (Sriprapat & Thiravetyan, 2016). The role of the plant microbiome and possibilities of its utilization have been described (Weyens et al., 2015; in chapter: Bio- and Phytoremediation of Pesticide Contaminated Environments: A Review by Eevers, White, Vangronsveld, & Weyens, 2017).

2. AEROSOL AIR POLLUTANTS

PM, nowadays the number one air pollutant in many regions of the world, is a complex mixture of chemical species within the solid or liquid particles suspended in the air defined as an aerosol. Solid aerosols can remain suspended in the air for minutes, hours to days or even months before deposition, depending mainly on size of the PM (Bidleman, 1988).

PM can be directly emitted into the air and comprise primary aerosols or can be formed in the atmosphere through particle coagulation of primary particles and incorporation of gaseous pollutants such as SO_2 and NO_x forming secondary aerosols ($PM_{2.5}$, <2.5 μm, fine particles). Larger particles than 2.5 μm in diameter are increasingly enriched with crustal elements due to the weathering process of rocks and soils (PM 2.5—10 μm, coarse particles). The smallest fraction of PM is generated almost exclusively by the anthropogenic sources, since 70% of $PM_{2.5}$ originates from combustion. A WHO report mentions that a major source of PM in urban areas of developed countries are fossil fuels and particularly internal engines combustion, while they are household cooking and heating in developing countries (WHO, 2015). Though the toxicity and health effects associated with the human exposure to PM are still a subject of major debate and extensive research, it is clear that different components of PM can contribute to the observed effects. These include not only native combustion borne particle components but also chemicals adsorbed on the PM when suspended in air. Recent data have shown that nonexhaust sources such as brakes, tires and asphalt dust contribute almost equally to the total PM emitted in highly urbanized areas (Grigoratos & Martini, 2015). If toxicity of noncombustion fractions of PM is equivalent to combustion fractions, the introduction of hybrid and electric cars will decrease the human risks associated with PM emissions only in part. A new discovery of Environmentally Persistent Free Radicals (EPFRs) has shed a new light on PM-associated toxicity (Cormier, Lomnicki, Backes, & Dellinger, 2006; Lomnicki, Truong, Vejerano, & Dellinger, 2008).

PM embedded in wax are more strongly associated with leaves in comparison to those located on the surface of the wax and that are more easily removed by rain or wind into the soil. In rainy regions, washing the surface of the leaves by precipitation is an efficient removal mechanism of PM-associated pollutants from the air. However, PM embedded in waxes and especially lipophilic organic compounds like PAHs, PCBs, PBDEs, dioxins and the charged electrostatic PM slowly penetrate into the leaf by diffusion and only after reaching the first living epidermal cells are displaced and deposited in the cell wall or the vacuole. Raking fallen leaves and young shoots from such sites and transporting them for disposal remove PM completely from the particular sites of urban areas. The best way to dispose PM-polluted litter is to burn it in strictly controlled conditions (backyard burning is absolutely not acceptable!), depending on amount of accumulated PM and associated pollutants they can be hazardous. The remaining ash (representing approximately 1% of the dry weight) must undoubtedly be handled as hazardous material, and disposed accordingly.

Black carbon (BC) particles comprise a significant fraction of PM originating from combustion and commonly are referred to as soot. Pure BC, commonly referred to as elemental carbon (EC), as a component of PM is not considered toxic to human health, but it plays a role of a carrier or vehicle for other toxic compounds on its surface. Their adverse role is strong light absorption and heat generation and therefore directly impacting the climate change through decreasing the Earth albedo (EEA, 2013). It is quite probable that BC affects plants similarly, i.e., increase of leaves temperature and reduction of the light transmittance to the photosynthetic apparatus. According to Naidoo and Chirkoot (2004) the inhibition of photosynthesis observed in plants growing in the vicinity of coal mines partially can be explained by the negative impact of BC deposited on leaves.

EPFRs are a recently discovered class of combustion-related pollutants. They are inherently associated with the PM, as they are formed at the gas–solid interphase and lead to the surface bound species. EPFRs are conjugated species between an organic fraction and metal centres embedded in the PM. The association between the metal and organic adsorbent (in general small aromatic compounds) results in an electron transfer from the organic molecules to the metal centre forming organic radical species, though bound to the surface. The organic forms of the EPFRs are generally of phenoxyl or semiquinone type structures, with the unpaired electron localized around oxygen, and stabilized by resonance. A unique characteristic of such species is their lifetime – while usually organic radicals have a lifetime of microseconds to milliseconds (and seconds in extreme cases), EPFRs are relatively stable with lifetime of hours, days or even months in ambient air (thus 'environmentally persistent') (Gehling & Dellinger, 2013; Vejerano, Lomnicki, & Dellinger, 2012). Recent studies implicate EPFRs to be responsible for observed health effects of PM in humans (Kelley et al., 2013; Reed, dela Cruz, Lomnicki, & Backes, 2015; Saravia, Lee, Lomnicki, Dellinger, & Cormier, 2013). In fact, it was suggested that based on the concentrations of EPFRs it is possible to compare the exposure of PM to the amount of smoked cigarettes (Gehling & Dellinger, 2013). The mechanism of their toxicity results from redox reactions with a net output of hydroxyl radicals which are extremely potent oxidizers and toxic to living organisms and resulting in oxidative stress (Khachatryan, Vejerano, Lomnicki, & Dellinger, 2011). So far the effects of the presence of EPFRs on either plants or their associated microbiome upon deposition of PM has not been investigated. However, based on the known chemistry and behaviour of EPFRs it is plausible that generation of hydroxyl radicals within thin water films on the surface of the leaves may result in the degradation of the protective wax layer and also be toxic to the microbiome. More studies are required in this field. It is also quite possible that through evolutionary processes some bacteria developed resistance to EPFRs or even can destroy them thus having a symbiotic protective function for their host plants. This will also contribute to the overall phytoremediation activity, as the PM collected by plants are neutralized in respect of the presence of EPFRs.

Semivolatile organic pollutants (SVOCs) are a large group of known toxicants formed during incomplete combustion of coal, oil and any other organic matter. Lower combustion temperatures and substoichiometric concentrations of oxygen increase formation of these products including PAH, biphenyls, phenolics, dioxins and many others. Compounds with low vapour pressure are partitioning to the surface of combustion formed particles such as carbon black or refractory oxide nanoparticles, forming soot and fly ashes.

Individual PAHs are different in environmental behaviour depending on the number of aromatic rings in the structure as well as their relative arrangement. Five and six rings are solid and very persistent in the environment and subject to environmental accumulation and bioaccumulation. Noteworthy are the two most carcinogenic five ring PAHs, i.e., benzo (*a*) pyrene and the more carcinogenic but fortunately present in lower quantities dibenzo (*a*, *h*) anthracene. Due to their stability, the fate of such PAHs is usually dependent on organisms capable to decompose them enzymatically such as by wood-rotting fungi; however, this is a very slow process. In Europe, Poland is one of the most polluted countries with benzo (*a*) pyrene (WHO, 2015) resulting from the dominant role of coal-based energy and the large number of old cars. However, similar situations are found in many places around the world.

Along with the progress in science the list of organic pollutants is increasing constantly, converting once praised new chemicals into banned and dangerous substances. A good example is the PCBs, originally introduced by Monsanto or PBDEs and recently diethyl phthalates. Though, in developed countries, the production of PBDEs was abandoned in the 1970s they are still released into the environment from old electronics, printed circuit boards, old furniture upholstery, construction and lighting equipment.

Besides SVOCs removal by PM deposition on the surface of leaves, many plants are capable to metabolize them making them perfect air remediation tools. PCBs from the air enter plants through the cuticle. Part of them remain trapped within the cuticle (Moeckel, Thomas, Barber, & Jones, 2008) and the remaining part penetrates into the cells and is metabolized possibly by cytochrome P-450 as shown for cells of pine needle and eucalyptus leaves by Chen et al. (2014). Winter melon (*Benincasa hispida*) is taking up di (2-ethykhexyl) phthalate very efficiently from the air through its leaves, shoots and even flowers, what makes this species much more effective than commonly recognized phytoremediation species like *Brassica chinensis* and *Brassica campestris*. These results indicate that *B. hispida* is a good candidate for phytoremediation of phthalates (Wu et al., 2013). Although it is a valuable vegetable in the warmer regions of the world, it cannot be used for consumption when cultivated as phytoremediation species.

HMs present in air PM mainly originate from combustion of fossil fuel. HMs are sometimes associated with larger, coarse particles in areas of active mining and where mine tailings are present. The list of HM is quite long, but lead (Pb), zinc (Zn), cadmium (Cd), arsenic (As), mercury (Hg), chromium (Cr) VI and copper (Cu) are very often dominating. Their presence in plant tissues should be interpreted with some caution because even metals that are very toxic and unnecessary to plants such as Cd and metalloids like As are very easily taken up from the soil and translocated to the above ground parts of plants. Also other elements like zinc (Zn) and slightly less copper (Cu) and manganese (Mn) are subject to similar translocation mechanisms and all of them can get toxic for plants at high levels. Other elements such as Pb, Ni, Cr, V and generally also Fe are poorly taken up from soils and generally do not translocate to the aerial plant parts, and their presence in leaves and twigs indicates absorption from air (Ny & Lee, 2011; Sæbo et al., 2015).

Increased concentrations of Pb and Cr in leaves confirm that they penetrate into these tissues via the epidermis as their uptake from the soil is minimal (Megremi, 2010). In cells, toxic forms of metal(loid)s like As V and Cr VI are reduced to less toxic forms, i.e., As III and Cr III. Plants have an even more effective defence mechanisms, i.e., methylation of As, Cr and Se and release of methylated forms to the air. With high levels of As in the soil, some plant species release significant amounts of As into the air; However, recent studies suggest that As methylation occurs already in the soil by rhizobacteria and that plants absorb methylated As (Lomax et al., 2011).

Both, organic compounds of high molecular weight (Wild, Dent, Thomas, & Jones, 2006) and HM (authors own data) slowly, through physical diffusion, penetrate through the wax into the living cells of the epidermis where they are sequestered in the vacuole or cell wall.

Aerosols moving near the ground encounter plants as 'obstacles', equipped with folded leaves, often covered with trichomes or hairs and possessing an electromagnetic charge, what together effectively increases the accumulation of PM on their surface (Gawronska & Bakera, 2015; Popek et al., 2013; Pough, MacKenzie, Whyatt, & Hevitt, 2012).

All plants accumulate PM on their surface, which means all are involved in air phytoremediation and act as biofilters (Bakker, Vorenhout, Sijm, &

Kolloeeffel, 1999; Barack, 2002; Farmer, 2002). However, research performed on a large number of trees and shrubs revealed that they can differ by more than 10-fold in their PM accumulation ability (Popek et al., 2013; Sæbo et al., 2012).

The following tree species are known as potentially good phytoremediation species: *Pinus sylvestris, Betula pendula, Fraxinus pennsylvanicus, Fraxinus excelsior, Pyrus calleryana, Sorbus intermedia, Populus* sp., *Alnus spaethi, Robinia pseudoacacia, Sophora japonicum, Elaeagnus angustifolia, Ligustrum lucidum, Quercus ilex, Tilia europaea* 'Pallida'. Equally excellent functions for phytoremediation perform shrubs that are growing closer to the ground like: *Pinus mugo*; *Syringa meyeri*; *Spireae* sp., *Stephanandra incisa, Taxus media, Taxus baccata, Hydrangea arborescens*; *Acer campestre, Physocarpus opulifolius*; *Sorbaria sorbifolia*; *Forsythia x intermedia* (Dzierzanowski, Popek, Gawronska, Sæbo, & Gawronski, 2011; Popek et al., 2013; Sæbo et al., 2012; Sgrigna, Sæbo, Gawronski, Popek, & Calfapietra, 2015; Wang, Shi, Li, Yu, & Zhang, 2013; Wang, Shi, & Wang, 2015). In the densely built-up city centres where the surface for growing plants is limited, following climbers can fulfil a role in phytoremediation: *Hedera helix*; *Parthenocissus tricuspidata, Parthenocissus quinquefolia, Vitis riparia* and *Polygonum aubertii* (Borowski, Loboda, & Pietkiewicz, 2009; Ottele, von Bohemen, & Fraaij, 2010 and own not yet published data). Also herbaceous plants deserve attention as phytoremediation plants: *Achillea millefolium, Berteroa incana, Polygonum aviculare, Brickellia veronicifolia, Flaveria trinervia* and *Aster gymnocephalus* (Weber, Kowarik, & Saumel, 2014).

Conifers plants are excellent for air phytoremediation due to the abundant wax layer on the needles and, as evergreens, they are active in pollutant accumulation during the whole year. These species keep needles for more than 1 year but being so efficient in retention of pollutants, they eventually can die because of too 'heavy loads' of pollutants. The only genera very tolerant and well-surviving contamination with PM in this group is the yew (*Taxus* sp.) with a very efficient self-cleaning mechanism.

For their effective remedial properties, conifers should be more frequently recommended for planting, however, at an adequate distance from the emission source and at moderate pollution intensities. When creating 'green filters', significant declines in PM were reported already at 100 m distance from the edge of a highway (Fuller, Bai, Eisinger, & Niemeier, 2009).

3. GASEOUS AIR POLLUTANTS

The main gaseous air pollutants are CO, NO_x, SO_2, O_3 and some organic compounds. One of the main greenhouse gases causing human concern is CO_2, which for the plant is the essential molecule in the process of photosynthesis and in an elevated concentrations, to certain extent, promotes the synthesis of plant biomass. Gaseous air pollutants, though at a marginal level, enter into the plants mainly through the stomata and through wax and cutin.

CO in the environment raises larger concerns for humans for whom it is acutely toxic already at 750 ppm. Plants, however, tolerate CO to a relatively high concentration. CO originates from the combustion of organic materials in conditions with limited oxygen.

Toxicity of CO in humans results from binding with the active site of heamoglobin, obstructing oxygen transport. CO is metabolized by plants either by further oxidation to CO_2 or is reduced and incorporated into amino acids. There exist large differences between species in the ability of uptake of this gas by plants (Bidwell & Bebee, 1974). From a list of 35 investigated woody plant species 17 were characterized by high ability to fix CO. Among trees these are: *Acer saccharum*, *Acer saccharinum*, *Gleditsia triacanthos*, *Pinus resinosa*, *Populus nigra*, *Fraxinus pennsylvanica* and two shrubs species *Syringa vulgaris* and *Hydrangea* sp. Two popular ornamental indoor plant species in moderate climate (and gardened in subtropical and tropical zones) *Ficus variegata* and *Phenix roebelenii*, also metabolize CO efficiently (Bidwell & Bebee, 1974).

Nitrogen (N), an important building block of proteins, is next to carbon the second most important element for plants. In nature it occurs in several different forms; however, plants can only absorb fixed nitrogen (i.e., other than in the molecular form N_2). The atmosphere is often contaminated with nitrogen-containing gasses such as NO_x and NH_3. In urban areas NO_x (primarily NO_2) are the dominant N-containing pollutants due to vehicular emissions. The negative environmental impacts of NO_x are associated with (1) active participation in the UV radiation assisted ozone formation cycle (however, the net increase of ozone requires also copresence of volatile organics), (2) acid rain formation and (3) recently discovered posttranslational modification of allergenic proteins as a result of their nitration (Reinmuth-Sezlet et al., 2014). Significant intensification of such processes is observed in polluted air (Bryce et al., 2010; Lu et al., 2014;

Zhao et al., 2016). However, plants are capable to utilize nitrogen in the form of NO_2. Some years ago Morikawa et al. (1998) investigated the ability of 217 herbaceous and woody species to take up NO_2 and discovered up to 600-fold differences in NO_2 uptake and assimilation between them. They reported the following woody species to be the most efficient: *Magnolia kobus*, *Eucalyptus viminalis*, *Eucalyptus grandis*, *Eucalyptus globulus*, *Populus nigra*, *Populus* sp., *R. pseudoacacia*, *S. japonicum*, *Prunus cerasoides* and as herbaceous species: *Nicotiana tabacum* and *Erechtites hieracifolia* calling them NO_2-filice (Morikawa et al.,1998). Further investigations of 70 species of woody plants recommended for the cultivation in the vicinity of heavy traffic roads revealed greater differences in the response to NO_2. Among the tested species *R. pseudoacacia*, *S. japonica*, *P. nigra* and *Prunus lannesiana* proved to be the most interesting species tolerant to high levels of NO_2 and efficiently assimilating this form of nitrogen. Therefore, they are the recommended for air phytoremediation (Takahashi et al., 2005).

Ammonia (NH_3) is also considered to be one of the primary N-containing air pollutants The major sources of atmospheric NH_3 are agricultural activities, animal feedlot operations, and domestic waste water effluents. All plants are able to take up ammonium directly from the air to a certain extent. Adverse effects on vegetation occur when the rate of foliar uptake of NH_3 is greater than their level of tolerance (Ghaly & Ramikrishnan, 2015; Krupa, 2003; Rogers & Aneja, 1980).

Ecosystem pollution with nitrogen gasses cannot be viewed from the amounts of NO_2 or NH_3 but must be seen in the context of total deposited N. Data suggest that a critical load of 5–10 kg/ha year of total N can be deposited on more sensitive terrestrial ecosystems (heaths, bogs, cryptogams) and 10–20 kg/ha year would be tolerated by forests. Crop plants could tolerate higher than the earlier listed values of nitrogen from air, but it is highly dependent on soil conditions (Krupa, 2003).

Biogenic volatile organic compounds (BVOCs) that are natural organic compounds should also be taken into account as substances polluting the air as they are released to the environment in considerable quantities. Plants as sessile organisms use such compounds to defend themselves against pathogens, herbivores and other stress factors. Typically, these are compounds of low molecular weight, including isoprenes, monoterpenes, sesquiterpenes and other C10–C15 chemical structures (Curtis, Helming, Baroch, Daly, & Davis, 2014). BVOCs are easily released to the atmosphere in gaseous form, and in the presence of NO_x and light, they contribute to photochemical reactions involved in the formation of secondary pollutants.

Introduction of biogenic and anthropogenic VOC shifts the equilibrium of the atmospheric processes in which NO and NO_2 are involved towards higher ozone contents (Calfapietra et al., 2013). Production of BVOC by the plants is still intensively investigated, also because of the possibility of EPFR formation when adsorbed on PM. The phenomenon of O_3 and EPFR creation should be taken into account in the current recommendations to increase greenery in urban areas. Studies allowed distinguishing between two groups: low BOVC and high BOVC emitting plants. The low BVOC emitting group of plants includes genera such as *Malus*, *Camphora*, *Citrus*, *Pyrus* and species as *Ginkgo biloba* and *Juglans nigra*. High-emitting BVOC plants belong the following genera: *Salix*, *Quercus*, *Populus*, *Pinus* and *Liqidamber* (Benjamin & Winer, 1998; Calfapietra et al., 2013; Curtis et al., 2014; Karl, Guenther, Koble, Leip, & Seufert, 2009).

Ozone contained in the troposphere is among the main threats to plants. Although all plants are susceptible to ozone, interspecies and intraspecies differences in ozone tolerance were found and depend on the existence of a genetically determined efficient antioxidant system. In recent years, studies have been undertaken on the active protection against the ozone damage in plants by the application of synthetic antioxidants, i.e., ethylene urea (EDU). European ash (*F. excelsior*), previously damaged by ozone during subsequent years, was successfully protected against ozone injury after application of EDU (Paoletti, Manning, Ferrara, & Tagliaferro, 2011).

To summarize, during evolution, plants colonized different environments by developing abilities to cope with all kind of stresses including air pollution. Plants with their large, conducive to pollutants accumulation surface are the most efficient and natural tool in the hands of a man for decreasing air pollution and with our knowledge will be effectively and widely used for this purpose.

4. PLANTS AS A BARRIER FOR AIR POLLUTANTS

The primary processes of plants are based on gas exchange to obtain CO_2. However, during this process, they inevitably also absorb contaminants present in the air. Consequently, they play a role as biological air filtering systems. The scale of this positive role of plants can vary, from large areas of forest, which can protect a city or even regions, to small scale, with only a few indoor plants in apartments or offices. It does not matter what is the scale; the intended use of plants in this context is to reduce air pollution and to protect human health and well-being as well as improving the environment for other organisms, improve climate and even enhance the monetary value of properties. The positive functions of trees and forests were object of many studies and a very good example is one conducted by Nowak, Hirabayashi, Bodine, and Greenfield (2014). They estimated that trees and forests in 2010 removed 17.4 million tons of air pollutants throughout the United States, with human health effects valued at 6.8 billion US dollars, including avoidance of more than 850 human deaths. The positive role of forests on climate and the environment, both globally and regionally is clearly large. As air pollution became one of the major threats in many parts of the world, forests may play a very positive role, which is a good argument to invest in afforestation policies. On smaller scales, parks and plant barriers can play a similar role to improve environmental goods and services. A fully overgrown stand was shown to stop most of the pollution within a distance of the first 20 m, with as dominant species *Tilia platyphyllos* (Popek, Gawronska, & Gawronski, 2015) and *Q. ilex* (Sgrigna et al., 2016). In case of pollutants emitted by vehicles, the concentration can be lower close to the road, because pollutants are blown further on where the particles are partially stopped, leading to increased pollutant concentrations (Popek et al., 2015; Sgrigna et al., 2016). To obtain a good protective role of parks in the cities, the belts probably need to be at least 20 m wide, with a canopy consisting of trees and shrubs. The mixture of species, age and the proper space for the air to move into the filter is beneficial for a good effect. In a belt of 50 m in width, a decrease in PM with more than 50% was observed (Popek et al., 2015). Especially fine and coarse PM decreased, with a much lower decline of the ultrafine fraction. The results presented indicate that forests and parks are sanctuaries with clean air. Especially the places where people live, play and work should have the best conditions. Particularly people with cardiovascular and respiratory problems are very sensitive to PM, and for young children, it is crucial for their health. City planners should include such aspects when planning parks, in which the outside belt of trees should possess a high capacity for air pollutant biofiltration.

Streets and roads with intensive traffic are sources of linear pollution that are difficult to control. Vegetation can be useful to reduce the pollutants emitted by transport vehicles at such places. The vegetation in the vicinity of transport routes can fulfil a useful role in all its growth forms, as trees (Popek et al., 2013; Sæbo et al., 2015; Tong, Baldauf, Isakov, Deshmukh,

& Zhang, 2016), shrubs (Popek et al., 2013; Sæbo et al., 2015) and herbaceous plants (Weber et al., 2014). One of the most air-polluting environments are highways, which also produce noise pollution. To combat the latter negative impact on the environment usually noise screens are used. Screens protect better against noise pollution than plants, but pollutants are dispersed. They are transported with the air streams, some passing over the screens, and they may be deposited on the screens or on the road, being easily resuspended by the next vehicles that are passing. Tong et al. (2016) evaluated the protective role of noise screens as a solid barrier only (1), solid barrier covered with vegetation as green wall (2), vegetation barrier (3) and the combination solid-vegetation barrier (4) as compared to free space in the vicinity of the road. Solid barriers (1) increased PM concentrations along the roadside though part of the PM was dispersed above the barrier. Solid barriers with vegetation cover (2) slightly reduced the PM concentration but the best result was obtained with a combination of a vegetation barrier (3) and a solid barrier with vegetation (4) since both significantly reduce PM concentration. The authors used evergreen conifer plant species with low BVOC emission and high LAI, to obtain a biofiltering function the whole year around. The results of this experiment indicated an additional positive effect for drivers. Long trips on motorways surrounded by solid noise screens become tedious, which increases the chance for accidents. Trees behind the screens both change the landscape and create a surface for the deposition of air pollutants.

When buildings are constructed close to the road they form canyons. In such canyons, trees may reduce the concentrations of NO_2 and PM by 40% and 60% respectively (Pough et al., 2012), although there exist also papers reporting the opposite (Jin, Guo, Wheeler, Kan, & Che, 2014). The latter might be due the fact that trees lower wind speed in and ventilation of the streets. In this respect, two factors are important: (1) the density of the 'phytobarrier' and (2) the location of the vegetation. The most frequently used method of measurement for density of the barrier is Leaf Area Density (Petroff, Zhang, Pryor, & Belot, 2009; Pough et al., 2012). A comprehensive overview of urban vegetation and air particle dispersion and deposition was made by Janhall (2015), highlighting several significant relationships. Phytobarriers need to be permeable enough to let air to penetrate through it but solid enough to keep moving air close to the surface of the vegetation. In developing the optimal solution for the best pollutant deposition in street canyons the model of plant distribution developed by Gromke, Jamarkattel, and Ruck (2016) can be helpful. An optimal situation can be achieved by using the appropriate species and pruning the barriers to the best height and porosity. Retention is higher in case the plants are equipped with trichomes, rough wax and an abundant microbiome (fungal mycelium) (Sanchez-Lopez, Carrillo–Gonzalez, Gonzalez-Chaves, Rosas-Saito, & Vangronsveld, 2015). PM deposition in street canyons can be increased, without the problems outlined earlier, if climbing vegetation is used to form green walls in the cityscape (Ottele et al., 2010). With wide streets, limited dispersion is not a problem. Usually, there is also more space for plants, so that deposition areas can be established for the pollution to improve air quality. Other groups of plants should also be used, such as shrubs and herbaceous plants. These plants grow lower than trees, and can be located closer to the edge of the road without disturbing neither the driver's view for traffic and road signs, nor air movements. Pollutants deposited on herbaceous plants are removed by mowing and removing the grass during the season. Herbaceous plants possess more shallow roots and also take up some pollutants (mainly HMs) from the top layer of the soil, thus the harvested biomass of these plants also removes some pollutants from the soil. Removal of air pollutants should take place on all possible levels both regional and local as well as inside our own homes and offices (see later).

5. INDOOR PHYTOREMEDIATION: MAIN POLLUTANTS AND RECOMMENDED PLANTS

In many parts of the world air pollution still continues and will remain a problem threatening the health of people and the environment for a long period of time. The purpose of this subsection is to illustrate the problem and to present measures that can be taken to protect people from negative effects in houses, apartments and workplaces. Indoor air pollution has been recognized as a public hazard globally. In developed countries, polluted air occurs from polluted outdoor air and products and materials, which are used in apartments and offices. In contrast, in underdeveloped countries it comes mainly from household cooking and heating and in some places partially also from transportation. Nowadays, many people spend most of their time indoors where the air is often more polluted than outside because there are additional sources of pollutants. The scale of the problem is indicated by the WHO's recognition as a 'sick building syndrome' (SBS) disease. Particular attention should be paid to this problem in newly constructed buildings because in these VOC emissions are the highest and only decrease over time. Usually, the emission of formaldehyde represents a significant risk factor for the SBS symptoms. At this point, it is important to monitor the indoor pollution and to increase awareness of residents (Takigawa et al., 2010). Heating systems and air conditioning equipment that are functioning in buildings and apartments need to be very well maintained because their use sometimes leads to increased ionization and ozone in the air (Britigen, Alshawa, & Nizkorodov, 2006). Plants biofiltrating indoor air, which in recent years was referred to as indoor phytoremediation, is of high interest because of (1) the low cost, (2) the ease of implementation and maintenance of this environmental friendly technology and (3) the fact that some species possess a surprisingly high efficiency of uptake of pollutants. The use of potted plants to remove VOCs from indoor air and a list of assessed species have been reviewed by Cruz, Christensen, Thomsen, and Muller (2014).

The list of indoor air pollutants is very long, starting from chemical impurities coming from outdoor sources: vehicle exhaust, plumbing vents and building exhaust, radon and combustion byproducts as well as from indoor sources: VOC from adhesives, upholster, carpeting, copy machines, cleaning agents, tobacco smoke, combustion byproducts from stove and fireplace, biological contaminants: pollen, bacteria, viruses, fungus, molds etc. Gaseous pollutants are the most difficult to remove from the air, which also not often achieved by air conditioners. Therefore, here the valuable ability of plants comes to the fore, because during gas exchange they simultaneously absorb and inactivate pollutants.

The following VOCs are most commonly found in homes and offices: formaldehyde, benzene, toluene, xylene, CO, naphthalene, nitrogen dioxide, ammonium, PAHs including benzo [*a*] pyrene, trichloroethylene, cigarette smoke and others. Sometimes the list contains even several hundreds of compounds. In the presently constructed buildings, formaldehyde is most often listed on the first place of pollutants and a great number of investigations among indoor phytoremediation are devoted to this compound. Despite the fact that it is already over 20 years old, the work of Giese, Bauer-Doranth, Langbartels, and Sandermann (1994) dedicated to the detoxification of formaldehyde is still very interesting. One of the investigated species was the spider plant (*Chlorophytum comosum*), which is a very common ornamental plant in apartments and offices. Plants were exposed to ^{14}C-formaldehyde for 24 h and analysed. Results showed that about 88% of the radioactivity was recovered in the plant and was incorporated into organic acids, amino acids, free sugars and lipids as well as cell-wall components. The authors identified also two dehydrogenases (including one glutathione dependent) involved in oxidative detoxification of formaldehyde (Giese et al., 1994). The role of the enzyme formaldehyde dehydrogenase in the degradation of formaldehyde was also endorsed by others (Xu, Wang, & Hou, 2011). The activity of formaldehyde dehydrogenase in three plant species: *C. comosum*, *Aloe vera* and *Epipremnum aureum* (7.8, 2.8 and 4.4 nmol/min mg protein respectively), confirms the role of *Chlorophytum* as a good air phytoremediator plant. Kim et al. (2010) evaluated capacity of 86 species of plants for removing formaldehyde from indoor air. The most efficient species are: *Osmunda japonica*, *Selaginella tamariscina*, *Davallia mariesii*, *Polypodium formosanum*, *Psidium guajava*, *Lavandula* spp., *Pteris dispar*, *Pteris multifida*, and *Pelargonium* spp., that are removing more than 1.87 $\mu g/m^3$ cm^2 of leaf area over 5 h. Based on efficacy of formaldehyde removal authors divided tested species into three groups: excellent, intermediate and poor removal. Species classified as excellent (removing more than 1.2 $\mu g/m^3$ cm^2 of leaf area over 5 h) are taken into account as the best candidates for phytoremediation in places where formaldehyde may appear as a health threatening gas. The results of this valuable work contribute to the development of indoor phytoremediation.

Indoor plants are cultivated in a prepared growing medium or in soil. Both substrates are rich in microorganisms which also participate in the degradation of VOC, using them as a source of carbon and energy. Gas exchange with soil is facilitated by watering plants which 'pushes' air out of the soil followed by a period of drying which allows 'new' batch air (with contaminants) to penetrate into the soil pores. The participation of both, the aerial parts of plants and the root zone in the formaldehyde removal during the day and night was assessed using two species, *Fatsia japonica* and *Ficus benjamina*. During the day both species were equally well removing formaldehyde by both the aerial parts and the root zone. However, during the night the above ground parts of the plants removed less than one tenth of what was achieved in the soil where in the process of removing of formaldehyde is performed by soil and rhizobacteria microorganisms (Kim et al., 2008). Xu et al. (2011) demonstrated that cultivation of plants like *C. comosum*, *A. vera* and *E. aureum* resulted in almost equal proportions of 50% formaldehyde removal by both microorganism and plants.

Aromatic hydrocarbons like benzene and its derivatives toluene, xylene ethylbenzene (known as BTEX) are common air pollutants in apartments and offices. They enter into indoor from the outside air but are also emitted by internal sources in the room. Applied ^{14}C-benzene is taken up by plants and in a first step hydroxylated to phenol or pyrocatechol. A further step is cleavage to muconic acid and in a next step fumaric acid. These metabolites were confirmed by the high radioactivity of these products. The research was conducted with three woody plant species: *A. campestre*, *Malus domestica* and *Vitis vinifera* (Ugrekhelidze, Korte, & Kveitdze, 1997). One of the first publications describing the removal of benzene from indoor air presented data for seven taxa: *Dracaena* 'Janet Criag', *Dracaena marginata*, *E. aureum*, *Schefflera* 'Amate', *Spathiphyllum* 'Petite', *Spathiphyllum* 'Sensation', *Howea forsteriana*. Tested dracaenas are from different species and differ in their efficiency of benzene removal. Differences between cultivars of *Spathiphyllum* show potential of improving efficiency during breeding processes (Orwell, Wood, Tarran, Torpy, & Burchett, 2004). During their search for plant species removing benzene from indoor air Liu, Mu, Zhu, Ding, and Arnes (2007) evaluated 73 species exposing them during 2 h to fumigation with this gas. Results showed high differences between tested plant species in their ability of benzene removal. Out of the 73 species 23 did not change the benzene level in the air, 13 species removed between 0.1% and 9.99%, 17 species removed between 10% and 20% and another 17 species removed 20%–49% of benzene; three species even were able to remove 60%–80% of the gas. Out of the 73 species 10 were selected for the second experiment, in which plants were exposed to fumigation for 2 days: *Crassula portulacea*, *Hydrangea macrophylla*, *Cymbidium* 'Golden Elf', *Ficus microcarpa* var. *fuyuensis*, *Dendranthema morifolium*, *Citrus medica* var. *sarcodactylis*, *Dieffenbachia amoena* 'Tropic Snow', *Spathiphyllum* 'Supreme', *Nephrolepis exaltata* 'Bostoniensis' and *Dracena deremensis* 'Variegata'. After standardization of the results to plant leaf area and extrapolation to room size, the authors concluded that for indoor phytoremediation of benzene the best were *C. portulacea*, *H. macrophylla*, and *Cymbidium* 'Golden Elf' while *D. deremensis* 'Variegata' was the least effective (Liu et al., 2007). Sriprapat and Thiravetyan (2016) investigated the ability of eight plant species for removing benzene from the air: *Syngonium podophyllum*, *Sansevieria trifasciata*, *Euphorbia milii*, *C. comosum*, *E. aureum*, *Dracena sanderiana*, *H. helix* and *Clitoria ternatea*. All species efficiently removed benzene, but the best was *C. comosum*. These authors also underlined the importance of the plant microbiome, both the endobacteria and phyllobacteria, in this process.

Another common air-polluting VOC is toluene, which is removed by several plant species commonly grown in homes and offices: *Schefflera elegantisima*, *Philodendron* spp. 'Sunlight' and *H. helix* (Kim et al., 2011) and *H. helix* 'Gitte' that was investigated in the work by Cruz, Muller, Svensmark, Pedersen, and Christensen (2014). To evaluate the removal of the aromatic hydrocarbon xylene from the air, Sriprapat, Boraphech, and Thiravetyan (2013) tested 15 species. The best species in this study was the recently very popular in public places, very resistant to drought and low light intensity *Zamioculcas zamiifolia* (Sriprapat et al., 2013). Tobata, Vangnai, and Thiravetyan (2016) demonstrated that this species can also successfully remove another VOC, ethylbenzene, from the air.

Yang, Pennisi, Son, and Kays (2009) assessed 28 ornamental plant species for their ability to remove five volatile indoor pollutants: benzene, toluene, octane, trichloroethylene (TCE) and α-pinene. Huge differences in the efficiency of removal of these compounds were found between the different plant species. Out of the 28 species only *Hemigraphis alternata*, *H. helix*, *Hoya carnosa* and *Asparagus densiflorus* were efficiently removing all five pollutants. Others as *Tradescantia pallida* displayed high removal efficiencies for four of the five VOCs. *Fittonia argyroneura* very effectively removed benzene, toluene and TCE, *F. benjamina* effectively removed octane and α-pinene while *Polyscias fruticosa* remove only octane.

In most laboratory tests plant species are assessed in relation to one chemical compound while in real life conditions plants usually have to cope with cocktails of several pollutants. Therefore, experiments using cocktails of pollutants are crucial. Plants exposed to a mixture of pollutants are removing them with varying efficiencies depending on plant species and type(s) of pollutant(s). Removal efficiency of a mixture of benzene, TCE and toluene from polluted air was evaluated using following species of ornamental plants: *Pelargonium domesticum*, *Ficus elastica*, *C. comosum* and *Kalanchoe blossfeldiana*. All species were good in removing benzene and toluene. TCE was preferably taken up by *C. comosum*, but the removal of benzene and toluene was decreasing (Cornejo, Munoz, Ma, & Stewart, 1999). Orwell, Wood, Burchett, Tarran, and Torpy (2006) reported an effect of synergy was observed when *Dracaena* 'Janet Criag' was exposed to a mixture of toluene and xylene in ranges between 1 and 10 ppm, these gases were more efficiently removed in comparison to when it was exposed to each of them individually.

6. FUTURE RESEARCH NEEDS

Because people in their homes and offices try to maintain optimal climatic conditions for life and since in our everyday life, we use fragrances and many other substances of plant origin (Rhind, 2014) there is chance that plants can improve human well-being. In many areas of the world the same ornamental plant species are grown. This has the advantage that the results of the investigations on the main pollutants removal can be applied at many places.

In case of indoor phytoremediation, the world can almost be considered as one global village. Despite the fact that over 20 years passed since the first book of the 'father' on indoor phytoremediation Dr. Wolverton (1996) was published and the knowledge already is quite extensive, still much is left as open questions and to be done. Indoor phytoremediation arouses growing interest of society, which often asks what species are recommended to maintain air in homes and offices safe and how many plants should be cultivated. Regarding plant species we already have at our disposal a significant amount of information, but knowledge on how many plants and what should be the leaf surface are needed for an adequate pollutants removal is still limited. One of the first real scale trials to reduce indoor air pollution was carried out with the very good indoor phytoremediation species *Dracaena deremensis* 'Janet Craig'. Three and six plants of this variety were placed in office rooms and it turned out that just three plants reduced pollution level by 75% below the threshold level of 100 ppb. These potted plants were equally effective in both air-conditioned and non—air-conditioned rooms (Wood et al., 2006).

In an attempt to solve the lack of adequate practical experience, Thomas, Kim, and Kay (2015) developed a mathematical model that takes into account the amounts of plant material, building air volume, VOC concentrations and air exchange. After a positive validation this model can become a very valuable tool in the implementation of the technology of indoor air phytoremediation.

Another area in which indoor phytoremediation can be a valuable ally to limit is to combat the negative effects of smoking tobacco. The reason for undertaking this study was the finding that nicotine is present in many foods of plants origin. The first clue was that the very common ornamental plant *E. aureum* takes up and transports nicotine to its above ground organs. This species does not metabolize the nicotine, and in higher concentrations it is even toxic for the plants resulting in an acceleration of the ageing process

(Weidner, Martins, Muller, Simon, & Schmitz, 2005). Innovative are the results of a study with peppermint (*Mentha x piperita*) as a model plant. The results demonstrated that peppermint takes up nicotine from a soil with plant debris and also absorbs it from air contaminated with tobacco smoke (Selmar et al., 2015). The presence of nicotine in many plant foods allows to suppose that many plant species possess the ability to take up this pollutant from the environment. This opens up new possibilities for indoor air phytoremediation research to find the most effective ornamental plants to protect nonsmoking family members from indirect smoking effects.

There is still a strong need to apply indoor phytoremediation for the immediate protection of human health. Already in 1996, Wolverton and Wolverton conducted a pioneering experiment of placing in the same house in the living room plants and treating the adjacent bedroom as controls without plant. The experiment was carried out for 3 months and was repeated at another season of the same year. In the room with plants humidity was higher than in the plant-free bedroom. Concurrently, microbial levels were more than 50% higher in a room without plants despite lower humidity. These results indicate that the plants generate a more friendly environment i.e., with lower numbers of microbes and a higher air humidity. With today's possibilities of molecular analyses, we should more concentrate on the role of the microbiomes associated with plants. In the aforementioned case, the plants' microbiomes might act as counterparts against pathogens within microbial ecosystems. In future, strategies can probably be developed of management of the environment in homes, apartments and offices that would be beneficial to human (Berg, Mahnert, & Moissi-Eichinger, 2014).

Phytoremediation of indoor air also opens up new possibilities for human well-being. A good example is provided by Boraphech, Suksabye, Kulinfra, Konsang, and Thiravetyan (2016) who used the popular plant *Sansevieria* spp. for removal of trimethylamine (TMO) (responsible for so-called fishy odor) from air. Both living plants and nonliving plant materials removed the smell. Massive secretion of TMO by the human body is called as fish odor syndrome, a rare metabolic disorder causing weak (or none) oxidation of trimethylamine to trimethylamine oxide. When the oxidizing enzyme is in disorder, TMO is accumulated in excess and released from the body not only through sweat and urine but also in the breath, causing a strong odor resembling the smell of rotting fish. Especially difficult experience with this disease is during childhood when peers do not accept such a person claiming that she/he does not care about hygiene. Currently, some drugs are available that reduce the disease state and also proper diet may be helpful, but it was demonstrated that help also might came from indoor phytoremediation, thus it can be said that phytoremediation marched with help into medicine (Boraphech et al., 2016).

Summarizing it can be concluded that with help of the properly chosen plant species there is a chance to create a sanctuary for a more safe and pleasant life.

REFERENCES

Bakker, M. I., Vorenhout, M., Sijm, D. T. H. M., & Kolloeeffel, C. (1999). Dry deposition of atmospheric polycyclic aromatic hydrocarbons in three *Plantago* species. *Environmental Toxicology and Chemistry, 18*, 2289–2294.

Barack, C. L. (2002). Pollution mitigation and carbon sequestration by an urban forest. *Environmental Pollution, 116*, 195–200.

Barthlott, W., Neinhuis, C., Cutler, D., Ditsch, F., Meusel, I., Theisen, I., & Wilhelmi, H. (1998). Classification and terminology of plant epicuticular waxes. *Botanical Journal of the Linnean Society, 126*, 237–260.

Benjamin, M. T., & Winer, A. M. (1998). Estimating the ozone-forming potential of urban trees and shrubs. *Atmospheric Environment, 32*(1), 53–68.

Berg, G., Mahnert, A., & Moissi-Eichinger, C. (2014). Beneficial effects of plant-associated microbes on indoor microbiomes and human health? *Frontiers in Microbiology, 5*, 15. http://dx.doi.org/10.3389/fmicb.2014.00015.

Bidleman, T. F. (1988). Atmospheric processes—wet and dry deposition of organic-compounds are controlled by their vapor particle partitioning. *Environmental Science & Technology, 22*(4), 361–367.

Bidwell, R. G. S., & Bebee, G. P. (1974). Carbon monoxide fixation by plant. *Canadian Journal of Botany, 174*(8), 1841–1847.

Boraphech, P., Suksabye, P., Kulinfra, N., Konsang, W., & Thiravetyan, P. (2016). Cleanup of trimethylamine (fishy odor) from contaminated air by various species of *Sansevieria* spp. and their leaf material. *International Journal of Phytoremediation, 18*(10), 1002–1013.

Borowski, J., Loboda, T., & Pietkiewicz, S. (2009). Photosynthetic rates and water efficiencies in three climber species grown in different exposures a turban and suburban sites. *Dendrobiology, 62*, 55–61.

Britigen, N., Alshawa, A., & Nizkorodov, S. A. (2006). Quantification of ozone in indoor environments generated by ionization and ozonolysis air purifiers. *Air & Waste Management Association, 56*, 601–610.

Bryce, M., Drews, O., Schenk, M. F., Menzel, A., Estrella, N., Weichenmeier, I., ... Traidl-Hoffmann, C. (2010). Impact of urbanization on the proteome of birch pollen and its chemotactic activity on human granulocytes. *International Archives of Allergy and Immunology, 151*(1), 46–55.

Calderon-Garciduenas, L., Solt, A. C., Henriquez-Roldan, C., Torres-Jardon, R., Nuse, B., Herritt, L., ... Reed, W. (2008). Long-term air pollution exposure is associated with neuroinflammation, an altered innate immune response, disruption of the blood–brain barrier, ultrafine particulate deposition, and accumulation of amyloid beta-42 and alpha-synuclein in children and young adults. *Toxicologic Pathology, 36*(2), 289–310.

Calfapietra, C., Fares, S., Manes, F., Morani, A., Sgrigna, G., & Loreto, F. (2013). Role of biogenic volatile organic compounds (BVOC) emitted by urban tree on ozone concentration in cities: A review. *Environmental Pollution, 183*, 71–80.

Chen, S. J., Tian, M., Zheng, J., Zhu, Z. C., Luo, Y., Luo, X. J., & Mai, B.-X. (2014). Elevated levels of enantioselective biotransformation of chiral PCBs in plants. *Environmental Science & Technology, 48*(7), 3847–3855.

Cormier, S. A., Lomnicki, S., Backes, W., & Dellinger, B. (2006). Origin and health impacts of emissions of toxic by-products and fine particles from combustion and thermal treatment of hazardous wastes and materials. *Environmental Health Perspectives, 114*(6), 810–817.

Cornejo, J. J., Munoz, F. G., Ma, C. Y., & Stewart, A. J. (1999). Studies on the decontamination of air by plants. *Ecotoxicology, 8*, 311–320.

Cruz, M. D., Christensen, J. H., Thomsen, J. D., & Muller, R. (2014). Can ornamentals potted plants remove volatile organic compounds from indoor air?–A review. *Environmental Science and Pollution Research*. http://dx.doi.org/10.1007/s1356-014-3240-x.

Cruz, M. D., Muller, R., Svensmark, B., Pedersen, J. S., & Christensen, J. H. (2014). Assessment of volatile organic compound removal by indoor plants- a novel experimental setup. *Environmental Science and Pollution Research, 21*(13), 7838–7846.

Curtis, A. J., Helming, D., Baroch, C., Daly, R., & Davis, S. (2014). Biogenic volatile organic compounds emissions from nine tree species used in an urban tree-planting program. *Atmospheric Environment, 95*, 634–664.

De Kempeneer, L., Sercu, B., Vanbrabant, W., Van Lanenhove, H., & Verstraete, W. (2004). Bioaugmentation of the phyllosphere for the removal of toluene from indoor air. *Applied Microbiology and Biotechnology, 64*, 284–288.

Dzierzanowski, K., Popek, R., Gawronska, H., Sæbo, A., & Gawronski, S. W. (2011). Deposition of particulate matter of different size fraction on leaf surfaces and in waxes of urban forest species. *International Journal of Phytoremediation, 13*, 1037–1046.

EEA. (2013). *Status of black carbon monitoring in ambient air in Europe*. Technical Report No 18.

Eevers, N., White, J. C., Vangronsveld, J., & Weyens, N. (2017). Bio- and phytoremediation of pesticide contaminated environments a review. In A. Cuypers, & J. Vangronsveld (Eds.), *Phytoremediation* (Vol. 83, pp. 277–318).

Fang, G. C., & Huang, C. S. (2013). Monitoring and modeling concentration and dry deposition of ambient air particulates and metallic elements Mn, Fe, Zn, Cr and Cu in central Taiwan. *Environmental Earth Sciences, 69*(3), 773–782.

Farmer, A. (2002). Effects of particulates. In J. N. B. Bell, & M. Treshow (Eds.), *Air pollution and plant life* (pp. 187–199). Hoboken, NJ: John Wiley and Sons Inc.

Feigin, V. L., Roth, G. A., Naghavi, M., Pamar, P., Krishnamurthi, R., Chugh, S., & Mensah, G. A. (2016). Global burden of stroke and risk fact ors in 188 countries, during 1990–2013: A systematic analysis for the global burden of diseases study 2013. *Lancet Neurology, 15*, 913–924.

Fuller, M., Bai, S., Eisinger, D., & Niemeier, D. (2009). *Practical mitigation measures for diesel particulate matter: Near-road vegetation barriers*. UC Davies – Caltrans Contract Report http://www.dot.ca.gov/hq/env/air/research/ucd_aqp/Documents/Mitigation-Measures-Package-Report-5-Micah-v3.pdf.

Gawronska, H., & Bakera, B. (2015). Phytoremediation of particulate matter from indoor air by *Chlorophytum comosum* L. plants. *Air Quality, Atmosphere, and Health*, (8), 265–272.

Gehling, W., & Dellinger, B. (2013). Environmentally persistent free radicals and their lifetimes in $PM_{2.5}$. *Environmental Science & Technology, 47*(15), 8172–8178.

Ghaly, A., & Ramikrishnan, V. V. (2015). Nitrogen sources and cycling in the ecosystem and its role in air, water and soil pollution: A critical review. *Journal of Pollution Effects & Control, 3*, 136. http://dx.doi.org/10.4172/2375-4397.1000136.

Giese, M., Bauer-Doranth, U., Langbartels, C., & Sandermann, H. (1994). Detoxification of formaldehyde by the spider plant (*Chlorophytum comosum* L.) and by soybean (*Glycine max* L.) cell-suspension culture. *Plant Physiology, 104*, 1301–1309.

Grigoratos, T., & Martini, G. (2015). Brake wear particles emissions: A review. *Environmental Science and Pollution Research International, 22*, 2491–2504.

Gromke, C., Jamarkattel, N., & Ruck, B. (2016). Influence of roadside hedgerows on air quality in urban street canyons. *Atmospheric Environment, 139*, 75–86.

Janhall, S. (2015). Review on urban vegetation and particle air pollution – deposition and dispersion. *Atmospheric Environment, 105*, 130–137.

Jin, S., Guo, J., Wheeler, S., Kan, L., & Che, S. (2014). Evaluation of impacts of tree on $PM_{2.5}$ dispersion in urban streets. *Atmospheric Environment, 99*, 277–287.

Karl, M., Guenther, A., Koble, R., Leip, A., & Seufert, G. (2009). A new European plant-specific emission inventory of biogenic volatile organic compounds for use in atmospheric transport models. *Biogeosciences, 6*, 1059–1087.

Kelley, M. A., Hebert, V. Y., Thibeaux, T. M., Orchard, M. A., Hasan, F., Cormier, S. A., … Dugas, T. R. (2013). Model combustion-generated particulate matter containing persistent free radicals redox cycle to produce reactive oxygen species. *Chemical Research in Toxicology, 26*(12), 1862–1871.

Khachatryan, L. E., Vejerano, S., Lomnicki, & Dellinger, B. (2011). Environmentally persistent free radicals (EPFRs). 1. Generation of reactive oxygen species in aqueous solutions. *Environmental Science & Technology, 45*(19), 8559–8566.

Kim, K. J., Jeong, M. I. I., Lee, D. W., Song, J. S., Kim, H. D., Yoo, E. H., … Han, S. W. (2010). Variation in formaldehyde removal efficiency among indoor plant species. *HortScience, 35*(10), 1489–1495.

Kim, K. J., Kil, M. J., Song, J. S., Yoo, E. H., Son, K., & Kays, S. J. (2008). Efficiency of volatile formaldehyde removal by indoor plants: Contribution of aerial plant parts versus the root zone. *Journal of the American Society for Horticultural Science, 133*(4), 521–526.

Kim, K. J., Yoo, E. H., Jeong, M. I., Song, J. S., Lee, S. Y., & Keys, S. J. (2011). Changes in the phytoremediation potential of indoor plants with exposure to toluene. *HortScience, 46*(12), 1646–1649.

Krupa, S. V. (2003). Effects of atmospheric ammonia (NH_3) on terrestrial vegetation: A review. *Environmental Pollution, 124*(2), 170–221.

Liu, Y. J., Mu, Y. J., Zhu, Y. G., Ding, H., & Arnes, N. C. (2007). Which ornamental plant species effectively remove benzene from indoor air? *Atmospheric Environment, 41*, 650–654.

Lomax, C., Liu, W. J., Wu, L., Xue, K., Zhou, S., McGrath, S. P., … Zhao, F. J. (2011). Methylated arsenic species in plant originated from soil microorganism. *New Phytologist, 193*(3), 665–672.

Lomnicki, S., Truong, S. H., Vejerano, E., & Dellinger, B. (2008). Copper oxide-based model of persistent free radical formation on combustion-derived particulate matter. *Environmental Science & Technology, 42*(13), 4982–4988.

Lu, S., Ren, J., Hao, X., Liu, D., Zung, R., Wu, M. Y. F., … Wang, Q. (2014). Characterization of protein expression of *Platanus* pollen following exposure to gaseous pollutants and vehicle exhaust particles. *Aerobiologia, 30*, 281–291.

McDonald, A. G., Bealey, W. J., Fowler, D., Dragosits, U., Skiba, U., Smith, R. I., … Nemitz, E. (2007). Quantifying the effect of urban tree planting on concentrations and depositions of PM10 in two UK conurbations. *Atmospheric Environment, 41*(38), 8455–8467.

Megremi, L. (2010). Distribution and bioavailability of Cr in central Europe, EOB Greece. *Central European Journal of Geosciences, 2*(2), 103–123.

Moeckel, C., Thomas, G. O., Barber, J. L., & Jones, K. C. (2008). Uptake and storage of PCBs by plant cuticles. *Environmental Science & Technology, 42*(1), 100–105.

Morikawa, H., Higaki, A., Nohno, M., Takahashi, M., Kamada, M., Nakata, M., … Goshima, N. (1998). More than a 600-fold variation in nitrogen dioxide assimilation among 217 plant taxa. *Plant, Cell & Environment, 21*, 180–190.

Naidoo, G., & Chirkoot, D. (2004). The effects of coal dust on photosynthesis performance of the mangrove *Avicennia marina* in Richards Bay, South Africa. *Environmental Pollution, 127*, 359–366.
Nowak, D. J., Crane, D. E., & Stevens, J. C. (2006). Air pollution removal by urban trees and shrubs in the United States. *Urban Forestry & Urban Greening, 4*(3–4), 115–123.
Nowak, D. J., Hirabayashi, S., Bodine, A., & Greenfield, E. (2014). Tree and forest effects on air quality and human health in the United States. *Environmental Pollution, 193*, 119–129.
Ny, M. T., & Lee, B. K. (2011). Size distribution of airborne particulate matter and associated elements in an urban area an industrial cities in Korea. *Aerosol and Air Quality Research, 11*, 643–653.
Orwell, R. L., Wood, R. A., Burchett, M. D., Tarran, J., & Torpy, F. (2006). The potted-plants microcosmos substantially reduces indoor air VOC pollution: II Laboratory study. *Water, Air, Soil Pollution, 177*, 59–80.
Orwell, R. L., Wood, R. L., Tarran, J., Torpy, F., & Burchett, M. D. (2004). Removal of benzene by the indoor plant/substrate microcosm and implications for air quality. *Water, Air, Soil Pollution, 157*, 193–207.
Ottele, M., von Bohemen, H. D., & Fraaij, A. L. A. (2010). Quantifying the deposition of particulate matter on climber vegetation on living walls. *Ecological Engineering, 36*, 154–163.
Paoletti, E., Manning, W. J., Ferrara, A. M., & Tagliaferro, F. (2011). Soil drench of ethylenediurea (EDU) protects sensitive trees from ozone injury. *iForest—Biogeosciences and Forestry, 4*, 66–68.
Petroff, A., Zhang, L., Pryor, S. C., & Belot, Y. (2009). An extended dry deposition model for aerosol onto broadleaf canopies. *Journal of Aerosol Science, 40*(3), 218–240.
Popek, R., Gawronska, H., & Gawronski, S. W. (2015). The level of particular matter on foliage depends on the distance from the source of emission. *International Journal of Phytoremediation, 17*, 1262–1268.
Popek, R., Gawronska, H., Wrochna, M., Gawronski, S. W., & Sæbo, A. (2013). Particulate matter on foliage of 13 wody species: Deposition on surfaces and phytostabilisation in Waxes-a 3-year study. *International Journal of Phytoremediation, 15*, 245–256.
Pough, T., MacKenzie, A. R., Whyatt, J. D., & Hevitt, C. N. (2012). Effectiveness of green infrastructure for improvement of air quality in urban street canyons. *Environmental Science & Technology, 46*, 7692–7699.
Reed, J. R., dela Cruz, A. L. N., Lomnicki, S., & Backes, W. L. (2015). Environmentally persistent free radical-containing particulate matter competitively inhibits metabolism by cytochrome P450 1A2. *Toxicology and Applied Pharmacology, 289*(2), 223–230.
Reinmuth-Sezlet, K., Ackaert, C., Kampf, C. J., Samonig, M., Shiraiwa, M., Kofler, S., ... Pöschl, U. (2014). Nitration of the birch pollen allergen Bet v 1.-1-: Efficiency and site-selectivity of liquid and gaseous nitrating agents. *Journal of Proteome Research, 13*(3), 1570–1577.
Rhind, J. P. (2014). *Fragrance and wellbeing. Plant aromatics and their influence on the psyche.* London: Singing Dragon.
Rogers, H. H., & Aneja, V. P. (1980). Uptake of atmospheric ammonia by selected plant species. *Environmental and Experimental Botany, 20*, 251–257.
Sæbo, A., Hanslin, H. M., Torp, T., Lierhagen, S., Gawronska, H., Dzierzanowski, K., & Gawronski, S. (2015). Chemical composition of vegetation along urbanization gradients in two European cities. *Environmental Pollution, 198*, 116–125.
Sæbo, A., Popek, R., Nawrot, B., Hanslin, H. M., Gawronska, H., & Gawronski, S. W. (2012). Plant species differences in particulate matter accumulation on leaf surfaces. *The Science of the Total Environment, 427–428*, 347–354.

Sanchez-Lopez, A. S., Carrillo −Gonzalez, R., Gonzalez-Chaves, M. C. A., Rosas-Saito, G. H., & Vangronsveld, J. (2015). Phytobarriers: Plants capture particles containing potentially toxic elements originating from mine tailings in semiarid region. *Environmental Pollution, 205*, 33−42.

Saravia, J., Lee, G. I., Lomnicki, S., Dellinger, B., & Cormier, S. A. (2013). Particulate matter containing environmentally persistent free radicals and adverse infant respiratory health effects: A review. *Journal of Biochemical and Molecular Toxicology, 27*(1), 56−68.

Selmar, D., Engelhardt, U., Hansel, S., Thrane, C., Nowak, M., & Kleiwachter, M. (2015). Nicotine uptake by peppermint plants as a possible source of nicotine in plant-derived products. *Agronomy for Sustainable Development, 35*, 1185−1190.

Sgrigna, G., Bldacchini, C., Eposito, R., Calanderlli, R., Tiwary, A., & Calfapietra, C. (2016). Characterization of leaf-level particulate matter for an industrial city using electron microscopy and X-ray microanalysis. *The Science of the Total Environment, 91−99*, 548−549.

Sgrigna, G., Sæbo, A., Gawronski, S., Popek, R., & Calfapietra, C. (2015). Particulate matter deposition on *Quercus ilex* leaves in an industrial city of central Italy. *Environmental Pollution, 197*, 187−194.

Sriprapat, W., Boraphech, P., & Thiravetyan, P. (2013). Factor affecting xylene-contaminated air removal by the ornamental plants *Zamioculcas zamiifolia*. *Environmental Science and Pollution Research, 21*, 2175. http://dx.doi.org/10.1007/s11356-013-2175-y.

Sriprapat, W., & Thiravetyan, P. (2016). Efficacy of ornamental plants for benzene removal from contaminated air and water: Effect of plants associated bacteria. *International Biodeterioration and Biodegradation*. http://dx.doi.org/10.1016/j.ibiod.2016.03.001.

Sun, F. B., Yin, Z., Lun, X. X., Zhao, Y., Li, R. N., Shi, F. T., & Yu, X. (2014). Deposition velocity of PM2.5 in the winter and spring above deciduous and coniferous forests in Beijing, China. *PLoS One, 9*(5). http://dx.doi.org/10.1371/journal.pone.0097723.

Takahashi, M., Higaki, A., Nohno, M., Kamada, M., Okamura, Y., Matsui, K., … Hiromichi, M. (2005). Differential assimilation of nitrogen dioxide by 70 taxa of roadside trees a tan urban pollution level. *Chemosphere, 61*(5), 633−639.

Takigawa, T., Wang, B. L., Saijo, Y., Morimoto, K., Nakayama, K., Tanaka, M., & Kishi, R. (2010). Relationship between indoor chemical concentrations and subjective symptoms associated with sick building syndrome in newly built houses in Japan. *International Archives of Occupational and Environmental Health, 83*, 225−235.

Thomas, C. K., Kim, K. J., & Kay, S. J. (2015). Phytoremediation of indoor air. *HortScience, 50*(5), 765−768.

Tobata, M., Vangnai, A. S., & Thiravetyan, P. (2016). Removal of ethylbenzene from contaminated air by *Zamioculcas zamiifolia* and microorganisms associated on *Z. zamiifolia* leaves. *Water, Air, Soil Pollution, 227*(4), 115. http://dx.doi.org/10.1007/s11270-016-2817-z.

Tong, Z., Baldauf, R. W., Isakov, V., Deshmukh, P., & Zhang, K. M. (2016). Roadside vegetation barrier designs to mitigate near-road air pollution impacts. *The Science of the Total Environment, 541*, 920−927.

Ugrekhelidze, D., Korte, F., & Kveitdze, G. (1997). Uptake and transformation of benzene and toluene by plant leaves. *Ecotoxicology and Environmental Safety, 37*, 24−29.

Varma, A., Abbott, L., Werner, D., & Hampp, R. (2008). *Plant surface microbiology*. Berlin: Springer, 628 pp.

Vejerano, E., Lomnicki, S., & Dellinger, B. (2012). Lifetime of combustion-generated environmentally persistent free radicals on Zn(II)O and other transition metal oxides. *Journal of Environmental Monitoring, 14*(10), 2803−2806.

Vorholt, J. A. (2012). Microbial life in the phyllosphere. *Nature Reviews. Microbiology, 10*(12), 828−840. http://dx.doi.org/10.1038/nrmicro2910.

Wang, H., Shi, H., Li, Y., Yu, Y., & Zhang, J. (2013). Seasonal variation in leaf capturing of particulate matter, surface wettability and micromorphology in urban tree species. *Frontiers of Environmental Science & Engineering, 7*(3), 579–588.
Wang, H., Shi, H., & Wang, Y. (2015). Effects of weather time and pollution level on the amount of particulate matter deposited on leaves of *Ligustrum lucidum. The Scientific World Journal*. http://dx.doi.org/10.1155/2015/935942.
Weber, F., Kowarik, I., & Saumel, I. (2014). Herbaceous plants as filters: Immobilization of particulates along urban street corridors. *Environmental Pollution, 186*, 234–240.
Weidner, M., Martins, R., Muller, A., Simon, J., & Schmitz, H. (2005). Uptake, transport and accumulation of nicotine by the Golden Potho (*Epipermnum aureum*): The central role of the root pressure. *Journal of Plant Physiology, 162*, 139–150.
Weyens, N., Thijs, S., Popek, R., Witter, N., Przybysz, A., Espenshade, J., … Gawronski, S. W. (2015). The role of plant-microbe interaction and their exploitation for phytoremediation of air pollutants. *International Journal of Molecular Sciences, 16*(10), 25576–25604.
WHO Report. (2015). *Reducing global health risks through mitigation of short-lived climate pollutants* (Report For Policy-maker).
Wild, E., Dent, J., Thomas, G. O., & Jones, K. C. (2006). Visualizing the air-to-leaf transfer and within-leaf movement and distribution of phenanthrene: Further studies utilizing two-photon excitation microscopy. *Environmental Science & Technology, 40*, 907–916.
Winter, K., & Smith, A. J. S. (1996). *Crassulacean acid metabolism*. Springer.
Wolverton, B. C. (1996). *Eco-friendly houseplant*. London: Weidenfeld & Nicolson.
Wood, R. A., Burchett, M. D., Alquezar, R., Orwell, R. L., Tarran, J., & Torpy, F. (2006). The potted-plant microcosm substantially reduces indoor air VOC pollution: I. Office field-study. *Water, Air, Soil Pollution, 175*, 163–180.
Wu, Z., Zhang, X., Wu, X., Shen, G., Du, Q., & Mo, C. (2013). Uptake of di(2-ethylhexyl) phthalate (DEHP) by plants *Benincasa hispida* and its use for lowering DEHP content in intercropped vegetables. *Journal of Agricultural and Food Chemistry, 6*(22), 5220–5225.
Xu, Z., Wang, L., & Hou, H. (2011). Formaldehyde removal by potted plant-soil system. *Journal of Hazardous Materials, 192*, 314–318.
Yang, D. S., Pennisi, S. V., Son, K. C., & Kays, S. J. (2009). Screening indoor plants for volatile organic pollutants removal efficiency. *HortScience, 44*(5), 1377–1381.
Zhao, F., Elkelish, A., Dumer, J., Lindermayer, C., Barbro, J., Rueff, F., … Frank, U. (2016). Common ragweed (*Ambrosia artemisiifolia* L.): Allergenicity and molecular characterization of pollen after plant exposure to elevated NO_2. *Plant, Cell & Environment, 39*, 147–164.

FURTHER READING

Baldacchini, C., Castanheiro, A., Maghakyan, N., Sgrigna, G., Verhelst, J., Alonso, R., & Samson, R. (2017). *How does the amount and composition of PM deposited on Platanus acerifolia leaves change across different cities in Europe? Environmental Science and Technology*. http://dx.doi.org/10.1021/acs.est.6b04052.
Buters, J., Ring, J., Gorg, A., Behrendt, H., & Traidl-Hoffmann, C. (2010). Impact of urbanization on the proteome of birch pollen and its chemotactic activity on human granulocytes. *International Archives of Allergy and Immunology, 151*(1), 46–55.
United Nations. (22 May 2001). *Stockholm Convention on Persistent Organic Pollutants, Stockholm*.
Wolverton, B. C., & Wolverton, J. D. (1996). Interior plants: Their influence on airborne microbes inside energy-efficient buildings. *Journal of the Mississippi Academy of Sciences, 41*(2), 99–105.

SUBJECT INDEX

'*Note*: Page numbers followed by "f" indicate figures and "t" indicate tables.'

D

E

I

N

O

P

R

T

U

V

W

X

Z

AUTHOR INDEX

Note: Page numbers followed by "t" indicate tables.

A

B

C

D

E

H

K

M

N

O

Q

R

U

V

W

X

FSC
www.fsc.org
MIX
Papier | Fördert
gute Waldnutzung
FSC® C083411

Kolibri 360 GmbH
Ferdinand-Jühlke-Straße 7
99095 Erfurt, Deutschland
produktsicherheit@kolibri360.de